Canada's Fifty Years in Space

The COSPAR Anniversary

By

Gordon Shepherd and Agnes Kruchio

An Apogee Books Publication

We acknowledge the financial support of the Government of Canada through the Book Publishing Industry Development Program for our publishing activities.

Published by Apogee Books, Box 62034, Burlington,
Ontario, Canada, L7R 4K2, http://www.apogeebooks.com
Tel: 905 637 5737

Printed and bound in Canada
Canada's Fifty Years In Space - The COSPAR Anniversary by Gordon Shepherd and Agnes Kruchio
©2008 Apogee Books
ISBN 978-1-894959-72-8

Canada's Fifty Years in Space

The COSPAR Anniversary

By

Gordon Shepherd and Agnes Kruchio

Dedicated to Canada's space pioneers

To all the members of my family, for their loving support

Gordon (Dad, Grandpa, Great-grandpa)

To my father, Paul Kruchio, whose passion for astronomy and knowledge inspired mine.

Agnes

CONTENTS

Preface   **11**
Foreword   **13**

Chapter 1: The Beginning   **17**

    The formation of COSPAR   17
    COSPAR's first meetings   19
    Space science in the Canadian context   25
    The National Research Council of Canada   28
    Postscript   31

Chapter 2: Post-war Foundations of Space Science   **33**

    How Canadian space science began in 1925   33
    Davies and Currie at Chesterfield Inlet   34
    Donald Charles Rose   42
    Henderson and Rose's 1932 ionospheric observations   45
    The NRC during World War II and its impact on space science 47
    Rose's wartime years   49
    A post-war NRC discovery – radio wave emission from the sun 51
    Currie's wartime years   53
    Davies' wartime years   55
    Formation of the Defence Research Telecommunications Establishment 57
    Postscript   60

Chapter 3: Space Science Takes Root in Canada   **61**

    Nate Gerson, a US benefactor, plants space science in Canada 61
    Auroral radar reflections observed in Saskatoon   66
    Optical studies and Gerson's AFCRL contract   69
    Nate Gerson and the legacy of John Arthur (Jack) Jacobs   75
        Sir Charles Seymour Wright   79
    Gerson finds the University of Western Ontario   80
    Colin Hines at Cambridge University   81
    Prince Albert Radar Laboratory (PARL)   82
    Postscript   84

Chapter 4: The IGY and its New Moons   **85**

    How the IGY began   85
    Canadian response to Sputnik   86
    What lay behind the IGY?   87
    Meteorology   92
    Geomagnetism   92
    Aurora   92
    Ionospheric physics   97

Solar activity   98
Cosmic rays   98
Rockets and satellites   99
What was learned from the IGY?   102
    The sun   102
    The atmosphere   103
    The Van Allen belts   104
    The magnetosphere   105
    What causes the aurora?   105
A personal reflection of the IGY   107
Postscript   111

Chapter 5: The Sixties – a Decade of Exponential Growth   **113**

The stimuli   113
The importance of the Defence Research Board   113
The Theoretical Studies Group at DRTE   115
The University of Saskatchewan   117
    Formation of the Institute of Space and Atmospheric Studies   117
    The creation of SED Systems   119
    The story of O_2 singlet delta   121
    Hunten and alkali metals in the upper atmosphere   124
    Currie steps down   126
Other universities   126
The Alouette/ISIS Program   128
A home-grown Canadian rocket program   140
Enhanced rocket instrumentation   146
Passive auroral plasma observations at the NRC   147
    Auroral electron precipitation and ionospheric electric fields 148
    Solar wind source for auroral ions   149
    Transverse ion acceleration (TIA)   149
    Polar cap aurora   149
Canada Centre for Remote Sensing   150
Postscript   154

Chapter 6: Decades of Transition   **155**

A Department of Communications   155
Impact of NASA's Space Shuttle on Canada   158
A Grass-Roots Space Science Initiative   159
The National Research Council takes charge   161
A maturing rocket program   163
Active experiments at the NRC   168
    "Waterhole" auroral perturbation experiments   169
    Echo auroral probes   169
    Firewheel   170
Termination of the Churchill Research Range   170
The Apollo-Soyuz Docking – Another kind of Active Experiment   170

Space Shuttle science initiatives 173
The WAMDII and WINDII missions 174
COSPAR in Canada – 1982 175
Nuclear winter and the Defence Research Establishment Valcartier 176
STRATOPROBE – High altitude balloon flights 177
Canada begins an Astronaut Program 178
A Canadian Auroral Imager on the Swedish Viking satellite 182
SMS on Akebono 184
The Swedish Freja and Russian Interball Missions 186
    Freja CPA 186
Freja and Interball Auroral Imagers 187
Postscript 188

Chapter 7: Birth of the Canadian Space Agency   **191**

The CSA moves to St. Hubert 191
The astronaut program within the CSA 194
Microgravity in Space 198
    Background 198
    STS-77 experiments 199
    Soret coefficient in crude oil (SCCO) 200
Space Life Sciences 201
    Aquatic Research Facility (ARF) 201
    e-OSTEO 202
    Canadian Protein Crystallization Experiment (CAPE) 202
    Perceptual Motor Deficits In Space (PMDIS) 203
RADARSAT 204
The International Space Station (ISS) 208
WINDII on UARS 209
Other Science missions under the CSA 213
    Observations of Electric-field Distributions in the Ionospheric 213
    Plasma – a Unified Study (OEDIPUS)
    Nozomi Thermal Plasma Analyser (TPA) 214
    Termination of Space Plasma studies at NRC 215
Continuation of the Space Plasma Studies at the Univ. of Calgary 216
    Nozomi Thermal Plasma Analyser (TPA) 216
    Suprathermal Ion Imager (SII) 216
    Imaging and Rapid scanning ion Mass spectrometer (IRM) 216
Rocket measurements of the Cosmic Background Radiation 217
Measurements of Pollution In The Troposphere (MOPITT) 219
Ozone and the Odin mission 220
The SCISAT mission 223
SWIFT on Chinook 225
Ground-based space science 226
Postscript 227

Chapter 8: Beyond the Earth   **229**

Very Long Baseline Interferometry – a Canadian invention   229
Space Astronomy   231
     Far Ultraviolet Spectroscopic Explorer (FUSE)   232
     Herschel   233
     The James Webb Space Telescope (JWST)   235
     ASTROSAT   236
     BLAST   237
     MOST – an all-Canadian astronomy mission   237
Gravity Probe B mission to test Einstein's general relativity   238
Canada joins the Phoenix mission to Mars   240
     Early background   240
     MITCH   242
     MATADOR   245
     Phoenix begins   247
Postscript   253

Chapter 9: Canada's Future in Space   **255**

Co-operation and competition in space   255
The Canadian space industry   256
How does one formulate a program?   258
Approaches by other countries   260
Canada's future   263
The benefits of space research   265
Steve MacLean and the International Space Station   267
Postscript   272

List of Acronyms   **273**

Index   **276**

PREFACE

The writing of this book took place in two phases. Phase 1 began with support from the National Research Council when Balfour Currie, Frank Davies, Don Rose and John Chapman were still with us, and able to give interviews. Much of the material used here came from these sources, as the authors have interviewed most of the Canadian space science pioneers who appear in this book. It's too late to thank them in person now, but we can and do remember them with admiration. This phase did not result in a published work. Phase 2 began with the idea of COSPAR's (and Canada's) Fifty Years in space.

The second phase required considerable updating because of the lapse in time; in this we received a great deal of input from the next generation space scientists. In particular Peter Forsyth and Ian McDiarmid were steadfast sources of information, advice and support. David Kendall also read all of the CSA material and made many helpful suggestions. Many others contributed material, much of which has been used as provided, shown in quotes. It would be impossible to create a scale of importance for contributions, so we refrain from identifying the "major" contributors. All of these are identified in the text and to them we are greatly indebted. Finally, many others provided advice, corrections, photos and other assistance. All of these, to the best of our knowledge, are listed below, in alphabetical order. To any who might have been inadvertently overlooked, we offer our sincere apologies.

Dennis Akins, John Brebner, Allan Carswell, Ruth Ann Chicoine, Leroy Cogger, Fokke Creutzberg, Wayne Evans, Barry Fowler, Rick Gerson, Sareen Gerson, Del Hansen, Donald Hunten, John Hutchings, Tim Hutchinson, Gordon James, Aaron Janofsky, Doris Jelly, Gloria Kavadas, Jim Koehler, Debbie Kowaliuk, Ted Llewellyn, Robert Lowe, Steve MacLean, Michael Maguire, Terry McCawley, Don McEwen, Joe McNally, Don Moorcroft, Al McNamara, Larry Morley, Bill Morrow, Don Muldrew, Ralph Nicholls, Atsuhiro Nishida, Gordon Rostoker, George Sofko, Boris Stoicheff, Ken Tapping, Alister Vallance Jones, Brian Whalen, Berton Woodward, Andrew Yau.

It has been both a pleasure and a humbling experience to work with these Canadian space pioneers. So much was accomplished in such a short time.

We thank Robert Godwin for his enthusiastic support for this endeavour and his energy in making it happen. It is really because of him that this work has finally appeared in print.

The final preparation of the text was carried out while GS was a visiting scientist at the Research Institute for Sustainable Humanosphere, Kyoto University, Uji, Japan. He is indebted to Professor Toshitaka Tsuda for providing such an inspiring environment for this work.

We close by again thanking all the contributors. If your material has in any way suffered from mid-interpretation, or actual errors, those remain the responsibility of the authors.

Gordon Shepherd   **Agnes Kruchio**   **March, 2008**

FOREWORD

Gordon Shepherd's active participation in Space Science began about half a century ago and is still going strong. Together with Agnes Kruchio, who has had a long standing interest in space research, they have produced an interesting, and in some cases detailed, story covering Canada's space programs from the early beginnings to the present day. Their book differs from previous books on the subject in that it not only describes what and how things were done, but also includes interesting information about the people who did them.

Studies of the upper atmosphere in Canada date back to the 1930s when ground-based measurements of the aurora were initiated by scientists at the University of Saskatchewan. Canada is an ideal place for this kind of research because electrically charged particles from the Sun, which are responsible for aurora, magnetic storms, and other high latitude phenomena, are funneled by the earth's magnetic field into ring-like regions surrounding the magnetic poles, one of which is located in northern Canada. To investigate these high-latitude and high-altitude effects, some of which have important practical implications, Canada in the 1950s and 60s developed a rocket program and, in collaboration with the US, a satellite program that resulted in Canada's first satellite, Alouette I. This early work not only produced some excellent science but also generated a new interest in space among university and government scientists, and at the same time it brought Canadian industry on board. In fact, it set the stage for Canada's future programs both in space applications and space science. Since Canada has no satellite launch capability, collaborative programs like Alouette/ISIS offer important advantages, particularly in terms of cost, but also by providing opportunities to interact with scientists and engineers from other countries and to have increased access to the technologies of these countries. The success of this early collaboration encouraged government agencies, and scientists, to continue this approach and Canada has had joint programs with the US, Sweden, Japan, France, the United Kingdom, Germany and Russia.

In 1967 the federal government began to redirect Canada's space objectives from purely scientific pursuits toward applications, particularly domestic communications and remote sensing of the earth's surface. With the launch of Anik A1 in 1972, Canada became the first nation on Earth to have a commercial domestic communications satellite in geostationary orbit. Later work in communications gave us the technology that is the basis for today's direct-to-home satellite broadcasts. Canada, with its abundance of natural resources, has had a long standing interest in developing ways to exploit and manage these natural assets. This interest led to the operational earth observation satellite Radarsat 1, another Canada-US project, which was launched in 1995. Built in Canada by Canadian industry, the satellite has on board an advanced radar system capable of taking observations in all weather conditions.

The flight of the US Space Shuttle Columbia in 1981 marked another major achievement in Canadian space technology. This was the first flight of the much photographed CANADARM, the remotely controlled space arm that has played an important role in Shuttle operations. It has been used for a variety of difficult tasks including the deployment and recovery of satellites. The technology developed for the 'arm' was further enhanced to create the Mobile Servicing System (MSS), a robotic system installed on the International Space Station (ISS) and which plays a vital role in the construction and servicing of the station. Canada is one of a number of countries participating in this international project whose assembly in orbit started in 1998. The ISS is a platform on which a low gravity environment is available for experiments and from which spacecraft with astronauts will likely be launched to the moon and possibly Mars.

To participate in the Shuttle and Space Station projects the Canadian Astronaut Program was started in 1983, again as a joint Canada/ US program. Initially, six astronauts were selected from a large number of applications, and in 1984 Marc Garneau, as a member of the crew of Shuttle flight 41-G, became the first Canadian in space. Today there is a permanent corps of astronauts and eight Canadians have flown on eleven separate missions.

The establishment in 1989 of the Canadian Space Agency (CSA) brought together the space programs that had previously been the responsibility of individual government departments, albeit with major participation from industry and universities. The creation of the CSA was a logical step in the evolution of Canada's varied and complex space program. It has provided the essential focus needed for the ongoing activities as well as providing a stronger voice in negotiations with partner countries. Since the establishment of the Agency many new projects have been initiated using both rockets and satellites. New collaborative projects are underway with a number of countries to further our understanding in several scientific areas, possibly including the weather on Mars. The US satellite Pheonix, which was launched in 2007, is expected to arrive at Mars in the spring of 2008. On board is a Canadian instrument to determine local weather conditions around the landing site.

Prior to 2003 the last Canadian scientific satellite was ISIS II, launched in 1971, but in 2003, no less than two scientific satellites were placed in orbit. The first was MOST, a suitcase-sized astronomical satellite designed to measure the age of stars in our galaxy. The second was SCISAT which is furthering our understanding of the ozone layer, particularly at high latitudes. Another satellite, CASSIOPE, is a small scientific satellite scheduled for launch in 2008 with instrumentation to study the effects of energetic particles produced by solar storms on communication satellites, satellite navigation systems, and other space based systems.

It is sometimes asked what have we gained from the considerable expenditures that have gone into the Space Program, a program that has evolved from a small government funded scientific endeavor into a complex array of government, industry, and university activities. We have certainly gained a wealth of new knowledge about what is going on in the space above us, ranging from our atmosphere to near earth space and beyond, knowledge that could not have been obtained without the use of rockets and satellites. Obviously, we have also acquired important new technologies that have been put to many uses. Telephone calls, computer data, medical services, and television programs are routinely sent to all parts of the country on Canadian satellite systems. Earth observation data are received and processed in government and industry facilities. The processed data are used in resource management, weather forecasting, and map making. Canadian space technology is playing a critical role in the construction of the International Space Station. About 25 years ago, an international search and rescue satellite system was established with important input from Canada, which expedites locating downed aircraft and ships in distress, a satellite application that has saved many lives. These are some of the uses to which space technology has been put, and there are many others.

Canada now has the knowledgeable personnel and the physical infrastructure in place to support an evolving and exciting space program which can be expected to yield future benefits that we can't yet even imagine.

Ian McDiarmid

Canadian space pioneers gathered at the Inaugural DASP Winter Workshop at Bristol Aerospace in 1974. Left to right First row (kneeling):Don Moorcroft, Ralph Nicholls, Alan Manson (with toque), Gordon Rostoker (sunglasses), Bob Horita, Ted Hartz. Second row (standing), Jim Koehler, John Stephenson, Bob Lowe (glasses) , Frank Palmer, Gordon James (sunglasses), Al McNamara (black beret), "unidentified", behind him is Don McEwen, and behind him Doraswamy Venkatesan, and just to the right, with white hair, John Craven. To the right of "unidentified" is Gordon Shepherd (white collar, dark scarf), Jim Yee in front of him, Jim Whittaker (glasses) is behind and to the right, Balfour Currie is further to the right (in dark fur hat), then John Olson. Behind them Harry Sullivan (mutton chops, centre of left door), Tomiya Watanabe (against the frames between the two doors), John Walker (to the right, in dark coat), and then Brian Whalen (light coat, looking off to one side). To the left of and behind Frank Palmer is Del Rice (against the door, to the right of Harry Sullivan). To the left of John Stephenson, partly hidden is Hugh Wood (glasses). Finally, behind Shepherd and Whittaker, in the dark cap is Peter Forsyth. Photo by courtesy of the Division of Aeronomy and Space Physics of the Canadian Association of Physicists.

A group of space pioneers gathered at the University of Western Ontario on November 25, 2002, hosted by Don Moorcroft and the Canadian Space Agency. From left to right, Colin Hines, Doris Jelly, Al McNamara, Peter Forsyth, Gordon Shepherd, Alister Vallance Jones, Ralph Nicholls, Ian McDiarmid, Don Moorcroft. Photo by courtesy of Don Moorcroft.

Chapter 1

The Beginning

The formation of COSPAR

The space age dawned with a shot literally heard all around the world, on October 4, 1957, with the launch of Sputnik 1 by the USSR. This 84 kg satellite, a sphere some 58 cm in diameter, launched into a 65° inclination orbit with a perigee of 227 km and an apogee of 945 km, broadcast its signals to radio receivers all over the globe. Scientists and radio amateurs alike listened to its "beep beep beep" and tried to determine its orbit during the 21 days that the signal was transmitted.

The launch of Sputnik 1 occurred just a few minutes after the closing ceremony for a memorable Canadian event, the rollout of the CF-105 Avro Arrow, Canada's new twin-engine, long-range, day-and-night supersonic interceptor. There are two important connections of this event to space. The first was that the onset of the space age raised questions as to the role and value of this type of aircraft in the military future—an argument that probably influenced John G. Diefenbaker, Prime Minster of Canada to cancel the program just 16 months and 16 days later. The second was that this decision resulted in the end of not just the aircraft, but also of its developer, Avro Canada, with the result that many of its engineers moved to the US to join a space program that was looking for experienced people. That story is not pursued here but is told in detail by Gainor (2004).

In the US, there were immediate repercussions. President Dwight Eisenhower called a meeting of the Science Advisory Committee (SAC) created by then President Harry S. Truman, but never used by him, likely because of his basic distrust of scientists. Eisenhower, on the other hand, related well to scientists, especially the committee chair, I.I. Rabi; and transformed the committee into the PSAC (President's Science Advisory Committee) with direct access to the president. This was one of many ways in which Sputnik opened new doors for scientists. One of the first actions of PSAC was to propose a treaty banning atmospheric nuclear tests, although an incident in which a US spy plane, was shot down over the USSR (the so-called U2 incident) delayed the signing of this treaty to the administration of John F. Kennedy. PSAC also enlisted Eisenhower in a campaign to change the attitude of the US public towards science and to strengthen the teaching of science in schools. A detailed account of this is given by Rigden (2007).

Shortly afterwards, on November 3, 1957, Sputnik 2 followed its predecessor into orbit, demonstrating that Sputnik 1 was not a one-time wonder, but that access to space was a major program of the USSR and proclaiming that this would become a whole new human endeavour. Not only that, but Sputnik 2 carried the first living creature into space, a dog named Laika, although for her, there was no return trip.

Sergei Pavlovich Korolev had first proposed the launch of an artificial satellite on May 26, 1954. On March 14, 1955, President Dwight Eisenhower announced that the US would launch such a satellite during the International Geophysical Year (IGY) which was to take place 1957-58. On August 8, 1955 the USSR satellite program was approved, paving the way for Sputnik 1 some two years later. The USSR had originally planned that its first launch would be a 1,327 kilogram scientific satellite with twelve instruments making many scientific measurements in space, led by Academician Keldysh, but after learning of the US plans for the tiny "grape-fruit" sized *Vanguard,* Kruschev gave his approval for a "Simplest Satellite."

Sputnik 1, in its low Earth orbit burned up in the atmosphere on January 4, 1958, but its orbital decay provided completely new and unique data on the atmospheric drag on satellites, yielding information about the atmospheric density of this region that was hitherto unknown. The pressure and temperature of the spacecraft itself were coded into the beep duration. After one failure the original Keldysh mission, Sputnik 3, was successfully launched on May 15, 1958, but a tape recorder failure prevented it from discovering the radiation belts that were found with the US Explorer satellites.

The Vanguard satellite launch attempt by the US Navy on December 6, 1957 failed spectacularly on the launch pad. On February 5, 1958, the US Navy failed again with Vanguard, but succeeded on March 17. Before that, however, the US Army launched Explorer 1 on January 31, 1958, from which the Van Allen belts were discovered, an aspect of the Earth's space environment that was not only unpredicted, but totally unanticipated by scientists. Space was not simply uncharted; it was unknown, a new cognitive horizon with enormous implications, a highly dangerous region for humans, stimulating a host of researchers to become space scientists. The US Air Force followed by launching a Thor-Able 1 lunar probe on August 17, 1958, but this failed. Once NASA was formed, very soon after, on October 1, 1958, all of these military space projects then came under civilian control.

While Sputnik 1 came as a momentous shock, there was already awareness that the exploration of space was just over the horizon, and in September, 1956, the International Astronautical Federation took up the question of the contamination by space vehicles of the moon and planets at its seventh Congress in Rome. The US National Academy of Sciences considered this important issue in 1957, and acted as the contact point for the US with other national and international scientific bodies. In February 1958 (after the launch of Sputnik 1 and 2) it established a Space Science Board to consider this serious challenge and it, in turn, formally transmitted its recommendations to the International Council of Scientific Unions (ICSU), the body responsible for all international scientific agreements. ICSU took this seriously and established a committee called CETEX (Contamination by Extraterrestrial Exploration) which met on May 12-13, 1958 in The Hague. At its meeting in Washington, October 2-4, 1958, (one day after the formation of NASA) ICSU formed COSPAR, the international Committee on Space Research. The list of attendees included one J.T. Wilson, representing the International Union of Geodesy and Geophysics (IUGG). This was Canada's John Tuzo Wilson, the renowned geophysicist who made incisive contributions to the concept of continental drift long before it became an accepted theory, who was then President of the IUGG. So there was a Canadian presence at this meeting. As an ex-officio member he likely did not vote, but as frequently in international affairs, the birth of COSPAR was witnessed by one of Canada's quiet sons – in this case a very distinguished and highly respected one.

John Tuzo Wilson (Jock to some, Tuzo to others; it's interesting that he began using the name Tuzo in order to avoid confusion with another J.T. Wilson, but there can surely be only one J.T. Wilson who was President of the IUGG) was born on October 24, 1908 in Ottawa, of remarkable parents. His father established civil aviation in Canada and chose the sites for the Montreal and Toronto airports. His mother, born Henrietta Tuzo, was an accomplished mountain climber and after she climbed it, Peak 7 in the Valley of the Ten Peaks near Banff was named Mount Tuzo. Wilson was the first student in Canada to study geophysics, which he did at the University of Toronto. He obtained a second B.A. at Cambridge and then a Ph.D. in geology at Princeton in1936. He worked for the Geological Survey of Canada for a time, with a break during World War II where he became a Colonel. He directed Exercise Musk Ox to test all-weather vehicles, which traveled from Churchill up to the Arctic Ocean and then to Great Bear Lake before returning to Edmonton. (The civilian consultant on the project was Armand Bombardier, who later put this experience to good use in developing Canada's first popular snowmobile and establishing Bombardier Inc. of Montréal, now known for manufacturing aircraft as well as railway and subway cars.)

Wilson's academic career came about through heading up the Geophysics Laboratory within the Department of Physics at the University of Toronto from 1946 to 1974. Early in his professorial career he was asked by the President of the National Research Council of Canada (NRC), C.J. Mackenzie, to chair and re-vitalize the Canadian National Committee (the Associate Committee) for the IUGG, and his active leadership led to his becoming Vice-President of the IUGG at the Rome General Assembly of 1954. The 1957 General Assembly, which would mark the beginning of the International Geophysical Year (IGY), was to have been held in South America, but it turned out to be impossible to carry out the arrangements there. Wilson then organized an invitation from Canada, and hosted the Assembly at the University of Toronto, with the largest attendance in IUGG history, and the first sizable delegation of scientists from the USSR at an international scientific event. He was rewarded by being elected President.

His Presidency carried through the International Geophysical Year (IGY), and he wrote two books about this afterwards: One Chinese Moon and The Year of the New Moons (the latter, of course, referring to artificial satellites). It was following the IGY – and likely using its results – that he developed and championed the theory of plate tectonics. Initially rejected, it introduced the idea of hot spots that produce island chains like Hawaii. He then identified trenches where the plates collide and rifts where they separate. He was the first foreign President of the American Geophysical Union (AGU), and hosted the AGU in Toronto in 1982. He served as Adviser on Earth Science for Expo 1967 in Montreal, and after his retirement from the University of Toronto, became Director of the Ontario Science Centre. He also made contributions to space science as is noted later in this volume.

COSPAR's first meetings

COSPAR held its first Plenary Meeting in London, England on November 14-15, 1958. The records (kindly provided by COSPAR) show that this meeting was attended by Donald Rose, of the National Research Council of Canada (NRC), and that John Chapman also attended as an adviser. Rose's participation was likely because the NRC is the body that represents Canadian scientists on the international scene, but in addition Rose had established himself as a space scientist, had guided Canada through its IGY program, and had a vision for a space science program in Canada. The second Plenary meeting was held at The Hague, March 12-14, 1959, and Rose attended and Rose attended again and he appears in the photo of the group. Rose's report at that meeting was brief and is reproduced below. It shows that Canada Rose's report at that meeting was brief, and is reproduced below. It shows that Canada was already well into space research in 1959.

Attendees at the COSPAR Plenary Meeting in The Hague, March, 1959.
(Front row left to right): H.Massey, M.Roy, E.K.Fedorov, H. Kallman-Bijl
(sec.), H.C. Van de Hulst, R.W.Porter, K. Maeda. *(Second row):* -, -, J. Bartels, D.
Rose, H.E. Newell, A. Wexler, A.C.B. Lovell, F.A. Kitchen, A. Dollfus, *(Third row):* -,
-, A.W. Frutkin, J. Kaplan, -, -, M. Nicolet, M. Florkin, H. Odishaw, M.O. Robins.
(Back row): -, L.G.H. Huxley, -, F. Hewitt, -.

Photo by courtesy of the COSPAR Secretariat.

CANADA

The delegate from Canada, Dr. Rose, discussed the close relation of the United States
Rocket Research Programme, which has been conducted at Ft. Churchill during the IGY, to that
of the present Canadian program, which will be continued. Canada is developing solid propellant rockets (more or less parallel to the British rocket) which are expected to be fired within a
year.

Dr. Rose remarked that tracking facilities were complete and also that an official
Canadian Committee will be formed which will report to COSPAR, and will conduct studies of
the aurora, the ionosphere, and studies of cosmic rays up to about 700 – 800 miles. Canada
has also been engaged in tracking U.S.S.R. satellites by camera and radio.

The First International Space Science Symposium organized by COSPAR was held
January 11-15, 1960 in Nice, France, with presentations of scientific results and their publication afterwards, creating Volume I of Space Research which has now evolved into a journal called *Advances in Space Research*. R. F. (Bob) Chinnick, of the Defence Research Board
(DRB) Canadian Armament Research and Development Establishment (CARDE) in

Valcartier, near Quebec City gave a paper entitled "Measurements on OH and Na in the upper atmosphere", the result of the first Canadian rocket flights. John Chapman gave one entitled "Two high altitude rocket experiments", also built by the Defence Research Board, but by the Defence Research Telecommunications Establishment (DRTE) at Shirley Bay, near Ottawa. Donald Rose did not give a scientific paper, but would have been acting as Canada's national representative. The report of Canada's Associate Committee on Space Research was presented by Rose and is reproduced below.

REPORT OF CANADIAN ACTIVITIES
I. The Canadian National Committee

Early in 1959 the National Research Council of Canada formed an Associate Committee on Space Research. This Committee is organised along parallel lines to associate committees in other scientific disciplines. These committees are national in character and form the connecting link between international unions or ICSU special committees and the activities in that discipline in Canada. The committee is, therefore, the Canadian National Committee on Space Research and coordinates non-defence space activities in Canada.

The membership in the Committee is as follows:

CHAIRMAN
Dr. D.C. Rose — National Research Council

MEMBERS

Dr. J. Auer	- National Research Council
Mr. C.M. Brant	- Department of Transport
Dr. J.H. Chapman	- Defence Research Board
Mr. R.F. Chinnick	- Defence Research Board
Mr. J.W. Cox	- Defence Research Board
Dr. P.A. Forsyth	- University of Saskatchewan
Professor C. Fremont	- Laval University
Dr. G.M. Griffiths	- University of British Columbia
Dr. A.D. MacDonald	- Dalhousie University
Mr. J.A. McCordick	- Department of External Affairs
Mr. P.D. McTaggart-Cowan	- Department of Transport (Meteorology Division)
Dr. D.W.R. McKinley	- National Research Council
Dr. A.D. Misener	- University of Western Ontario
Dr. G.N. Patterson	- University of Toronto
Mr. M.M. Thompson	- Dominion Observatory
Mr. F.R. Thurston	- National Aeronautical Est.
Dr. B.G. Wilson	- University of Alberta
Dr. C. Winkler	- McGill University

SECRETARY

Mr. B.D. Leddy   - National Research Council

An executive and coordinating committee has been formed consisting of :

Dr. D.C. Rose   - Chairman of the Associate Committee
Dr. J.H. Chapman
Dr. J.W. Cox
Dr. D.W.R McKinley
Mr. B.D. Leddy   - Secretary of the Committee

II. The Fort Churchill Rocket Launching Facility

The rocket launching facility at Fort Churchill was established as an IGY activity by the United States on the invitation of the Canadian Government. Though the facility was financed by the United States, and United States personnel operated the facility, the Canadian contribution was considerable in that it was established at a defence base where housing and messing of personnel were already available. Laboratory space was provided by the Canadian Defence Research Board, the required meteorological services were partly provided by the Department of Transport and aurora and ionosphere observations supporting the rocket work were supplied by Canadian teams. The rocket launching facility operated under the jurisdiction of the Canadian Commanding Officer at Fort Churchill. Although the contribution of the United States authorities was great and, in fact, made the much smaller Canadian rocket program possible, the project was considered as a joint effort. Canadian research workers in this field appreciate the advantages in experience and opportunity given us by our United States colleagues in establishing this facility in Northern Canada.

At the close of the IGY the agreement between the Canadian and United States Governments terminated and the launching facility was closed for several months. When the post-IGY program was prepared and the formation of NASA was completed, a long-term program was planned. This is in hand at present. It involves the joint use of the Fort Churchill launching facility by United States and Canadian scientific teams. It is anticipated that the United States program at Fort Churchill will be considerably larger than Canadian activities for some time.

The launching facility at Fort Churchill is rather unique in that it was designed for scientific purposes, only a few launchings have been carried out in the area having a defence objective. The facility includes two launching towers, one for liquid fuel rockets 15 inches in diameter, particularly designed for launching of the Aerobee type of rocket. The second is a launching rail suitable for smaller rockets of the Nike-Cajun type. In addition, a simple rail launcher was built during 1959 by the Canadian Armament Research and Development Establishment for launching at elevation angles up to about 70° a series of 17 inch solid fuel rockets designed as propellant test vehicles (P.T.V.)

The facility is equipped with preparation buildings for the two vertical launchers and connecting corridors. A Doppler velocity and position indicator (DOVAP) is included with appropriate ground stations in an array up to about 10 miles from the launching site. A VHF communication system is installed connecting all stations. A radar tracking system is used, a beacon being carried in the rocket, for range safety control. Telemetering systems for recording signals from the rocket are available.

III I.G.Y. Rocket Firings

Approximately 120 rockets were launched at Fort Churchill during the pre-IGY and IGY period. The objectives covered measurements on pressure, temperature, density, winds, magnetic field, cosmic rays, various types of radiation and electron density in the ionosphere. Canadian scientists took some part in the whole program and the instruments in two IGY Nike-Cajun rockets were completely built by Canadian scientists at the Canadian Armament Research and Development Establishment. These were launched successfully in the autumn of 1958. The objectives will be discussed in Section V of this report.

IV Post-IGY Launchings

A continued program of launchings for scientific purposes has been arranged during the latter months of 1959. Plans are in hand for launchings during 1960 and for an indefinite period thereafter.

V Canadian IGY Launchings at Churchill

The two rockets launched at Churchill carrying Canadian built instruments were designed for studies in the sodium layer, and of radiation in certain infra-red bands in the upper atmosphere. The instruments were carried in Nike-Cajun rockets. They included photo-meters of spectral sensitivity selected to detect sodium and infra-red detectors sensitive to selected bands in the OH emission spectrum. Valuable data were obtained on the height of the sodium layer and the height at which the OH emissions were strong. The details will not be included in this report since they are being presented in a paper by Mr. R.F. Chinnick at the symposium being held at the time of this meeting of COSPAR.

VI Canadian Rocket Experiments during 1959

(a) Ionosophere and Cosmic Ray Measurements

Two Aerobee-Hi rockets were launched at Fort Churchill in September 1959 containing Canadian experiments. The primary experiment in each rocket was for the measurement of electron densities in the absorbing layers in the lower ionosphere, using the well known two frequency technique. Cosmic ray measuring instruments were included and auxiliary instruments included an aspect indicator based on the· direction of the sun and a magnetometer. One of the rockets was launched successfully and good data were obtained up to over 200 Km. The second rocket failed after a few seconds of flight. The details of these experiments will be presented by Dr. J.H. Chapman at this symposium.

(b) Propellent Test Vehicle (PTV)

Research on propellants for large rockets has been carried out for some time at the Canadian Armament Research and Development Establishment. Four propellant test vehicles (PTV) were launched at Fort Churchill in September 1959. Since investigations into rockets and propulsion are not in the terms of reference of COSPAR no details will be presented here except to say that modifications of this vehicle give promise of being a very useful and economical rocket for scientific purposes capable of carrying a payload of 100 Kg to a height of 250 Km.

VII The Canadian Satellite Experiment

Arrangements have been in hand for some time for the construction of instruments in Canada by our Defence Research Telecommunications Establishment to be launched in a polar orbit using a United States satellite launching rocket sometime in 1961. The experiment is in effect an upside down vertical incidence ionosphere sounder (topside sounder). Pulses from the satellite will be transmitted and the return echo from the top of the ionosphere will be observed. Preliminary experiments involve the measurements of cosmic noise in the region of 3 Mc/sec above the ionosphere. Equipment to do this has already been prepared and may be demonstrated at the symposium held at the time of this meeting. Dr. J.H. Chapman is presenting some details of this experiment at the symposium.

VIII Future Plans

It is anticipated that the growth of space research in Canada will be slow. Our United States colleagues will continue to operate the launching facility at Fort Churchill for the next few years and are planning for the launching of a number of rockets from these during 1960. We will contain the close cooperation that was carried out during the IGY.

During the year 1960-61 we expect to launch a few rockets from Fort Churchill in a Canadian program, but using the joint range facilities. The Canadian program will probably include between five and ten rockets, some of which will be the 17 inch solid fuel rockets mentioned above. These will carry instruments to measure electron density, atmospheric density, magnetic field and cosmic rays.

IX Satellite Tracking

A "Minitrack" satellite tracking station is being constructed near St. John's, Newfoundland, that is, near the most easterly of the Canadian coast. This will form one of the networks of Minitrack stations operated by NASA. The station will be established by NASA but its operation will be a Canadian project.

— END OF REPORT—

Space science in the Canadian context

The influence of space science spread rapidly through the Canadian research community. There had been a Canadian Aeronautics Institute since 1954. With the onset of the space age, astronautical societies sprung up in Toronto and Montreal, but eventually all of these bodies joined together in 1961 to become the Canadian Aeronautics and Space Institute under the influence of two men – Gordon Patterson of the University of Toronto Institute for Aerospace Studies, and Phil Lapp (of de Havilland).

Canada was able to be an active participant in COSPAR because of its long earlier history in space research, with a strong interest in the Arctic. A significant part of the image and history of Canada is wrapped up in the north: its Arctic explorers, from Henry Hudson to John Franklin (with fatal results in both cases) to Roald Amundsen's finally successful transit of the North-west Passage in 1903-06.

Although traversing the Northwest Passage was driven by commercial motivation, the early explorers all had a scientific bent, and devoted a significant fraction of their human resource effort to science. They measured the Earth's magnetic field and its variations, and observed its Aurora Borealis (Northern Lights), both closely related to Van Allen's radiation belts whose recognition only came with the space age. This was space science, in a form that was still nascent. These explorers did make important contributions to scientific observation that was little recognized in their day. But they overlooked the fact that the Arctic was already populated, if thinly, by the Inuit people, who had learned how to live in this harsh and unforgiving environment over many thousand of years. Almost everything that can be said about the Arctic environment can be said about space. Both are vast, cold, isolated and unforgiving. Failure to create adequate protection, whether it be by a fur parka, or a space suit, an igloo or a space capsule, leads rapidly to death.

Canada's first serious attack on the space environment came with the Second International Polar Year (IPY-2), 1932-33, in the form of an ambitious campaign to study the Arctic from a number of stations. This highly coordinated activity, conducted during the height of the Great Depression, was remarkable in many ways. The site that produced the most advanced data, pushing at the frontiers of the space age was Chesterfield Inlet, 63.3° N, north of Churchill, Manitoba. The spectacular results that came from Chesterfield were borne of the scientific and entrepreneurial spirit and great ability of its scientific team. The Chesterfield team was fortunate to have Frank Davies, an established explorer as its leader and Balfour Currie, a professor from the University of Saskatchewan, as his co-investigator. Atmospheric science was less structured then, with no hard lines between meteorology and what we now call space science. Davies and Currie became experts in both, and carried with them a panoply of equipment donated by various groups, all wanting their favourite measurements to be made in this unknown region. The data acquired provided for decades of scientific study for future research students and meant that at the time of the first COSPAR plenary session, there were knowledgeable Canadians at the table (although neither Currie nor Davies were present).

There were two other scientific developments that were important to Canada's future role in COSPAR. The first was one brought about by Don Rose himself. At about the time of the Second International Polar Year, there was a solar eclipse on August 31, 1932, near Kingston, Ontario. Rose became involved in the scientific question as to what the impact of this would be on the ionosphere. He acquired an ionosonde and made the measurements, and ended up addressing a fundamental question about the source of the ionization of the layer: his results showed, that what was known as the F-region ionization was a result of ionization by solar ultraviolet photons. Later, at the NRC, he worked on the applied topic of aerial photography, but became interested in the static electricity "flashes" that he found in the images, when the film was advanced to the next frame. This led him to wonder about the sources of static electricity and ionization at these altitudes and this, in turn, led him to cosmic rays—extremely high energy particles from the far reaches of outer space that had been discovered not long before.

The effects of these super-energetic particles striking the Earth's atmosphere were very evident at aircraft altitudes, but could also be seen from mountain tops, or even the sea level. After the Second World War, this led Rose to create a Cosmic Ray section at the NRC. From cosmic rays it was just a small jump to the Van Allen belt particles, one that was easy to make. For that reason Donald Rose would have felt very comfortable at the London COSPAR plenary session in 1958. He later went on to create a Canadian rocket program, as is described in succeeding chapters.

The second element, and in some ways the most decisive, arose from the Second World War. Canada's involvement had many aspects, one of which was the Battle of the Atlantic, the success of which hinged almost entirely on the detection of submarines, and, more generally, communications. Although much pioneering research on the ionosphere had been laid down in England, and Marconi himself had made the first transatlantic radio transmission from Signal Hill, outside St. John's in Newfoundland, the ionospheric paths taken from Canada to the UK, and from ships in the North Atlantic were not well known. In fact, these radio wave paths traversed the Arctic, with the aurora and its associated electrical effects, which were capable of disrupting radio communications, or even causing blackouts (from solar cosmic rays).

The problem became one of determining which radio frequencies would provide the best communications at a given location, time of day, and during solar-terrestrial events, the stuff of the Chesterfield Inlet observations. (It is interesting that Japan also saw the value of ionospheric knowledge and operated 14 ionospheric stations across South Asia during the war. They discovered the Appleton anomaly (near the geomagnetic equator) before it was recognized by Appleton and named for him. After the war the operators were able to get the data home by wrapping it around their bodies. (This information was provided by Atsuhiro Nishida during a talk in Kyoto)).

At the end of the war, it was clear that Canada had to have control of its northern communications, that it was important and, in the context of the Cold War, had security implications. Accordingly, the Defence Research Board with Frank Davies as superintendent of the Defence Research Telecommunications Establishment. Here was a highly experienced Welshman-turned-Canadian who could also sit at the COSPAR table. The story of the fifty years of Canadian space activity that correspond to the first fifty years of COSPAR is one in which Currie, Rose and Davies acted as an unofficial triumvirate that guided the University, NRC and DRB sectors respectively in a highly effective way. Much of what happened during these fifty years, 1958-2008, can be traced back to these individuals.

Donald Charles Rose, photo by courtesy of Ian McDiarmid.

Don Rose's (1960) impressions of the formation of COSPAR were recorded in an article in *Physics in Canada*. They are taken directly from that article.

"COSPAR is a special committee of the I.C.S.U. and was first formed in 1958, its first organization meeting being held in London in November of that year. A constitution had been prepared and was approved tentatively at that meeting. To go into the details of the constitution would be too involved for this report but important features were, that there were members representing national academies of countries taking part in space science activities, and there was representation of International Scientific Unions, all these members being on the same voting level. In scope it covered all scientific activities relating to space including tracking and telemetering of satellites, but excluded the more technological aspects such as propulsion and guidance.

The second meeting was in March 1959. In the interval between the first and second meetings the constitution was to have been considered by I.C.S.U. and the various national academies or national research councils. When the second meeting took place it was apparent that the representatives of the international unions were chosen by accident in such a way that the voting membership included only one Iron Curtain country, namely, the U.S.S.R., while U.K. and U.S.A. had three or four members each. The Soviet delegate objected to the membership clauses in the constitution which made a situation possible that there were 17 voting members of which only one was from the group which launched the first satellite. Most of the members felt that the Soviet delegate had a case, but the proposal he submitted to try and rectify the situation was not acceptable to anyone. A compromise was attempted and a new constitution was proposed which was referred back to the I.C.S.U. and national academies. In the course of the following ten months a great effort was made to find the right constitution of members which would meet everyone's point of view. The effort was successful and a new constitution was finally accepted wherein the membership was in effect open to any national academy (or equivalent) recognized by the I.C.S.U. which took part in any form of space activity or supporting research."

As has been noted, three Canadian institutions played a major role in the space activity that became part of COSPAR. The University of Saskatchewan was one, and its story is developed later. The Defence Research Telecommunications Establishment was another—it grew out of the Second World War and that emergence is also described later. The third was the National Research Council—it had a long and tortured birth that deserves a summary in this introductory chapter. Two of a number of books that describe this history in detail are by Thistle (1966) and by Eggleston (1978).

The National Research Council of Canada

What was later named the National Research Council (NRC) was created in 1916, at a very trying time for Canada. Sir Robert Borden was Prime Minister, the war was going badly, and the Parliament Buildings had just burned down. Nevertheless, the needs of the war, along with Canadian industry in general, prompted Sir George Foster, Minister of Trade and Commerce to put forward the Act creating a nine-members volunteer Advisory Council to the Privy Council. Another stimulus was the creation of a "Department of Scientific and Industrial Research", in Britain that was accompanied by suggestions that it might be expanded to the colonies – this was another stimulus. The first nine members of the Canadian Advisory Council included prominent financiers, industrialists and scientists – names like Walter Murray, President of the University of Saskatchewan and John McLennan, who had created an eminent Department of Physics at the University of Toronto and other distinguished academics from across the country. By 1917 it had developed the concept of Associate Committees. The Associate Committee on Space Research and the Associate Committee on Geodesy and Geophysics of the NRC have already been mentioned. Associate committees then were seen as a way to tackle specific research problems by groups that had essentially no budget, and they were remarkably effective at bringing relevant parties together.

But the first task of the Advisory Council was to determine its future shape. Should it be a centralized laboratory, located in Ottawa, or should it be a suite of laboratories spread across the country? As will be seen, it first became the former, and much later the latter as well. One argument in favour of the centralized model was the recognized need for a Canadian standards laboratory which did not yet exist – Canada was totally dependent on other countries for this. The Ottawa location, however, created the potential of overlap and competition with laboratories in government departments.

A.B. Macallum, a biochemistry professor from the University of Toronto who had already established a research laboratory, was elected the first Advisory Council chair and this decision fell under his leadership. He was given a salary of $10,000 per year; the other members got a per diem. By 1918 they had agreed on a central laboratory in Ottawa to be located at the Experimental Farm. Macallum had a clear vision of the research to be done, and how it should be distinguished from the more specific research tasks to be undertaken in government departments. The request for a half-million dollars to create this laboratory was however, refused by the government, who decided a parliamentary study was required first. This was led by Hume Cronyn, member from London, who did a very thorough job, examining the way research was conducted in many other countries. He also identified the "brain drain" as a problem that was already becoming serious. His report was tabled in 1920, but delayed for consideration by one year because of a change in Prime Ministers. The bill for what was now called the Research Council was passed by the House of Commons but the Senate voted to defer the capital cost section of the bill, killing the construction of the laboratories. Macallum then left the Council to take up a position at McGill University. However, there was now operating money for a General Clerk, two stenographers and a librarian, for a budget of $23,000. From the beginning, Council had from the beginning established scholarships for graduate students, as well as post-doctoral fellowships. There were few takers during the war, but after ten years they had awarded 344 scholarships and fellowships to some 199 students, including Gordon Shrum, who had a major impact on the University of British Columbia, and Donald Rose, who has been introduced already, and of whom more will be written. By using associate committees (who now received grant funding) they had already tackled four major problems: 1) the briquetting of lignite coal in Saskatchewan; 2) wheat rust on the prairies; 3) the deterioration of concrete – also a prairie problem – this was solved by Professor Thorbergur Thorvaldson from the University of Saskatchewan and 4) the discoloration of lobster in the canned lobster industry. While the agency cost of about $100,000 per year, benefits exceeding $30 million per year accrued from its work, a clear success for this fledgling research body.

Meighan was defeated by Mackenzie King in late 1921, so new relationships had to be developed. Henry Marshall Tory, President of the University of Alberta had been appointed to the Research Council. He was an impressive individual who originally intended to enter the clergy, but his skill at mathematics was identified at an early age, and after studying at McGill University in 1892 he managed to find funds to support him for five months at Cambridge University. He returned to a faculty position at McGill, where he became a research colleague of Rutherford, discoverer of the atomic nucleus who would later return to Cambridge to lead its physics research. Tory, who was greatly influenced by Rutherford, later became the first President of the University of Alberta. He was 60 years of age when he was appointed to the Research Council. Assessing the situation, he soon agreed to become its chair. He made rapid connections with the government and found a strong ally in Charles Stewart, Minister of the Interior, whom he had known in Alberta. They reviewed and revised the Research Council bill, which made Tory President rather than Chair, of what was now to be called the National Research Council. They decided that building was premature, so Tory rented space and tackled the magnesite problem. This was a major natural resource in Quebec, but the steel industry had been buying magnesite from Austria because the Quebec product needed some as yet undetermined type of processing to make it useful for the lining of furnaces. In 1929 the NRC started the Canadian Journal of Research.

After Mackenzie King won a majority government in 1926 and James Malcolm from Kincardine became the Minister of Trade and Commerce, fortune began to shine on the Research Council. Malcolm and Stewart supported buying the "Edwards property", a beautiful site on the Ottawa River, now 100 Sussex Drive, with a waterfall producing electric power and some mill buildings that could be put to use immediately. A sub-committee of Cabinet was created after which Tory resigned from the University of Alberta and moved to Ottawa. They put forward a bill in 1928 asking for a $750,000 installment for the building, against a total cost of $3 million, and $300,000 for an annual budget. This bill was supported by the Conservatives, with Hume Cronyn getting some credit. The building was to be 18 meters in height, 135 meters long and 52 meters deep, with a beautiful auditorium and impressive library. Constructed of Wallace marble quarried in Nova Scotia, it became known for a time as the "temple of science." At the same time, Tory began constructing a model for its staff, and was beginning to identify individuals. One of these was Robert William Boyle for physics (also from the University of Alberta) and soon after this Donald Rose's name appeared as a staff member in physics.

The official opening was held in August, 1932 with the Governor General, the Earl of Bessborough (whose names survives with the Bessborough Hotel in Saskatoon) present. Mackenzie King had been defeated in 1930 so he was not there, but it was the inscription that he selected that was on the plaque at the entrance to 100 Sussex Drive where it may be seen today on this graceful structure: "Great is truth and mighty above all things; it endureth and is always strong: it liveth and conquereth for ever more; the more thou searchest, the more shalt thou marvel", from the apocryphal book of Esdras.

The Prime Minister at this time was Richard Bedford (R.B.) Bennett, who had the misfortune to lead Canada into the Great Depression. The costs of the building had been already approved, so it was completed as planned, but the budget paring created havoc with the hiring of new staff, caused salary reductions for existing staff, and led to the "hiring" of non-paid employees. Daily expenses were scrutinized in detail by Henry Herbert Stevens, the new Minister of Trade and Commerce, who even reprimanded Tory for hiring a temporary secretary for one month without consulting with him first.

The relations between Tory and the new government were frosty, but it can be understood that a budget-sensitive government wanted no apparent inconsistency between a "temple of science" and the severe budget constraints running across the entire government, and the country. When Tory's appointment came to the end of its term, he requested a re-appointment, but it was not granted. This individual, who gave so much to the early structure of the NRC left it with much sadness, but he left the foundation of an organization with a future.

Enter Andrew George Latta (Andy) McNaughton, born in 1887 in Moosomin, in what is now Saskatchewan, but was then the North-West Territories. He completed his science education while in the Militia and the 4th Battery of the Canadian Expeditionary Corps, obtaining a B.S. (1912) and a M.S. (1914) degree from McGill University. As a scientist, able to apply his knowledge to artillery, he was rapidly promoted, so that by the end of the war he was head of the Canadian Artillery Corps. In 1920 he enlisted in the Permanent Force, becoming Chief of the General Staff in 1929. He continued to modernize the armed forces; the cathode ray direction finder, the forerunner of radar, was one of his inventions. During the same period he established a place for Canada in civil aviation and created a network of air fields that later enabled the creation of Trans Canada Airlines. But when the second World War broke out, he became commanding officer of the First Canadian Infantry Division. Under McNaughton's direction, the Division grew to be reorganized as a corps (1940), and then as an army (1942). McNaughton's contribution to the development of new techniques is outstanding, especially in the field of detection and weaponry. In 1944 he was appointed Minister of Defence, and later became Canada's Representative to the United Nations' Atomic Energy Commission while also the President of the Atomic Energy Control Board of Canada. At the UN he was Permanent Representative from 1948 to 1949, and a representative of Canada on the Security Council, chairing it several times (a chair was appointed one month at a time). In 1950 he became commissioner of the International Joint Commission and then President of its Canadian section. During this period the agreements of the St. Lawrence seaway and the Columbia River were worked out.

During the depression, he operated the labour camps for the unemployed. This made sense from a military perspective, as it was better to have men housed, fed and kept busy rather than to be fighting riots with the unemployed homeless on the streets. He performed well at this, but there was nevertheless considerable criticism from the opposition at these 20 cents per day "slave camps." Bennett decided that he didn't want McNaughton in this position during his next election, and so offered him the Presidency of the NRC. McNaughton was reluctant, proposing instead C. J. Mackenzie, Dean of Engineering at the University of Saskatchewan, but Bennett was firm, and finally proposed that McNaughton take a secondment, so that he would be able to return to the military if the need arose. On one point they both agreed – that the next war in Europe was not far off.

Although this bizarre approach to his appointment offended a number of people, including many staff at the NRC, he was in fact an excellent choice on his own merit, especially given the timing. With financial backing from Bennett he was able to get the staff deficiencies of the depression corrected and he would immediately begin to include defence research as part of the NRC mandate. When war did come the NRC was a strong organization, and able to meet its new responsibilities. That development led naturally into the space science story that is told in this volume.

Postscript

Canadian scientists had been doing space science for a long time, without realizing that one day this was what it would be called. In their framework they were simply studying the Canadian environment, at and above the Earth's surface. The launch of Sputnik 1, and the formation of COSPAR provided a new context for their work, and they were able to participate from the very beginning.

References:

Eggleston, Wilfrid, National Research in Canada: The NRC 1916-1966. Clarke, Irwin and Company Limited, Toronto. 1978.

Gainor, Chris, Arrows to the Moon, Apogee Books Space Series 19, 2004.

Rigden, John S., Eisenhower, scientists and Sputnik, Physics Today, page 47, June, 2007.

Rose, D.C., Space Research – Internationally and in Canada, Physics in Canada, Vol. 16, No. 2, pp 22-27, 1960.

Thistle, Mel, The Inner Ring: The early history of the National Research Council of Canada. University of Toronto Press, Toronto. 1966.

Wilson J.T. with J.A. Jacobs and R.D. Russell, The Year of the New Moons, Alfred A. Knopf, 1961.

Chapter 2

Post-war Foundations of Space Science

How Canadian space science began in 1925

This chapter is about how space science in Canada grew out of the work done before and during the Second World War. But more than that, it develops the stories of three individuals introduced in Chapter 1. Visionary scientists who laid the foundation for space science in Canada during that time: Donald Rose, born in 1901 in Ontario; Balfour Currie, born in 1902, in Montana; and Frank Davies, born in 1904 in Wales. The lives of Frank Davies and Balfour Currie first became intertwined in 1925. This is how that came about.

Frank Davies was born in 1904 in Merthyr Tydfil, a coal mining town in Wales. He got a Bachelor's degree (Honours) in physics at the University of Aberstwyth and in 1925 he came to Canada on a harvest excursion. There, in the area near Melfort, he worked on a farm as a labourer, waiting for a chance to buy his own land. He had a great time spinning yarns and getting to know the local farmers—but that was quite different from being one. It is said that being chased around the barn by his boss, who was yielding a pitchfork at the time, helped him decide to move to the city. Peter Forsyth claims that this event determined the origins of space science in Canada. Soon after Davies' move to Saskatoon for the winter, the University of Saskatchewan had an unexpected need for a demonstrator in physics and he was asked to fill the post on a temporary basis. During the 1925-26 year he also enrolled as a graduate student. Davies, who had not thought of continuing his studies in physics, was suddenly back in full swing in the career that would not only shape the rest of his life, but would take him as close to the two poles of the Earth as almost any other human had ever gone, as well as to the magnetic equator high in the Andes, and ultimately to assemble the team that would build Canada's first satellite, Alouette I.

In 1926 Davis went to McGill as a research assistant, completing his M.Sc. In 1928 he signed up with Richard Byrd to go to the Antarctic as a magnetician and auroral physicist, and spent the next two years on that expedition. After that he went to work with the Carnegie Institute in Washington for a year. Then, in 1932, he received a call from John Patterson, Director of the Dominion Meteorological Service of Canada, inviting him to lead Canada's expedition to the Arctic for the Second International Polar Year. Heading the team to be located at Chesterfield Inlet; the Carnegie Institute gave him a two-year leave of absence in order to accept this posting. Patterson thought he was perfect for the job. While Davies was in Saskatoon, he met and became close friends with a local farmer's son who had, like himself, been acting as a demonstrator in the physics laboratories. Balfour Currie had subsequently gone to McGill for his Ph.D. where he shared a room with Davies, and was now an instructor back in Saskatoon. Davies accepted the assignment on the condition that he would be able to take Currie along as second-in-command.

Balfour Watson Currie was born in Montana, in 1902, but his family moved to a homestead in Saskatchewan in 1910. Saskatchewan had just become a province, in 1905, and at this time was growing rapidly with a flood of immigrants seeking their own land. In those days it was difficult for rural students to achieve Junior Matriculation (Grade 11), which was sufficient at that time for university entrance. So it was not until four years later that his parents accumulated adequate funds for him to enter the University of Saskatchewan, in 1921, but perhaps this was just as well, as he was only 14 years of age when he finished high school. He intended to take a combined course in mathematics and economics as well as something related to his agricultural background and the marketing of wheat; but high marks in mathematics and physics (and a B in economics) as well as the impression left on him by the new Physics Building, changed his direction. He accepted a demonstratorship in physics in his third year, and took the three available senior courses in Physics, obtaining a B.Sc. degree with High Honours in 1925. Much later, Currie said that he didn't realize until he graduated that he had actually received an Honours Degree in Physics.

With no employment in sight he took a summer job on a farm, enhancing his skills in driving six, eight and even ten-horse teams. While so engaged, Ertle Leslie Harrington, head of the physics department sent him a letter, offering him a half-time job, instructing agriculture students and demonstrating in the physics laboratories. All of this was on the condition that he enroll as an M.Sc. student during his time there, a condition he accepted and this ultimately led to an M.Sc. degree in 1927. A National Research Council bursary then took him to McGill, but he received job offers from both Queen's and Saskatchewan in 1928, and so returned to the University of Saskatchewan as an Instructor in 1928. At the same time he completed his Ph.D requirements for McGill University in 1930, upon with this he became an Assistant Professor. He became a Professor in 1943 and head of the department in 1952. A rising star, he became Dean of Graduate Studies (1959-70), and Vice-President (Research) from 1967-74. As will be seen later, during his career, Balfour Currie established a flourishing Institute for Space and Atmospheric Studies at the University of Saskatchewan and produced a large number of space scientists who served other institutions in Canada and around the world. He provided the academic pillar for space research in Canada.

Davies and Currie at Chesterfield Inlet

A Second International Polar Year (IPY-2) was organized to take place 50 years after IPY-1, during which observations had been taken between August 1, 1882 and September 1, 1883. The idea of International Polar Years was the inspiration of the Austrian explorer and naval officer Lt. Karl Weyprecht who was a scientist and co-commander of the Austro-Hungarian Polar Expedition of 1872-74. One of the twelve Arctic stations during IPY-1 was at Fort Rae, on the northern shores of Great Slave Lake; this was the British station, sometimes shown in documents as "UK (Canada)." The federal government decided that Canada should mount a significant program for IPY-2. The search for staff began in early 1932, only months before the expedition got underway. Some $25,000 still had to be found and there was to be a vote in parliament on the subject in the summer of 1932. Frank Davies was an obvious choice. He was now a scientist on the staff of the Carnegie Institution of Magnetism, and had been recognized by a Special United States Congressional Gold Medal which was presented to him by President Herbert Hoover in 1930. The one condition that Davies set up was that Currie would come along on the expedition. Currie later said that the only reason he was taken along was because he was a farmer's son, a jack-of-all-trades, but Davies even much later insisted that it was because of Currie's scientific background and superior training. Currie hesitated for a moment before accepting the assignment. His mother had cancer and it was not likely that she would survive the year until her son would return. However, the university where he was teaching as an Assistant Professor gave him a two-year leave of absence and an annual salary of $2,500 as an inducement. The money made it possible for him to take care of the expenses connected with his mother's illness. And, as it turned out, $2,500 was to be a great deal of money for a university teacher, whose salary was frozen the year after his return. Because of the difficulties caused by the depression, all unmarried university teachers at the University of Saskatchewan were asked to take a year's absence without pay. Things became difficult enough during the depression to prevent all travel by airplane to the Arctic. When Currie did finally accept, he and the crew for IPY-2 would have to travel by the much riskier land-and-sea route to his destination.

By early May all the participants had arrived at the 315 Bloor Street West offices of the Meteorological Service. The other two team members were Stuart McVeigh and John Rae. Organizing for IPY-2 was led out of this handsome turreted building, which exists today as the Office of Admissions and Awards of the University of Toronto, with a plaque indicating its earlier history. There would not be very much time to allow everyone to learn all the ropes; in only three months they would have to be out in the frozen expanses of the Arctic, even if they were less than fully informed about all the equipment that was sent along with them. The Chesterfield party with all its equipment would have to be in Churchill by early August to connect with the ocean-going government tug which would take them to Chesterfield. Two sets of LaCour magnetometers arrived at the weather office, and one was tried out while the other was packed away at the Agincourt Magnetic Observatory. Størmer auroral cameras (named after Carl Størmer, the Norwegian physicist who had developed them and pioneered in height-finding of the aurora) also came in from overseas, and the Carnegie Institute of Magnetism provided a unit to measure Earth currents at Chesterfield. Two kite outfits were lent by the US weather bureau. These kites were equipped with hand-cranked reels to observe the temperature, humidity and pressure changes at different altitudes. A new invention called the Molchanoff Radiosonde was also provided by the International Polar Year Commission, equipped with radio receivers to make upper atmospheric soundings.

A range finder was borrowed from the University of Toronto to determine the height and movement of clouds. The University of Toronto also lent the expedition a spectrograph once used by Professor McLennan to observe the 557.7 nm green airglow emission, but one that could be used for photographing auroral spectra. The green airglow is a weak emission of light from the upper atmosphere, near 100 km, produced by chemical reactions there. McLennan and his student, Gordon Shrum, had discovered in the laboratory that the source of the green light was atomic oxygen, showing for the first time that atomic oxygen was a constituent of the high atmosphere. It is the same emission as emitted by the aurora, giving it the well-known green colour during normal conditions. This was another early Canadian contribution to space science. Shrum was mentioned earlier as an early NRC scholarship holder, but he went on to far greater things after obtaining his Ph.D. As head of the physics department at UBC from 1938-61, he created a forefront research organization. After his retirement he became head of BC Electric and at about the same time became the first Chancellor of Simon Fraser University.

An electrostatic galvanometer, designed by Patterson in about 1910 was included for making atmospheric electricity measurements. Never used before, Patterson insisted that the Chesterfield team try out his invention. McGill University provided kata thermometers to measure the so called wind-chill factor, simply by exposing one to the atmosphere. More importantly for their comfort, Burberry's, the prestigious English firm, came forward with six complete Shackleton gabardine outfits, snow and windproof, consisting of a blouse, trousers, overalls and head covering.

Crucial to the success of the mission was a small portable radio receiver equipped with batteries and earphones, which was provided by the Meteorological Service. This was to be used to take parallactic photographs at Chesterfield and at its substation to determine the height of the aurora. Once in the field, however, the crew discovered that it was a receiver only, and did not contain a transmitter. Whoever would be at the substation, some 32 km from Chesterfield would be cut off from the outside world, able only to listen to what was happening at Chesterfield. Currie would be left entirely on his own resources.

Chesterfield was a small settlement with a radio station, an Oblate mission (an order that was doing much work in the north), an RCMP detachment and a Hudson's Bay trading post. In the summer the Inuit would gather to deliver their furs to the Hudson's Bay post, then would leave again by the end of the summer to go to their fishing, trapping and hunting grounds. For most of the winter there would only be three RCMP constables, Catholic friars, three sisters in the settlement's hospitals who were never seen, and a population of only a few Inuit who would come and go at the hospital or local mission. The buildings of Chesterfield were all grouped around a shallow bay called Spurrel Harbour, a mile away from the mouth of Chesterfield Inlet, with four little islands in the inlet. At low tide, large stretches of rocks and mud-flats were exposed in the harbour. The polar party had to unload their heavy boxes, originally packed at the meteorological office in Toronto. The odd choice of putting the scientific instruments in well-marked discarded liquor boxes gave the locals on the shore cause for speculation about the true purpose of the expedition. The tug, ironically named the Ocean Eagle, arrived at Chesterfield on August 13, in the middle of a squall. In the rush to unload everyone's supplies for posts as far away as Baker Lake, 280 km to the west, things got hopelessly muddled. At convenient locations along the shore, every able-bodied person was out helping to bring the supplies ashore. The arrival of food and fuel supplies was the most important event of the year. From the Eagle, supplies for the party, the radio station and the RCMP had been towed in on an old vessel with another ironic name, Neophyte. The badly mixed supplies would take a long time to sort out.

The building that was the base for the Chesterfield Inlet operations. The dog team is being prepared to take Currie to the secondary auroral station. Photo by courtesy of the University of Saskatchewan Archives, Department of Physics fonds, photograph A-44G.

The "empty old hut" provided to them that was to be home for the next thirteen months, had two rooms, but a partition soon made that into two large and one small area. There was also a shed, but the total available space in the house and shed was small for all four men, especially considering the chores that had to be done. There was a small staircase leading to the attic, a space barely big enough for an adult to stand up. The attic was where personal belongings were stored and where the party, in shifts of two – two by day and two by night – would sleep. With a warm sleeping bag on top of a spring or a mattress on the floor it was warm enough, though hardly comfortable. The attic was freezing at times and on occasions the four scientists discovered that water had frozen solid in the cold of the night. Developing photographic plates would also be done in the attic. Currie developed the auroral plates; some twelve hundred plates, with six exposures on each, were taken during the course of the year, in addition to two hundred pairs of double station photographs, twenty-two hundred in all. During the winter this developing would be a "bit of a problem", Currie, who was in charge of developing the auroral plates, would recall. Since the temperature in the attic was often near the freezing point, the developer would cool down rapidly. Currie would heat the developer on the coal stove that warmed the large rooms downstairs, then would rush with it to the attic at breakneck speeds, in order to develop his plates with the time of development estimated to be the appropriate one at the temperature of the developer; he eventually became quite good at this process, with impressive results.

The tiny shed also had to serve as a housing for some of the instruments. It was also the only place where the observation balloons could be filled with hydrogen. The two kites (0.9 x 2.4 meters) would be broken if left outside, so they, too, found a home in the shed. The earth-current recorder had to rest on solid foundations and a concrete pier was poured into the ground on which the recorder could be placed; the recording was on photographic paper, so the shed had to be made light-proof as well.

A hut still had to be put up for the magnetometers. There were plans drawn up for this in Toronto, but problems arose once it had to be actually built. The "magnetic" hut had in fact to be non-magnetic, so it contained no iron and had to be far away from disturbances caused by the settlement. It also had to have a solid base provided to hold the magnetometers themselves (to keep them rigidly fixed with respect to the Earth's magnetic field) and the hut itself had to be insulated to protect the delicate instruments from rapid changes in temperature. Currie was the only one with any experience with a saw and hammer, so he was nominated as the foreman of the few carpenters. He recollects struggling with the materials:

"The frame was easy enough, since the roof was flat and there was only one door and one small, window-like opening with a door on hinges that could be opened for sighting on a fixed object required for absolute magnetic observations (i.e. for orientation purposes). Cement was mixed and poured into box-like forms on the rock substrata between floor joists, anchoring six inch bakelite cylinders to hold the aluminum plates on which the instruments would be placed." But after the sheathing was nailed on the framework (with copper nails), the actual insulation of the building became a miserable job. Eel-grass, a sea weed from the Maritimes was supplied for the purpose, and the building had to be covered with 7-10 cm of this material. It contained so much salt and sand from the sea shore that before long Currie became seriously ill from it. Davies became nurse with the only medication he knew about, the two bottles of precious liquor, the entire year's supply of alcohol. Currie recalls: *"I crawled into my bunk and Davies dosed me frequently from one of the two bottles of liquor*

for medicinal use. For most of a day all I recall is apparently floating trance-like in the attic." The treatment was effective and he was back on the job the next day. But he would not live down the incident. *"The claim was made for months and years thereafter that I had consumed half the liquor supplies for the group."*

But if building the magnetic hut was trying, building the auroral sub-station, some 40 km south of Chesterfield, became a veritable adventure. Again Currie was assigned the task of building the station. He made a shelter by digging several feet into the sand and gravel, surrounding the cellar-like hole with a low wall made of peat-like material available on the shore, cut into sod blocks and pinned together by two foot lengths of wood split from the boards especially imported for the purpose. The roof was made of two by fours, covered with boards, and these in turn insulated with more sod blocks. In a word, it was a real burrow, made of mud and muskeg. In the sub-zero temperatures encountered during winter, the muskeg blocks would freeze dry, and even a small bench made of this strange material would prove suitable to sleep on, provided one was equipped with a warm sleeping bag. Inside it was just big enough for a bench made of sod, a ten gallon tank for fuel for the lamp and the few supplies which the observer brought with him.

Balfour Currie in front of the outlying auroral station, named Fort Sik-Sik, in summer.
Photo by courtesy of the University of Saskatchewan Archives,
Department of Physics fonds, photograph A-45G

The shelter was soon built and it did truly resemble the burrow of some gigantic ground squirrel. Hence the name Fort Sik-Sik, Sik-Sik being the onomatopaeic Inuit word for the Franklin squirrel which was abundant in the region. The entire area of the enclosure was only 2 x 3 meters. A lamp was used to heat the shelter, melt ice for tea, cook bannock and heat other food brought from the main station. The entrance, a panel 1.2 by 0.9 meters, could be pulled in behind one after climbing inside and fitted into a framework embedded in the earth-wall. Getting to the site with the help of "Foxy" Brown, a trapper who was a local character hired for the purpose, was an ordeal. Currie made five separate trips to Fort Sik-Sik by dog team and would spend nearly a month altogether in the solitude of the earth-warren taking the necessary photographs. Years later he would still remember living for a week to ten days alone in the solitude of the Arctic wasteland: *"It was an experience not easy to forget - living for a week or more each time in a dungeon-like enclosure where temperatures were always below zero Celsius. The nearest settlement, Chesterfield, was some 40 km away. Sounds from Arctic hares occasionally scampering across the roof were easy to imagine as the arrival of a polar bear. The odd bear was known to wander along the coast in search of seals at the open water between the land fast ice and the ice sheet covering Hudson Bay. As a safety measure the RCMP had loaned me a Ross rifle* [the ineffective rifle used by the Canadian army during World War I], *and cartridges. In retrospect, I wonder what use it would have been to me if a bear had come and broken into the enclosure. Seen at a distance, a few shots would have scared them away."*

Auroral height measurements were taken using two-station parallactic auroral photography. The principle behind this was similar to calculating the altitude of a triangle in geometry. Knowing the length of the base of a triangle (the distance between the two stations) and the angles of the triangle it is possible to determine the altitude, the location of the highest point in the triangle, By picking out the star formation with an aurora before it, and by taking pictures simultaneously at two locations, knowing the positions of the stars, they were able to determine the angles of the triangle. The photographs had to be taken simultaneously, and against the same star background. Had there been a direct telephone line between Chesterfield and Fort Sik-Sik, there would have been no problem. But the only method available was to use the radio transmitter at Chesterfield. The equipment and the personnel of the station had been put at the disposal of the scientists by the Department of Marine. The camera station at Chesterfield was connected by telephone to the radio station, a fairly short telephone line that was laid down by the polar party. The person recording the information, usually McVeigh, would go to the radio station. Davies was in charge of the Chesterfield Station. He would pick an auroral formation, determine the name of the constellation against which it appeared in the sky, then telephone McVeigh at the radio station. Currie, it would be assumed, was outside of the hut at Fort Sik-Sik, ready to point his camera at the same auroral formation against the same star background. A supply of plates would be in his hand, and the marine radio donated to the group was tuned to Chesterfield. After selecting the appropriate auroral formation, Davies would tell McVeigh at the radio station, McVeigh would then relay this information to Currie. After a minute or so of waiting, to give Currie time to get organized, Davies would call "On", McVeigh then would call this out on the radio to Currie. This was heard simultaneously by both cameramen, the first one heard it via the telephone line, the second, on the air. Then Davies would call off the shot, saying "Off" and McVeigh would repeat the message, again both over the telephone and over the air. Meanwhile McVeigh recorded the auroral formation, constellation, times, etc. in the logbook. This ingenious method proved quite successful except for the rapidly moving types of aurora. Davies and Currie became so adept at this method that they took over 1000 pairs of photographs in this way, of which 200 were finally selected for auroral height studies, mostly because the auroral forms were pictured against well known constellations.

Much later Currie discovered that the photographing sessions became quite a popular radio show in the neighbouring settlements, giving rise to endless speculations. Currie recalls: *"A person, searching for programs on his radio, could pick up our broadcasts—names incorrectly pronounced of stars interspersed with ON's and OFF's. Ballpoint pens were many years in the future. McVeigh used a bottle of ink and a pen with a steel nib. Occasionally, ink on the nib would be used up or fail to run properly. In the rush to record data, there would come over the radio a piece of King's English....... all this would continue for hours, only interrupted when the aurora became weak or we had to stop long enough to remove the exposed plates from the plate holder and replace them with unexposed ones."*

During the periods that Fort Sik-Sik was unattended, single-station photographs were taken from Chesterfield. Regular visual observations of the aurora began early in September. This meant that every quarter-hour Davies and Currie would check for aurora. If there was an auroral display, they would take "single station photographs" on glass plates holding six exposures each.

McVeigh was in charge of the meteorological observations. Four times a day the team recorded observations from the meteorological instruments: atmospheric pressure, air temperature, humidity, rainfall, sunshine, solar radiation, wind direction and velocity, earth temperatures and minimum temperatures on the glass were observed at 1 am, 7 am, 1 pm and 7 pm. Hourly observations of the cloud and weather were taken during the International Cloud Months of October, 1932, and January and July, 1933. Kites and balloons were flown several times a day for the purpose of studying the conditions of the atmosphere. Pilot balloon flights were taken twice a day throughout the year, except when double station auroral photography interfered with the regular routine. During the three International Cloud Months, three observations by balloon were taken each day. Altogether, the polar party flew 496 balloons, the longest of which was in the air for more than 138 minutes. A theodolite was set up near the hut for observing the rate of ascent of the balloons. This was a telescope fitted with spirit levels and scales to measure vertical angles in order to measure altitude and azimuth, giving the position of the balloon. It was a common surveying instrument, but in the cold weather, it would inevitably misbehave and by December of 1932, very few flights were taken. Then it was discovered that if they took out the grease that kept the theodolite lubricated, replaced it with kerosene and then left the instrument continuously out of doors, it would not freeze. The details of the equipment and observations are available at the University of Saskatchewan archives (2002).

While the atmospheric potential measurements did not get started until December 1, 1933, the earth current potential reading was begun by the beginning of November. Electrodes for earth currents had to be located and this was a difficult task in the rocky ground. There were two pairs of electrodes, with those for each pair about one kilometre apart. One pair was in a north-south direction, the other east-west. Each electrode consisted of one hundred feet of 6 mm lead wire, arranged in a grid of 1.2 by 1.8 meters and buried 90 cm below the surface.

There were some fluctuations in declination (the difference between the magnetic field direction and true north) of as much as ten degrees. When, much later, the results were analyzed, correlations between the magnetic measurements and the earth current flows became extremely striking. The earth current observations showed that the direction of these so-called telluric currents changed in accordance with the magnetic fluctuations, and were possibly caused by them. Earth currents changed in direction during the day; in the morning and evening, they tended to run in a north-south direction and at noon in an east-west direction. Currie also noted a very high correlation between magnetic field differences, earth-currents and auroral observations, hinting at some common mechanism for all three. Today the significance of this is known, and how all three fit into the current picture of the magnetosphere. Scientifically, the expedition was successful beyond anyone's expectations.

Balfour Currie measuring the atmospheric potential gradient. Photo by courtesy of the University of Saskatchewan Archives, Department of Physics fonds, photograph A-22G

As for the auroral observations themselves, they noticed that the auroral activity at Chesterfield peaked at midnight; later Frank Davies compared the results obtained at Chesterfield with the results he had obtained at Little America in the Antarctic. In Little America the auroral activity peaked at six o'clock in the evening. Davies was never able to explain the difference, but we know now this was the first inkling of the global configuration of the aurora, later called the auroral oval. Because Little America was two degrees higher in magnetic latitude than Chesterfield, it passed under the oval ring of aurora much earlier than Chesterfield.

Currie had also measured the tilt (from an east-west direction) of auroral arcs from his photographs and found the tilt was greater in the early evening at Chesterfield than it was at midnight. When he repeated the measurements later at Saskatoon and failed to find the same effect, his scientific scepticism held him back from drawing any conclusions. It is now known that the tilt (oval) effect is more pronounced at high latitudes and that Currie and Davies were very close to discovering the auroral oval in 1932, long before Feldstein achieved that much later in 1968, using more extensive IGY data. Spectrograms were taken with great frequency. They obtained too much data to analyze on the spot, or to modify their procedures. Rather, they maintained the same observing pattern throughout the year so as to generate a consistent data base. Currie found that certain emissions did not go away when the aurora went away. This phenomenon had already been discovered by Anders Ångström of Sweden in 1868 and was called airglow. As noted earlier, the auroral green line also appears in the night airglow (also called the nightglow), where it is excited by energy released in the recombination of atomic oxygen atoms to form molecules. In aurora the same excited state of atomic oxygen is responsible, but the energy source is energetic electrons from the magnetosphere, bombarding the atmosphere. Currie built up a fine auroral and airglow spectral collection from which they found emissions that occurred only in the aurora, or only in the airglow, or both. They also found the airglow greenline to be weaker at Chesterfield than at Toronto.

Currie and Davies found a yellow emission that had puzzled some earlier observers. They noted that it was a feature of airglow and not aurora, and that it was especially bright just after sunset, as though it were in some way enhanced by sunlight. This feature was in the 589 nm region of the spectrum, but their spectrograph had insufficient wavelength precision for an accurate determination. It had been designed for maximum sensitivity to weak emissions, at the cost of reduced wavelength precision. Currie and Davies knew of the prominent sodium emissions at 589 and 589.6 nm, but again, their scientific honesty precluded their jumping to a premature conclusion: that sodium exists high in the atmosphere. And it was a rather crazy conclusion anyway - how could that much sodium get up there? It was as if nature had provided weak but permanent sodium streetlight illumination. Later, using an instrument of higher resolution, the French scientist R. Bernard identified the emission as sodium in 1939 and the IPY-2 polar party did not get the credit. Still, Canadians can be satisfied that Balfour Currie and Canadian Frank Davies observed it first at Chesterfield.

While Currie and Davies did not reach a full understanding of any of the phenomena they grappled with, they laid the foundations for auroral research that would continue in Canada for many decades. First of all, they provided a wealth of data for current and future study. But more important, their own experience, which could have been gained in no other way, put them in a unique position in Canada. Auroral physicists were oddities in 1932. But during and after World War II, when auroral and ionospheric scientists were suddenly needed, and space science was at hand, these men had the experience to create whole areas of activity in Canadian space science, which would carve out institutions and set the pattern of space science until at least 1980.

Donald Charles Rose

In Chapter 1, Donald Rose was introduced as the Canadian participant in the earliest COSPAR meetings. While Davies and Currie were at Chesterfield Inlet, Rose was engaged in making ionospheric measurements during a solar eclipse, also a space science topic. His life becomes meshed with that of Currie and Davies only after World War II.

As a young scientist, Don Rose had been seconded from the National Research

Council of Canada to the Department of Defence as a scientific advisor. His role as a scientist during World War II greatly influenced his post-war research and ultimately his leadership in Canadian space science. On the advice of General McNaughton, commander-in-chief of the Canadian Armed Forces, Don Rose became Scientific Advisor to the General Chief of Staff. As noted earlier, General McNaughton was head of the National Research Council between 1935 and 1939, and recognised early the significant role science and scientists would play in deciding the outcome of the Second World War.

Rose was born in 1901 of a cultured, established southern Ontario family. His father encouraged him towards study and a professional life and after obtaining a Bachelor's and a Master's degree in physics at Queen's University, young Rose went to obtain his doctorate in Cambridge, England on an 1851 Exhibition Science Scholarship under the illustrious Ernest Rutherford. While still at McGill, Rutherford also had a great influence on Henry Marshall Tory, the first president of the NRC. Don Rose was an example of the old-style scientist who rubbed shoulders with the best in the field. Born in 1871, Ernest Rutherford was a New Zealander who had earlier held an 1851 Scholarship at Cambridge, working under J. J. Thomson, then taking a faculty position at McGill for nine years, before returning to England, first at Manchester, then at Cambridge again. Of his many accomplishments he is famous for demonstrating that almost all the mass of an atom is concentrated in a tiny region at its center, the nucleus.

Rose was a shy, quiet, unassuming man who worked alone and quietly behind the scenes for what he thought to be the best policy for Canadian science to follow. He retained his timidity to the end of an illustrious career which saw him establish, almost single-handedly, an active scientific rocket research program in Canada and during which he laid the foundation for Canada to become involved in space research at the precise historical moment that made its future development possible. As described in Chapter 1, he ensured that Canada was involved in COSPAR from the very beginning.

A clear-eyed, clear-headed, supremely rational individual, Don Rose was a rare man of his time who saw his country as a potentially powerful nation with much to contribute, if only it could climb out of the morass of its provincialism. All his life he rejected the colonial attitudes that many of his countrymen had adopted: Rose was determined to conduct in Canada the scientific work that was both a responsibility and an opportunity for Canadians. He saw that Canada had capability to do space and upper atmospheric research, and research on the Earth's magnetic field, even while the United States set up a rocket research range at Churchill, Manitoba during the International Geophysical Year and brought in its own rockets to do this. The phenomena the American rockets explored were unique and occurred only in a few places on earth, and Canada had a natural advantage for exploring them. Rose, who was chairman of the Canadian Committee for the IGY, chafed that Canada had only a supporting role in the spectacular research program that took place at the Churchill rocket range. Working in the background without much fanfare and much ado, he quietly worked to convince government departments to support the development of Canadian sounding rockets, and also to support the type of expensive research these rockets made possible, so that Canadian scientists could live up to the obligations Canada had in studying what were both Canadian and global scientific problems. Through quiet persuasion Rose managed to obtain additional funds for university scientists to do research using rockets, and stimulated the establishment of serious rocket research of the upper atmosphere and the near-earth space environment. But

Rose was no empire builder and sought little credit for himself. As quietly as he had begun and operated it, he left the operations of the rocket program to others, once he was convinced that it was capable of surviving without him, and quietly stepped aside.

A recent photo of the National Research Council building (the "Temple of Science") at 100 Sussex Drive, Ottawa.

Rose was the first physicist hired by the NRC in 1930, at a time when the research council lacked even a basic laboratory. Henry Marshall Tory had resigned as President of the University of Alberta to become President of the NRC in 1928. One of his first tasks was to oversee the design and construction of the beautiful building to be located at 100 Sussex Drive, then often called "the temple of science." Major appointments began in 1929 including Robert Boyle as Director of the Department of Physics and Engineering Physics. Unlike most of his peers, Rose took a vivid interest in the fledgling National Research Council, but there were no positions there when he returned from England, and he spent the 1929-30 school year teaching at Queen's University. Boyle, who had met Rose in England and thought highly of him, appointed him in 1930 at a step above the minimum salary and later consulted him on new appointments. Rose had studied engineering physics at Queen's, and his scientific horizons had been stretched wide during the time he was at the Cavendish Laboratory getting his Ph.D., and in two further postdoctoral years at the University of Bristol. On his hiring, Boyle told him that he would have time to pursue his own interests, but otherwise he would have to study the problems that were sent to the division from the outside. These were unusual, often applied problems, and in studying them, Rose developed a wide spectrum of skills and interests that would stand him in good stead and would influence his development as a scientist in later years. The very first problem that Rose had to tackle was the development of a special light for the grading of grain. Later problems had similar practical importance.

During the 1930s, Rose studied a range of problems. Two of the most significant for his future development were 1) the study of atmospheric electricity and 2) the study of ballistics. (Even before the war, the NRC did some of the scientific research and development work that was referred to it by the military, but this was reorganised after the war, as is seen later.) A large program was under way at the NRC in the 1930s to map the vast uncharted territories of Canada by aerial photography. The photographs were taken from high-flying planes, in a rarefied atmosphere, and as the aircraft were not pressurised, a problem developed with the camera. As the film moved forward in the camera, static electricity built up and eventually a spark would jump across the film, making ugly streaks. In studying this problem, Rose learned a lot about atmospheric electricity. He wondered just how air at these high altitudes became electrified. Many years later this experience would prompt him to study the cosmic ray sources of atmospheric electricity. Cosmic rays proved to be his scientific life-long interest, and eventually he became the cosmic ray expert in Canada. The other line of research Rose pursued that proved to be significant for his later life as well as for Canada was the study of ballistics. Studying the trajectories and speed of gun projectiles, and a research project into the manufacture of explosives piqued Rose's future interest in ballistics, propellants and rocketry, an interest that was rekindled during the war years and in the years following the war, especially during IGY when Rose saw the potential of rocket power, both as a weapon and as a possible research tool.

General Andy McNaughton was always on the lookout for improvements in the equipment of the Canadian army, and purchased a new German high-speed sports rifle. It was to determine the speed of the bullets from the German gun that McNaughton had asked Rose over to DND headquarters in 1932; this visit resulted in the establishment of a ballistics laboratory in the sub-basement of the NRC building at 100 Sussex Drive that included a short rifle range. Rose decided that the only way to measure the speed of the bullets accurately was to use four photodetectors down the length of the range, and as the bullet passed each detector, it interrupted the circuits in turn. This was a new technique, full of difficulties, but one that was later perfected by Rose, who developed a portable version of this system that could be taken out into the field by the army. The velocity of 8 kilogram field gun shells was measured with an accuracy of about 0.6 meters per second by this technique. This must have delighted Gen. McNaughton, by this time (1937) President of the NRC, whose experiences in the war of 1914-18 made him acutely aware of the military value of such precision.

Henderson and Rose's 1932 ionospheric observations

In the early 1930s the United Kingdom was still a world leader in scientific research, and since Canadian universities offered few research opportunities, it was natural for Canadian scientists to further their education in England. Don Rose had gone to Cambridge in 1925 (the year Frank Davies arrived in Saskatoon), and in 1932 John T. Henderson of the National Research Council was at the newly formed Radio Research Station at Slough, created by Sir Edward Appleton, the pioneer of British ionospheric research, and co-discoverer of the ionosphere. (He received a Nobel Prize for this in 1947.) Henderson was a McGill graduate who had gone to the UK to work with Appleton, first in London and then at the Radio Research Station in Slough on a Province of Quebec Bursary. Following this he went to the

Sorbonne and Munich. Shortly before Henderson's return to Canada with his Ph.D., Appleton recognized that there was to be a total eclipse of the sun in Canada on August 31, 1932, and that this would provide a golden opportunity to determine which of two existing theories for the formation of the ionosphere's E-layer was correct. So with Appleton's guidance and support, Henderson began to prepare equipment that would be brought back to Canada with him for the experiment. This fitted well with the plans of the NRC, which had opened a Radio Section in 1931, and so Henderson was positioned with this group.

The dominant part of the ionosphere is the F-region (then referred to by Henderson as the Appleton layer), extending from roughly 150 km to 400 km. It was believed that it was produced by solar ultraviolet radiation. Below this, at about 100 km, there was another weaker layer, which required for its production more energetic particles than for the F-region, since these source particles would have to penetrate deeper into the atmosphere. Henderson called this the Kennelly-Heaviside layer, but it is now called the E-region. Appleton had proposed that these E-region producing particles were simply solar ultraviolet photons, of still shorter wavelength and thus greater energy—approaching the X-ray part of the spectrum. Sidney Chapman, on the other hand, who was much concerned with auroral phenomena, proposed that the source was charged particles shot out from the sun. Being charged, these particles would be guided along trajectories determined by the Earth's magnetic field, and so would approach the earth from a direction different from that of the ultraviolet light, which travels in a straight line.

This was the basis of the method. The "ultraviolet eclipse" would occur along a track on lines directly connecting the sun to the earth while the "particle eclipse" could occur on a different trajectory, which had been calculated by Chapman. For the 1932 eclipse, Vankleek Hill, halfway between Ottawa and Montreal, was selected as a site on the "ultraviolet eclipse" path at the 100 km level, and Corner Brook, Newfoundland was chosen as a site near the "particle eclipse" path. The plan, which was financed by the NRC, was to set up ionosondes at the two stations and then see whether the actual eclipse, as seen in the disappearance of the E-region, corresponded better in time and space to the hypothetical ultraviolet eclipse, or the hypothetical particle eclipse. Henderson did the preparatory work in England, with the assistance of R.A. Watson Watt (later the inventor of radar), and then brought it to McGill University for preliminary tests and observations. The equipment was then moved to Vankleek Hill and Corner Brook, where measurements were made from August 28 to September 3. W.B. Ross and J.C. Stadler went to Corner Brook while R.H. Smyth and R.H. Prissick assisted Henderson at Vankleek Hill. The result for the eclipse day of August 31 shows a sharp bite-out of the critical frequency versus time curve, with a minimum coinciding precisely with the time of optical (and thus ultraviolet) totality. At Corner Brook the data obtained were not as high quality, but no evidence of a "particle eclipse" was seen. Thus Henderson was able to confirm the hypothesis of his mentor, and the E-region source particles had been identified. It was an auspicious beginning for ionospheric research in Canada, which was later to grow to a dominant role in Canadian space science. His paper was published in the new NRC *Canadian Journal of Research* in 1933.

The NRC provided support for another ionosonde set up near Kingston, which was the responsibility of Don Rose. While Vankleek was under the E-layer at the time of totality, a point under the F-region at the same instant would be displaced on earth because of the angle the sunlight made with the earth at that particular time. Don Rose's intention was to study eclipse effects on the F-region and so he set up his transmitter at Queen's University and the receiver at the Royal Military College (RMC). His ionosonde construction was done independently of Henderson but it was also based on the pulsed method of the Americans, Breit and Tuve. The transmitter was borrowed from the Royal Canadian Signals and the aerial used with it belonged to the Queen's University broadcasting station. The receiver was a purchased unit (but modified) and Rose hung an aerial between two buildings at RMC. The originally intended method of sweeping the cathode ray tube – a method recommended by Appleton – did not work, and at the last minute they improvised a method of using the line voltage directly to sweep the CRT horizontally, which was then automatically synchronized to the transmitter pulses, generated by a mechanical keying unit driven by a synchronous motor.

A strong eclipse effect was also observed in the F-region, with about a 30% reduction in ionization. As the F-region is more sluggish than the E-region and thus has a greater capacity to maintain itself, this result was important. It also coincided with the optical eclipse, confirming the ultraviolet source for the F-region. This was the expected result and so Rose's results were not so dramatic as Henderson's. The most significant observation that Rose made, from the perspective of space science, was more in exasperation than in satisfaction. His attempts to establish baseline values for the F-region ionization in the few days prior to the eclipse were frustrated by the occurrence of a magnetic storm, which severely perturbed the ionization. This major effect, noted clearly by Rose (1933) in his paper in the *Canadian Journal of Research*, immediately following Henderson's, was later to become a major theme in Canadian studies of space. Following this paper there was a third, authored by Henderson and Rose together, describing the impact on commercial radio communications during the eclipse. All together, these three papers occupied the first 36 pages of Volume 8, Issue No. 1, January, 1933 of this new NRC journal.

From 1933 onwards, much of the work at the Radio Section at NRC was directed towards the location of electrical disturbances by radio direction finding. McNaughton's invention of a cathode-ray tube direction finder was pursued by Henderson (see the next section) who developed this for use between 1936 and 1938, and then installed it in Nova Scotia. The same year, McNaughton was consulting with DND about electrical detection of aircraft. Radio noise bursts generated by lightning and other atmospheric electrical sources called "atmospherics" were commonly detected and it was thought that radio direction finding would thus provide a useful tool for detecting low pressure weather centres by radio means. Henderson led the research in this area, and after initially encouraging results, found that intervening interference from minor storms made the method unreliable. In 1938, C.W. (Bill) McLeish, under Henderson's direction and with an $800 budget, built and operated a 1 kilowatt pulse ionosonde that produced continuous film recording at a number of fixed frequencies. This made automatic and untended operation possible. One of the phenomena they detected with this system was E-layer reflections during auroral disturbances. They then planned to operate the system at Coppermine, near the North magnetic pole, but the project was cancelled in the spring of 1939 when all the effort of the Radio Section was turned to radio navigation and location problems.

The NRC during World War II and its impact on space science

Although McNaughton had assumed the Presidency of the NRC with promises from Bennett for support, the support did not materialize immediately under Mackenzie King, who had come to power in 1935 with a very frugal approach to government. Nevertheless, McNaughton was able to gradually win support and his budget of $510,000 in 1936 had risen to $900,000 by 1939. Although he had inherited a large building (the "temple of science") with few staff, but by 1939 100 Sussex Drive was becoming overcrowded. During the pre-war years McNaughton had carefully prepared the way, with the following initiatives. John T. Henderson was pursuing radar investigations; Sir Frederick Banting, known as a co-discoverer of insulin had laid the foundations for aviation medicine; Donald Rose had developed a technique for the measurement of muzzle velocities of weapons; George C. Laurence had established a radiography laboratory and L.E. Howlett had established expertise in optics.

When war broke out and McNaughton was selected as Commander of the Expeditionary Force, he suggested that Chalmers Jack Mackenzie be named Acting President during his absence. Mackenzie was a graduate of Dalhousie and Harvard who had served during World War I and subsequently became a Professor and then Dean of Engineering (in 1921) at the University of Saskatchewan. He inherited at the NRC 300 staff and a budget of $900,000. The new laboratories on Montreal Road had been designed, with the first priority being wind tunnels. Realizing that it was far behind Germany in military power, the UK was well advanced in war research, recognizing that the only way to close the gap was through technological development. At first reluctant to draw on the scientific capability available in the US and in Canada, as war conditions worsened and the research environment deteriorated, the necessity for this became clear for the English war effort, and Sir Henry Tizard came to North America on a fact-finding mission.

At this time Henderson was head of the Radio Section, with one assistant. He seemed most able to appreciate the research activities in the UK and so was sent there for a tour. Here he was shown the secret British Aerial Defence System, based on radar. Henderson returned and submitted a radar plan to the Canadian government – it was turned down. So were subsequent plans. But in October, 1939 things began to move when two army personnel were added to his group. In November there were four NRC and four army staff, by June, 1940 he had a staff of 23. The NRC acquired 20 acres on Metcalfe Road for a radar facility, and in November the first Canadian radar echoes were obtained. Sir Charles Wright appealed to McNaughton to send Canadian graduates over to England to work on radar for the Navy, because as Wright said, the Air Force was soaking up all the available people. Peter Forsyth from the University of Saskatchewan was one of those who were sent over.

In April, 1940 some concerned citizens, including Sir John Eaton, put up $1 million to assist NRC research. The UK had invented the cavity magnetron, which operated at high frequencies, bringing the radar wavelength from 1.5 meters down to 10 cm. This improved the accuracy and target information and allowed radar to be carried on board aircraft. The problem now was how to mass produce the device and this became Canada's role. A crown corporation, Research Enterprises Limited (REL), to be located in Toronto, was formed to manufacture radar and optics. By July, 1940 the Montreal Road laboratories opened; there were now 500 people at the NRC, and Henderson had a group of 80. During the war, Mackenzie kept McNaughton fully informed but one particularly long letter he gave to Banting to carry with him on his flight overseas in February, 1941, never arrived. Sadly, Banting's crashed near the Canadian coast and he died soon after, a great loss to Canada. By this time, 1941, the NRC budget was $5-6 million and there were 1000 employees. Henderson's group produced the microwave mass-produced radars, for $100,000 each, requiring two trucks to transport one complete system. REL made 665 of these, the G.L. Mark III C, and another 4500 radars for aircraft, for attack purposes. REL had orders totaling $26 million for the G.L. and $15 million for the aircraft radars. Henderson now had 160 people (later 300) and a budget of $1 million.

McNaughton very much wanted a higher velocity projectile for anti-aircraft guns, and Rose did the muzzle velocity measurements for a Bofors 40 mm weapon. He was now considered Canada's "fire control expert" and in July, 1942 the NRC transferred him to the Proving Grounds at Valcartier, near Quebec City. One year later Rose became scientific advisor to Army Headquarters in Ottawa.

By 1944 it was clear the war was coming to an end and Mackenzie began to think of NRC's future after the war. A Committee on Research for Defence was appointed by the government, but it met only once. Some considered making defence a Division of NRC, but Mackenzie resisted this. He was proud of what had been accomplished during the war, but wanted to consolidate research at the international forefront. According to him the NRC "had emerged from the war, not the struggling infant it was in 1939, but a strong, experienced scientific establishment that had won international recognition as a competent and effective organization." In 1945 the NRC had a fair-sized budget of $5.2 million, and employed 1,400 persons; by 1950 this had expanded to $12.9 million and 3,000 staff. The emphasis on radar continued with the demands of the Korean war and the formation of NATO and NORAD.

Early in 1952, Mackenzie became President of Atomic Energy of Canada Limited, a new enterprise that had grown out of Laurence's work on nuclear fission that began in 1940. Mackenzie was replaced at the NRC by Edgar William Richard Steacie, formerly Vice-President (Scientific). He stressed fundamental research, and his first challenge was to increase university funding. This ran head-on into the new Diefenbaker government, which was now operating with austerity in mind. The 1957 launch of Sputnik 1, however, initiated a change of thought. Among all that flowed from this momentous event was a recognition that science in the West was not as advanced as it should be, and this stimulated the increased funding that Steacie had been advocating. New areas of fundamental research opened up, including cosmic rays studies, led by Rose. His group built a cosmic ray laboratory – a neutron monitor in Ottawa; then another in Resolute Bay at 74° latitude, on Cornwallis Island in the Arctic, and then another on the top of Sulphur Mountain in Banff National Park. As chair of the Canadian IGY committee, Rose's leadership won him praise from many quarters. On May 13, 1959, Rose sent a memo to Steacie, recommending the creation of an Associate Committee on Space Research; in this he predicted that Canada would be launching 20 rockets per year. In 1962, Steacie died in office and was replaced by B.G. (Guy) Ballard. From its uncertain origins, the NRC had become a formidable research organization, widely respected both inside and outside Canada, through the wise guidance of its presidents–Tory, McNaughton, Mackenzie and Steacie.

Rose's Wartime Years

During the war the NRC was the major military research body, its scientists and laboratories made equally available for the support of the war effort. In 1941, the NRC became the scientific research and development establishment of the Royal Canadian Navy. Rose was first appointed liaison officer, and later the Deputy Director of Scientific Research. By this time he was the head of NRC's Physics Division. In his new role he was responsible for the research being done at the new Scientific Research and Development Establishment of the Royal Canadian Navy on both coasts as well as in Ottawa. In 1943, he was appointed Scientific Advisor to the Chiefs of General Staff (Army), seconded from NRC for this duty. In August, 1945, still on loan from the research council, he was appointed Chief Superintendent of the Army Research Establishment, which was later to become the Canadian Armament Research and Development Establishment (CARDE) at Valcartier, Quebec.

During the war years Rose continued his work in ballistics, becoming the acknowledged expert in Canada. As to the secret work being done in the RCN's Scientific Research and Development Establishment, the correspondence between General McNaughton and C.J. McKenzie, the director of the RCN's Scientific Research and Development Establishment (SRDE), and Rose's immediate superior, contains frequent mention of Rose's "good work", and "speedy progress", although the nature of this work is not revealed. The appointment of Rose as the Scientific Advisor to the Chief of Staff reflected the sentiments of the Allied leaders in general, and of General McNaughton in particular, that the only way the Allies would win the war was by making major scientific advances and using such discoveries to military advantage. Rose was often sent to London on missions during the war and there he could see for himself the awesome advantage the V-2 rockets gave the Germans in their offensive against Great Britain, an experience that impacted on his mind the importance of this new technology.

During the war years, a new military establishment grew up in the blueberry swamps near Valcartier, Quebec, which was to test and proof (maintain quality control) rifles and was to conduct research in explosives and propellants; and investigate the possible improvement of old ones. Valcartier had a ballistics laboratory that was already operating and was isolated, near to the Dominion Arsenals, with factories in Quebec City and Valcartier. There the NRC set up labs for the chemical and physical study of explosives. According to a tri-partite agreement between the UK, the US and Canada, the three parties were to exchange technical and scientific information to aid the war effort. In the context of this agreement, the UK gave ample support to the setting up and direction of this new establishment, including the building of a pilot plant for the manufacture of explosives and propellants. Britain was well advanced in propellant research, and thus the pilot plant in Canada was set up with the best information available at the time, and eventually proved to be very advanced and efficient. Originally the purpose of this establishment was the development of manufacturing processes for explosives, of analytical methods for process and material control, and for the material control of explosives manufactured in Canadian factories. It was also used to study the stability of explosives, as well as their compatibility with other materials with which they might come in contact. All phases of organic chemistry were to be studied here, along with manufacturing processes, so propellants could be produced on an economical scale. Under the direction of a British scientist, Harold Poole, this new Explosives Experimental Establishment (EEE) made many advances in the production of propellants which greatly improved Canada's industrial position as a supplier of armaments. Indeed, by the end of the war, Canada was in fourth place in the world in the manufacture and sale of arms. At the end of the war, the NRC, intending to return to peacetime scientific research, wanted to divest itself of the explosives lab. But there was still a need for a permanent establishment so that Canada could keep up with developments in equipment and the technological phases of warfare. The various establishments that were operating at Valcartier—the ballistics lab, the propellant and explosives groups, inspection groups and operations divisions were amalgamated into one entity, which then was called the Canadian Armament Research and Development Establishment (CARDE). It was to serve all of the armed forces and was to be operated by the army.

NRC's Explosives Experimental Establishment, the ballistics laboratory, and the Army's field trials wings were combined, and a design wing was added for projectiles and weapons. Many of the new establishment's members were sent over to the UK during the war to get specialised training, and now this group of people was put to good use. There were five wings in the new establishment: the administrative ("A") wing and four technical wings. The "B" wing was responsible for ballistics; "C" wing dealt with chemistry; "D" wing took care of design and development of weapons up to the production of prototypes; "E" wing was the experimental wing and worked on field trials and the proof of field equipment. Eventually [one presumes] there was an "F" wing; and finally a "G" wing, which worked on the Velvet Glove missile in the early 1950s. An Order-in-Council placed CARDE under the Army's command in March, 1945, and two years later, when the Defence Research Board was created as a scientific research body responsible for basic research that might be of interest to the military, CARDE came under its umbrella. Don Rose became CARDE's first Chief Superintendent.

Rose remained at CARDE after the Defence Research Board took over control, and he "did much to help the establishment over its difficult transitory period", and set it up on the road to do more basic research into propellants. And even then, as early as 1946, he recalls he wanted to see the development of a rocket that was unique to Canada, but this suggestion fell on fairly deaf ears. As early as 1947, however, CARDE began the testing and improvement of propellants and explosives. Soon after, DRB began to work on the highly secret Velvet Glove air-to-air guided missile. By the time that work started, however, Don Rose had decided to return to civilian research and rejoined NRC to pursue his interest in the basic science that the war years had prevented him from pursuing.

The decision to develop a Canadian guided missile began in 1948, with sophisticated design facilities and advanced features such as a computer. Both DRTE and NAE (the National Aeronautics Establishment) at NRC supported the effort. The planned Velvet Glove was inspired by the experiences during the Korean war, and meeting Canada's NATO obligations. An air-to-air missile was chosen because this area was lacking, and because of its feasibility. It was to be used on the CF-100, built by A.V. Roe Canada Ltd. and on the F-86, built by Canadair Ltd. By 1951 the first prototype was launched from the Canadian army test range at Picton, Ontario and over four years some 300 missiles were tested, mostly at Cold Lake, Alberta. But by 1954 the project had to be abandoned in the light of supersonic bombers and because the missile had not caught up to such US counterparts as the Sparrow. By this time the focus of the cold war had shifted from aircraft to the ICBM (Inter Continental Ballistic Missile). This reinforced the interest in the upper atmosphere and in rocket propellants. There were also other post-war signs of "space" activity at the NRC.

A post-war NRC discovery—radio wave emission from the sun

The Radio and Electrical Engineering Division (REED) that NRC had a in place by 1948 reflected the division of responsibility between the NRC and DRTE. Chalmers Jack

(C.J. as he was often called) MacKenzie, the President at that time, didn't want classified work there after the war, thus this element was moved to DRTE, as described earlier. REED was to develop microwave expertise and in 1947, Arthur Covington had already built a radio telescope to work at 10.7 cm wavelength. The following story, which had great impact on the international space community, is adapted from its history as written by Ken Tapping, and available on the Dominion Radio Astrophysical Observatory (DRAO) website.

Immediately after the Second World War, Arthur Covington and his colleagues at the NRC in Ottawa used bits of military surplus radar and test equipment to make a radio telescope. The antenna was a 1.2 meter paraboloid from a Type IIIC Gun Laying Radar, mounted on a prototype mount casting for a Model 268 radar. By leaning the mount so that the azimuth axis was pointed at the Pole Star, it was converted into a simple polar mount, which made tracking celestial objects much easier. The radar system operated at a frequency of 2800 MHz, that is a wavelength of 10.7 cm.

The instrument was pointed in the direction of various celestial objects, including Jupiter, the Milky Way, the aurora borealis, and the Sun. It was too insensitive to pick up any cosmic source apart from the Sun. However, as time passed, Covington and his colleagues realized that the Sun's emission at 10.7 cm wavelength was varying. They did not expect this, as the thinking at that time was that the solar emission at centimeter wavelengths would be simply black body emission from a ball of hot gas. This led to the question of whether this was a variation in the emission from the whole disc or that smaller, variable sources were present, perhaps associated with active regions and sunspot groups.

The poor angular resolution of the radio telescope (a few degrees) made it impossible to distinguish between these two possibilities. However, an opportunity to address the question offered itself on 23 November, 1946, when an eclipse of the Sun occurred in the Ottawa area. The observation showed convincing proof that strong contributions to the total emission at 10.7cm originated in the vicinity of sunspots. The eclipse record showed a strong dip in signal strength after 11:40 local time, when the moon covered a large sunspot on the solar disc.

Covington then showed that the 10.7cm Solar Flux correlates with indices of solar activity such as sunspot number and total sunspot area, with the advantage over those indices that the measurements are completely objective, and can be made under almost any weather conditions. Since it is closely correlated with magnetic activity, it correlates closely with other activity indices and, since magnetic activity modulates the Sun's energy output, with solar irradiance. The 10.7cm Solar Flux is currently one of the best indices of solar activity we have. It now forms a consistent, uninterrupted database covering more than 50 years. Only sunspot number counts cover a longer period, going back to at least the 17th Century. However, these data are subject to subjective effects in observation and evaluation, and are affected by the weather.

Between 1946 and 1990, the measurements were made in the Ottawa area, first at Goth Hill, a site south of Ottawa, and then at the Algonquin Radio Observatory. In 1990, following the closure of that observatory, the Solar Radio Monitoring was relocated to the Dominion Radio Astrophysical Observatory, near Penticton, British

Columbia. At that point, what had been known for decades as the "Ottawa Flux", became the "Penticton Flux." In 2003, the Solar Radio Monitoring Programme became a joint program with the Canadian Space Agency and a component of the Canadian Geospace Monitoring Programme, which brings together various geospace monitoring programs in one entity. The data are distributed worldwide through the Boulder National Geophysical Data Center and are used by thousands of space scientists around the globe.

Balfour Currie during a lecture at the University of Saskatchewan. Photo by courtesy of the University of Saskatchewan Archives, Alumni & Development #2023, Star Phoenix photo by Peter Wilson (January 6, 1979).

Currie's wartime years

Following IPY-2, Balfour Currie returned to the University of Saskatchewan before carrying on to Toronto and the Meteorological Service to join up with the other Canadian expeditions from the International Polar Year in the fall of 1933. Here, with other members of the Canadian effort, Currie spent a year analyzing the information garnered during their thirteen months at Chesterfield Inlet. Davies worked on the magnetic data, while Currie was assigned the auroral height measurements, including the auroral log book, and the photographic plates. McVeigh was also in the same research group that worked the weather information gathered at Churchill, which he was to complete before his untimely death in a car accident.

The Meteorological Service of Canada provided funds for this one year of effort, and for the publication of a book containing the results. However, this was the limit of their support, so Currie was allowed to take the auroral photographic plates back to Saskatoon. Here, during the years between 1934 and the war, he taught his graduate students the technique of auroral height determination from parallactic photographs, and how to interpret the results. At long last the enormous effort expended by the polar party began to bear fruit in publications in the scientific literature. Currie himself greatly enhanced his knowledge of the aurora through this process, becoming an expert in the field. This was not all that he did, however. His interest in meteorology and atmospheric science generally had been aroused through all those observations, taken round the clock and he developed two undergraduate courses, one in meteorology and the other in climatology, which were open to all students, not just those in physics. These courses proved to be very popular.

Radio and radar technicians were urgently needed by the Armed forces during World War II. Along with other universities, the University of Saskatchewan provided the basic training that was needed to prepare recruits for the more specialized training provided by the Services. The first group of 150 students enrolled in the Air Crew school were billeted in the university residences, but were marched to the classrooms and laboratories and back again to the residence by non-commissioned officers. For succeeding groups the procedures became more relaxed and the men were allowed to come to classes and spend time in laboratories without supervision. The school stimulated much interest in radio science by the regular students.

Although research at the University of Saskatchewan Physics Department existed almost from the beginning, it was greatly enhanced when, finding themselves no longer able to work in Germany, Gerhard Herzberg and his wife Louise, both spectroscopists, arrived in 1935. The contact with the University of Saskatchewan was made by John W.T. Spinks, who had been appointed Assistant Professor of Chemistry in 1930, just in time to take an enforced leave during the depression, which he spent in Germany. While Herzberg's impact on space science was somewhat indirect, his influence on the Canadian scientific community and the NRC was profound and merits a very brief summary here. Herzberg fitted well into the department, becoming an excellent lecturer, while continuing his research and meanwhile writing his books on atomic and molecular spectroscopy. These books became foundational in the field. In 1940 and 1941 his undergraduate spectroscopy courses were crammed with students. During the war, Herzberg undertook research on explosives, conducted in a bunker beside the physics building. The explosions were photographed through a small tunnel, using a camera made in the physics shop. It included an airbearing cube of glass rotating at 800 rpm, allowing time-lapse photography of the explosion. The explosions were heard by most students on the campus. The program ended in 1949, when the remaining 40 kilograms of explosives were transported by Currie in his car to the river bank and detonated, but the results were reportedly a disappointment for the six carloads of students that followed behind. After the war Cecil (Cec) Costain and Peter Forsyth came back from their terms in the British Royal Navy with radio expertise. To Herzberg this offered an opportunity to move spectroscopy into the radio region (specifically microwaves) with two M.Sc. students. Costain then went on to Cambridge for his Ph.D. and eventually to work in Herzberg's lab at NRC. Later he established Canada's time service based on atomic clocks, removing it from the complications of the movements of astronomical bodies.

The Herzbergs left Saskatoon in 1945 for a position at Yerkes Observatory of the University of Chicago. His intent here was to work on spectroscopic problems of astrophysical interest. This move, however, did not provide the scope that he desired, and in 1948 he was attracted back to Canada, to the National Research Council in Ottawa, in the Pure Physics Division, where he assembled a stellar group of researchers, some of whom had been his students at the University of Saskatchewan. The result of this work was a Nobel prize, in 1971.

Herzberg's position in the Physics Department in Saskatoon was filled by William (Bill) Petrie, who had received a B.A. degree from the University of British Columbia and a Ph.D. in astrophysics from Harvard. He was able to fill the gap left by Herzberg for the teaching of spectroscopy, but for his research there was no scope in astrophysics, as the university telescope was far too small for research purposes. He therefore turned his attention to the spectroscopy of the aurora, building an auroral spectrograph of his own design. The spectra that he acquired were later to become important in the sudden growth of space science in Canada.

A recent photo of the Physics Building at the University of Saskatchewan. The penthouse that housed the spectrographs is located in the centre of the roof.

Davies' wartime years

As the war progressed, the importance of communications increased dramatically. The British Admiralty was having trouble communicating with its ships in the North Atlantic, and recognized that this was happening on polar radio communications paths, where the behaviour was much more erratic than at low latitudes. In 1941 they asked the Canadian Navy (RCN) which had access to lots of high latitude territory, to operate two ionosondes which the Admiralty would provide at northern sites to study the nature of the ionosphere in this difficult region. The Royal Canadian Navy had no scientific staff at this time and turned to the National Research Council for advice. Since Frank Davies had been in charge of a geophysical station in Huancayo, Peru, that included a "Berkner" ionosonde, he was asked to take on the job. Seconded from the NRC, Davies headed up an inter-service Section of the Royal Canadian Navy, concerned with the application of ionospheric data to communications, detection and direction finding in the high frequency (HF) band. He also served in both British and

Canadian Naval Intelligence during the war. The sites selected were Churchill and Pond Inlet on Baffin Island. In early 1942 the RCN borrowed McLeish's ionosonde, and began to train a group of radio operators at a temporary station set up in a farmer's field near Chelsea, Quebec, just 11 km northwest of Ottawa. Under Lt. Graydon Lloyd a 24 hour observing program was started in mid-1942 and the results sent to the Admiralty in the United Kingdom. Frank Davies felt that a scientist would still be needed to interpret the Churchill results, and Ross Freeman of NRC was placed in charge of the five RCN operators. The Admiralty ionosonde arrived in early 1943 and was set up at Chelsea. Lt. Jack Meek arrived at Naval Headquarters in 1943 to work with Frank Davies, who now had been seconded full-time to the RCN. Jack Meek was given the responsibility for the interpretation and dissemination of the observations.

By now, the value of continued observations at Chelsea was recognized and so McLeish was asked to build two new and better ionosondes. One would replace the Admiralty model at Chelsea, and the second would be a back-up to the Admiralty system moved to Churchill. The Baffin Island operation called for Arctic expertise and so the Carnegie Institute of Washington was called in to help. They set out in 1944 with the second Admiralty ionosonde (Model 249) in the Hudson's Bay Company northern supply ship Nascopie. The ice was unusually heavy that year and the Nascopie was unable to reach the top of Baffin Island—instead they disembarked at Clyde River, and so the most northerly station was established here. However, the Carnegie Institute had personnel instability problems during the winter, and asked the Canadian Department of Transport to take over. J. Armour Warwich went in to Clyde River by dog team in early 1945 to help out, and when the supply ship went in that summer, a full Canadian crew took over.

To appreciate this work, we need to know that the ionosphere splits radio waves into two polarization components, just as polaroid sunglasses do with ordinary light. These travel separately in the ionosphere in what are called ordinary and extraordinary modes. These were already known to radio scientists, but Jack Meek found that at Churchill, where radio waves can propagate parallel to the magnetic field direction, a third mode, called the Z-mode could be observed, and Jim Scott worked out the theory for this. Auroral scattering from the highly variable (called sporadic) E-region was found to be exceedingly common at Churchill, sometimes blanketing the ionospheric reflections completely. This was a first inkling of the amount of energy deposited by the aurora and its dominant effect on the high latitude ionosphere. Unusual reflections were also seen from the D-region, lying at 90 km, just below the E-region. During major magnetic disturbances, radio blackouts would occur, with a total absorption of radio waves in the D-region. It was recognized that any solar particles responsible for this would have to have a lot of energy to penetrate this deeply into the atmosphere.

These were important scientific discoveries, but what the Admiralty wanted was a method of predicting what frequencies would give most reliable communications for a given time of day, location, season and level of disturbance. This had already been established for communications with the rest of the Empire, but Canada presented a special problem and the new data were crucial. It was found that 27-day recurrence patterns in auroral disturbances, caused by the rotation of the sun, could be used in frequency predictions, and that about one and a half days after a solar flare (observed by telescope) that radio blackouts would occur. The problem was never fully solved, but it was possible to reach Newfoundland and Halifax with some reliability. Submarine cables and now communications satellites provide incredibly reliable telephone and TV links, which we now take for granted. It was not so during the war. The German Navy was communicating in the North Atlantic too, with its submarines. In fact, unlike the British and Canadian Navy requirements of radio silence at crucial times, the German submarines were required to report all sightings and to receive orders before attacking. This was done on the assumption by the German Navy that the ionosphere was too erratic for consistent monitoring and that their codes would never be broken. As is well known now, the Canadian William Stephenson, otherwise known as Intrepid, did break that code secretly and early in the war. The ability to pick communications frequencies reliably could then be used to monitor much of the submarine activity. This was done in a coordinated effort between Canada, the United States and the United Kingdom.

Frank Davies, Explorer, and Chief Superintendent of DRTE. Photo by courtesy of the Communications Research Centre, Ottawa.

Frank (Taffy) Davies received many honours. He was a Fellow of the Royal Society of Canada; a Governor of the Arctic Institute of North America, a member of the Antarctican Society of Washington, DC, a former president of both the Arctic Circle Club of Ottawa and the Ottawa Welsh Society. Mr. Davies was physicist of the First Byrd Antarctic Expedition of 1928-30. He was awarded a Special United States Congressional Gold Medal which was presented by President Herbert Hoover in 1930. He was also awarded the Queen's Coronation Medal and Jubilee Medal for service to Canada.

Formation of the Defence Research Telecommunications Establishment (DRTE)

The coordination of the radio work during the war was done by the Canadian Radio Wave Propagation Committee. After the war it was recognized that this work was too important to be stopped, so it was continued by a small unit in Naval Headquarters, called the Radio Propagation Laboratory. The Dept. of Transport gradually took over the operation of ionospheric stations in Canada, while the Radio Propagation Laboratory provided supervision and interpretation of results, this now consisted only of Frank Davies, its Superintendent from 1948 to 1951, Jack Meek and Jim Scott. A little later they were joined by J.W. (Jim) Cox and R.C. (Bob) Langille. At that time Cox was a member of the Marconi group in England but he immigrated to Canada after the war to join his wartime scientific colleagues. Meek started an experimental laboratory with six summer students, of which two, Fred Green and John Chapman, remained in radio research the rest of their careers. John Chapman was later to have a very prominent role in the Canadian space science story.

This is Frank Davies' own account of the beginnings of DRTE, transcribed from hand-written notes and recorded on the Friends of CRC website. He was Chief Superintendent of DRTE, except for a period when he was seconded to DRB headquarters, from 1951 until he retired in 1969.

"We were in fact orphans after WW II. DRB hadn't planned to include this research and did so only because a) the Services could no longer support the set of stations, b) NRC declined to accept an RCN request to take this responsibility, and c) DOT agreed to run the field stations but not the training nor data utilization.

"It had become necessary to continue an ionospheric service as part of the rapidly growing international effort which in a relatively short time included more than one hundred ionosondes around the globe. In this the Canadian ionosondes were much more important than their number because they spanned the northern auroral zone in which ionospheric disturbance was a maximum. (At that time there was no equivalent ionosonde net across the southern auroral zone). Scott, Meek and I, with Miss Richard as secretary, were transferred to DRB in August 1946 and were joined by Bob Langille who brought with him from Army Operational Research the continuing radar program called 'Stormy Weather', a study of local weather by radar. This program, with equipment and DRB grants, was later transferred to McGill University where, under the direction of Dr. Stewart Marshall, it has proven a very creditable scientific success.

"DRB did not decide to build a full Telecommunications Establishment until some four years later when the two early laboratories, RPL (in a surplus RCN signal station at the Experimental Farm), which developed from the original ionospheric group, and EL (at Rockcliffe), were combined in a single DRTE. This was located at Shirleys Bay, but for a decade EL continued to grow on the Montreal Road, closely linked with CSRDE (Army). As superintendent of both wings, I spent alternate days east and west of Ottawa and in the process got to know people in both wings rather better I think than any of my colleagues. Jim Scott (RPL) and Jim Cox (EL) were the first superintendents of these separate laboratories when these had grown to fill two large new buildings.

"The term 'superintendent' was not customary in Canada in the sense of laboratory superintendent. I remember being drafted to the jurors' panel in the Carleton Fall Assizes and was interviewed in my turn as panel members registered. Two men before me described their jobs as superintendents which meant apartment house supervisors. I simplified explanations by describing myself as a laboratory director.

"The early history of DRTE is not fully understood unless it is realized that NRC retained the military research fields of Radar, Antennas (Radar and Communications), Direction Finding, and Electronic Tube development after WW II. NRC's large division of Radio and Electrical Engineering had done a remarkable job for the Services in these fields during WW II and had a very experienced staff to continue. It was only in later years when NRC interests broadened into space developments that radar, electronics and D/F were added to the main role of DRTE—i.e. communications.

"The Glassco Commission in 1962 recommended that ionospheric and communication research in DRTE should be transferred to NRC, perhaps in ignorance of the RCN request to NRC to do this in 1945."

With the end of Frank Davies handwritten notes, the story continues as follows.

The Radio Propagation Laboratory wanted to explore in more detail the auroral effects on the ionosphere that had been observed in Churchill during the war. So in what was then, and still is, a rather outlandish idea, they commandeered a CN railway car, and fitted it out as a mobile geophysical observatory, including an ionosonde, a photoelectric auroral intensity recorder (provided by Bill Penn, one of Balfour Currie's students from the University of Saskatchewan), a fluxgate magnetometer provided by Harold Serson (of what is now Natural Resources Canada—NRCan), and a 2 MHz Loran receiver. This observatory was operated for a week at Portage la Prairie, Manitoba, and then pulled by regular passenger train to The Pas, Waboden, Pikwitonei, Gillam, Herchmer and Churchill, stopping for a week of observations at each site.

Seven such traverses were made during the year 1948-49. The observations showed that during this year of sunspot maximum the auroral zone was displaced 100 miles south of the Churchill location Herman Fritz, a 19th century auroral authority, ascribed to it, that magnetic disturbances were strong and nearly always present, and that sporadic E reflections had several origins and were also difficult to understand. They mapped the patterns of visual aurora, blanketing sporadic E and F-region frequency depressions, as well as the region of auroral absorption. At the end of this work it was clear that the University of Saskatchewan was becoming a major centre of auroral research, and Jack Meek was posted there. Here he set up an ionosonde (an LG17 manufactured by Canadian Marconi to DRB standards), a magnetometer of the Serson design, and he designed an all-sky camera, using a simple convex mirror and a modified movie camera, mounted on legs above the mirror into which it viewed the entire sky. Pictures were taken automatically once per minute and this technique later became a major cornerstone of the International Geophysical Year. The all-sky camera was invented independently by Carl Gartlein in the US and Willy Stoffregen in Sweden, but Meek's conception was entirely his own. With this assembly of ionospheric, magnetic and auroral equipment he made the first truly coordinated measurements of these phenomena, for which he received his Ph.D. before returning to Ottawa.

The Defence Research Telecommunications Establishment at Shirley Bay.
Photo by courtesy of the Communications Research Centre.

When DRB was first formed in 1946, RPL became an integral part of it, housed at its headquarters. In 1949 RPL moved to a site on the Dominion Experimental Farm, as already noted by Frank Davies, where there was room to erect antennas. In 1950 the Defence Research Electronics Laboratory was formed and it was housed in army quarters at Rockcliffe. But much more space was needed and after a personal search by Frank Davies some sheep-grazing land next to the Connaught Rifle Range was found 18 km west of Ottawa, at Shirley Bay. This had lots of room for radio antennas and was free from electrical interference from the city. A huge copper shield was buried in the ground around the site to improve this shielding and in 1952 the Radio Propagation Laboratory (now called the Radio Physics Laboratory) and the Defence Research Electronics Laboratory were eventually united into the Defence Research Telecommunications Establishment (DRTE) at this site, with Frank Davies as Chief Superintendent. The Electronics Lab, however, did not move here for some time. The second lab on the Shirley Bay site was the Defence Chemical Laboratory and the third was the Communications Laboratory under John Chapman (as part of DRTE, but separate from RPL). Jim Scott was head of RPL and Jim Cox head of the Electronics Laboratory. The RPL developed into six sections, the Theoretical Section, the Atmospheric Studies Section, the Radio Prediction Section and three radio propagation sections—for low frequencies, high frequencies and microwaves. The enthusiasm and drive released into DRTE from the cessation of the war was enormous and led to a rapid buildup of high quality research.

Postscript

At the end of World War II Currie (and Petrie) were at the University of Saskatchewan with a remarkable expertise in auroral behavior and its spectrum. Rose had returned to NRC with expertise on propellants and rockets; and Davies was Superintendent of a large organization (DRTE) in Ottawa, one with enormous capacity from which the first Canadian satellite would later emerge. The ground was ready, but nothing had taken root as yet. How would this come about? As is seen in the next chapter, the answer to this leads back to Saskatchewan.

References:

Henderson, John T., Measurements of ionization in the Kennelly-Heaviside layer during the solar eclipse of 1932. Canadian Journal of Research, Volume 8, No. 1, January, 1933.

NRC Herzberg Institute of Astrophysics Dominion Radio Astrophysical Observatory website: http://www.drao-ofr.hia-iha.nrc-cnrc.gc.ca

Rose, D.C., Radio observations on the upper ionized layer of the atmosphere at the time of the total solar eclipse of August 31, 1932, Canadian Journal of Research, Volume 8, No. 1, page 15, January, 1933.

University of Saskatchewan Archives, 2002. http://scaa.usask.ca/gallery/northern/currie/
Friends of CRC website: http://www.friendsofcrc.ca/

Chapter 3

Space Science Takes Root in Canada

Nate Gerson, a US benefactor, plants space science in Canada

Following the war the major strategic importance of the polar cap became clear as the US and USSR glared at each other across the ice and snow of the Canadian north. They used radar as the technique with the establishment of the mid-Canada line, and the Distant Early Warning system; both were set up to detect hostile aircraft. The USSR no doubt built similar systems. With the newly developed Inter Continental Ballistic Missiles, it became crucial to develop long range over-the-horizon radars and the problems of immediate detection and interpretation of radar echoes came to the fore. But echoes can be obtained from many other things. With the abundance of radars after the war, the British scientists Lovell, Clegg and Ellyett soon discovered that they could get radar echoes off meteor trails. While studying these, they got a persistent echo that they attributed to the presence of aurora. Here was one of those accidental discoveries, even more so since the aurora they saw visually was overhead, and it is now known that this aurora could not produce the echoes they saw; this must have come from unseen aurora at their northern horizon. At about the same time, Leiv Harang and Willy Stoffregen were experimenting with radars at Tromsø in Norway, and were certainly getting auroral echoes.

There arose a simultaneous need for the military agencies to identify auroral radar echoes, and a strong desire by scientists to understand what caused them. Balfour Currie made a trip to Ottawa in early 1948, and while he didn't leave a record of what transpired, two other people did. One of these was N.C. (Nate) Gerson. His story begins very shortly, but in his article in Physics in Canada he stated that "In the spring of 1948 I had contacted Dr. Cullwick then at the Defence Research Board, about sponsoring auroral research (at the University of Saskatchewan) and radio meteorological investigations (at McGill University). Cullwick arranged a meeting at the old National Defence headquarters building to include himself, Don McKinley of the National Research Council, Balfour Currie of Saskatoon, Stuart Marshall of Montreal and myself." His initial reaction was that "Currie thought that the radar observations might (could?) fail and seemed reluctant to agree." Next day however, proposals were agreed to by all parties.

The other record of Currie's visit to Ottawa is a note from Don McKinley to Peter Forsyth hand-written in 1977 on a copy of a scientific paper by McKinley and Peter Millman. In this note McKinley indicates that he had already published auroral echoes in "this paper", writing that "Balfour visited us at the Radio Field Station, Ottawa, about late April 1948 (or May) and saw some of our early 33 Mc/s auroral echoes (see p. 172 of this paper). He departed full of enthusiasm for the possibilities of radar detection of aurora – and you know the rest. At the time Peter Millman and I agreed to leave the auroral radar field to Balfour, as we had our hands full with meteor echoes." [The authors are grateful to Peter Forsyth for providing this information]. Whether these were indeed auroral echoes or not, Currie arrived back in Saskatoon with the enthusiasm noted by McKinley, determined to begin such experiments at Saskatoon.

Peter Forsyth had just finished his Master's degree and Currie, knowing of his background in radar during the war, interrupted his move into the study of medicine by hiring him to set up a radar system. They borrowed the first radar they could get their hands on, from the Canadian Navy, sight unseen, so to speak. It turned out to have a very high frequency, 3000 MHz and a totally unsuitable antenna. Forsyth constructed a new antenna and set the system up on a specially built cubicle on the north-sloping roof of the physics building, where it could be directed towards the aurora. A camera was mounted on the antenna so the aurora could be photographed as well. After a cold winter's work, scrambling around on that cubicle, with no guard rail, Forsyth obtained a lot of auroral photographs, but no echoes at all. They then concluded that the frequency was just too high to get any reflections. To continue, they would need another radar, but where would they get it? The answer was rather evident, since Currie had already met Gerson the previous year. Balfour Currie was not the only North American thinking about the auroral ionosphere, the United States Air Force was too, and they had commissioned Nate Gerson to establish the study of auroral physics in North America. Up until then it had existed only in Europe and mainly in Scandinavia.

The cubicle installed on the roof of the Physics Building at the University of Saskatchewan to accommodate the first "auroral" radar, and later used for many auroral observations. The picture is recent; the structure remains to this day. The elegant new Agriculture Building can be seen in the upper left corner.

In 1940, President Franklin D. Roosevelt created the Office of Scientific Research and Development (OSRD) to direct civilian industry conducting research and development (R&D) efforts required to develop weaponry and to support war-related requirements. Unprecedented in history, the US mobilized the civilian science and technology community for employment in the war effort. Breaking new ground, OSRD was the first federal agency to provide large-scale government support for university-based scientific research.

Dr. von Karman, one of the world's leading aerodynamicists, stated in a report to the Scientific Advisory Group, of which he was chair, entitled Toward New Horizons, that a national program in basic research was a "necessary adjunct" to the maintenance of a strong military posture. In 1948, the Air Force (AF) established the Office of Air Research within the Air Materiel Command. In 1950, the AF established the Air Research and Development Command (ARDC) devoted entirely to R&D efforts. In 1951, due in large part to the intercession of the AF's first Chief Scientist, Dr. Louis Ridenour, the AF created the Office of Scientific Research (OSR) as a small staff office in ARDC headquarters.

By establishing ARDC and OSR, the AF acknowledged that R&D was vital to the AF mission. In 1955, the AF separated OSR from ARDC and established a new organization designated the Air Force Office of Scientific Research (AFOSR) to emphasize the importance of basic science and technology research. The mission was basically to plan, formulate, initiate, and manage a basic research program.

Post-war budgets were lean, but AFOSR's budget tripled within two years of the 1957 Sputnik launch. In 1958, the AF made a major change in the way basic research was funded – a grant program – which remains as the basic instrument to sponsor extramural basic research. In the 1950s, AFOSR contributed to many notable scientific advancements.

In 1961, the AF had a major reorganization leading to the Office of Aerospace Research (OAR). The AF designated OAR as a separate operating agency under Headquarters, United States Air Force, with the functions and responsibilities of a major air command. The AF placed four research organizations under OAR: the Air Force Cambridge Research Laboratories, the Aeronautical Research Laboratory, the Frank J. Seiler Research Laboratory, and AFOSR. The Air Force Cambridge Research Laboratories (AFCRL) were located at Hanscom Field, near Boston, and it was from here that research funds began to flow to Canada.

Nathaniel (Nate) C. Gerson was born in Boston in 1915. He was a meteorologist with the US Weather Bureau between 1938 to 1946, and assigned to Puerto Rico. Here he attended the University of Puerto Rico by day and was a meteorologist at night, in this way attaining a B.Sc. in Physics in 1944. He was transferred from the Weather Bureau in 1946 to the USAF Watson Labs in 1946, and enrolled for his M.S. at New York University, completing this in 1948. During this time he traveled across Canada and the Arctic in order to evaluate a new experimental low frequency Loran Chain for aircraft navigation. When the Geophysical Directorate of the AFCRL was formed in 1948 he became Chief of its Ionospheric Research Laboratory. It was during this period that he made his contributions in establishing space science in Canada; he did this until 1956, when he joined the US National Security Agency. His claim that he had seen more of Canada than most Canadians, is well founded; he even suggested the establishment of the station at Alert on Ellesmere Island, Canada's most northern station, and one that is popular today as a jumping-off point for North Pole expeditions (but fundamentally it is a military base).

Nathaniel C. (Nate) Gerson. Photo by courtesy of Sareen R. Gerson.

Gerson's philosophy was strikingly simple, but effective. "Find a good man and back him", he was fond of saying. But he credits this approach to someone else, Frank Davies, who had told him, "Nat, in physics if you want something done well, find a bright young lad and back him. You can always find plenty of mediocre chaps, but they won't get you anywhere." Gerson's corollary to this was "you do not fish in the desert." He also thought that at least three years would be required to accomplish something and evaluate it, so he insisted on three-year contracts. During that time he did not pester the research group, although he maintained a close interest. From anyone else, the close scrutiny would have seemed to be pressure, but no one resented the barely five-foot tall, energetic little man who could barely sit still, yet could get so involved in discussing the arcane details of the research that he would forget to eat or move in his chair. His self-effacement, modesty and alertness as well as continued child-like interest in his science made him the closest thing a human can be to a leprechaun. In contrast with the traditional (and highly successful in its own right) way of pursuing scientific research by dividing the world of physical phenomena into smaller units for more detailed analysis, he was determined to see the whole global picture as it related to electromagnetic phenomena. Soon after joining the USAF, he was charged with setting up a laboratory at Cambridge, Mass. to study electromagnetic propagation. But not liking a title that was so specific, he asked his superiors to change the aim of the research lab to ionospheric research. Then he stewed for six months, trying to sort out what should be done. He had only a secretary, but he said later that he did not want to hire anyone either on staff or on contract before he knew what a good research project would have to cover. Having recently finished a thesis on the ionization of gases, he thought he knew all there was to know about the field and where the gaps were in the knowledge of ionospheric physics. "I was young and arrogant and thought I knew it all", Gerson chuckled many years later.

He did, however, develop a vision of a total research program. The four areas of electromagnetic phenomena—what he called the four electro-magnetic sciences—ionospheric physics, geomagnetism, cosmic radiation and auroral physics - were inseparable and had to be studied in parallel. It is interesting that three of these corresponded to the expertise of Canada's pioneers: Davies in ionospheric physics, Rose in cosmic radiation, Currie in auroral physics, and Gerson succeeded in establishing the fourth, in geomagnetism in Canada with Jack Jacobs. Although his superiors wanted him to set up a facility for the study of electromagnetic wave propagation (radio wave transmission) he was impatient with routine measurements, and wanted to have nothing to do with what he considered technician's work. He persuaded his superiors to allow him to study the medium—that is, the physics of the ionosphere.

Gerson's search for "good men" took him across the North American continent. At Rice University, the President of the University spent a whole morning with him and Gerson concluded that any University where the President had an entire half-day to spend with him couldn't be very active. He also considered the University of Alaska, Penn State University and New York University, but decided against all of them.

Arriving at Saskatoon in 1948, he was delighted to find a highly motivated group already deeply involved in auroral research. There was Balfour Currie with nearly twenty years of experience in auroral photography, height finding, auroral morphology, magnetic variations, earth currents and auroral spectroscopy. This depth of auroral knowledge was unique in North America. There was Peter Forsyth, a fledgling scientist, but with as much experience in radar as anyone could have then, who had set up a radar by himself and attempted to obtain auroral echoes. In addition there was William Petrie, Herzberg's replacement, who had been appointed to the Physics Department in 1946 as described earlier. Petrie was an astrophysicist, trained at Harvard University and came to Saskatchewan by way of Manitoba. Rather than pursue astronomy, he decided to study the spectrum of the aurora prevalent in Saskatoon, which has many similarities to stellar spectra. Both are produced in the plasmas of atmospheres, and although stars have much hotter plasmas than the Earth does, and the constituents of their atmospheres are different, the same basic physics applies. Astronomy was in the family since Petrie's brother Robert was to later become Canada's Dominion Astronomer, housed at the major Canadian observatory in Victoria, B.C. Petrie knew a great deal about spectrographs, and recognized that one designed for the extended auroral light would be different from ones designed for the point-like stellar source. He set about building an auroral spectrograph and was (no doubt) taking spectral measurements by the time Gerson arrived. Nate Gerson's intuition told him that this was the place to invest his money. The institution was small, but it had three unusually knowledgeable and motivated people, whose knowledge spanned the range of techniques and basic physics needed for a comprehensive study of the aurora. Currie's long experience and broad range of interests formed a perfect background for the two young radar and spectroscopic workers, Forsyth and Petrie, respectively.

Before leaving Saskatoon, Gerson had arranged for a $50,000 annual contract for basic research on aurora. The message from this, that productive experimental research by universities required far more substantial funding than had previously been the case in Canada, especially where new technology was being implemented, sent shock waves across the Canadian research community. Significantly, grants from the NRC and the DRB increased rapidly from that point on, in some cases by a factor of ten.

Currie and Forsyth knew by now that they wanted a lower frequency radar; Gerson had access to one, and was happy to provide it. But he couldn't do this without envious reaction—from both the University of Alaska, and the NRC in Ottawa. His Solomon-like solution was to give radars to all three. One 106 MHz radar was shipped directly to Saskatoon and the other went to NRC, to have its frequency halved to 56 MHz before then being shipped to Saskatoon. Another political aspect Gerson had to deal with was the developing Defence Research Telecommunications Establishment (DRTE) in Ottawa, under Frank Davies' direction. The auroral ionosphere was to be their theme also, and it wouldn't do for them to be in competition with a USAF-supported Saskatchewan operation. After some negotiations in Ottawa it was agreed that DRTE would financially support the auroral radar work in Saskatoon, using Gerson's gift radar, and that the USAF would fund the optical spectroscopic work.

This was entirely satisfactory to Gerson, who was not an empire builder—he only wanted to see the research done. The DRTE financial commitment to Saskatoon formed a collaborative link between Davies and Currie that ensured the optimization of the Canadian ionospheric work. The NRC was not entirely happy, as it was their business to give research grants to universities, and the USAF contract provided about ten times as much funding as an NRC grant. This, NRC officials worried, would throw the whole grant system out of kilter, and Saskatchewan shouldn't expect to be rescued by NRC when the American money disappeared. This turned out not to be a problem, as by the time the USAF contract did phase out, the NRC funds were reaching reasonable levels. In fact, the USAF example may have helped stimulate this growth to realistic national funding.

Auroral Radar Reflections Observed in Saskatoon

Balfour Currie was no doubt well satisfied with his USAF contract to begin optical auroral studies and the promise of a new radar from Gerson. His equanimity was shattered when the CPR freight office telephoned to say that they had two railway flat cars of equipment for him, that the freight bill was five thousand dollars and that there were customs charges as well. It was only then that he realized that this mammoth radar bore little resemblance to the tiny 3 GHZ radar they had already worked with. It was an AN/SCR 270 Westinghouse radar, of which 800 were fabricated by Westinghouse during the war. The radar was completely self-contained, in four trucks, one for the antenna, one for the diesel electric power plant, one for spare parts and one a repair shop. There was also a jeep. Gerson was unable to solve the administrative problem of having the USAF pay for the freight and DRB came to the rescue. The instruction manuals referred to a crew of fifteen, so Currie hired Floyd Vawter on a part-time basis to assist Forsyth. Vawter was an assistant in the physics department and his technical skills and energy were an important aspect during this period. During the summer of 1949 the first echoes were obtained with the 106 MHz system, along with simultaneous auroral photographs. It was immediately evident that the echoes were coming only from very low elevation and so would have nothing to do with the visual auroras which could take place anywhere in the sky. The problem of determining the relationship between optical and radar aurora thus turned out to be much more complicated than expected.

In 1990 it was discovered that in 1941 this very radar had been in operation at Kauai and had detected reflections from the incoming Japanese aircraft that were about to bomb Pearl Harbour, but that these had been ignored. Apparently the operators were told that American aircraft were arriving, but the detected ones were coming from the wrong direction—from the west. Furthermore, the Saskatoon radars were the two remaining SCR270 units of the original 800; by now they have been refurbished by Westinghouse and returned to the Historical Electronics Museum near Baltimore Washington International airport.

Shortly after the radar, modified to accept long range echoes, was set up, it received echoes from targets more than 800 km away. Simultaneous photographs of the aurora suggested the echoes could be arising from aurora close to the horizon. There seemed to be no echoes coming from high angles. This was not conclusive, since the radar antenna had been designed to be most sensitive near the horizon (to detect incoming aircraft as soon as they appeared above the horizon). The vertical sensitivity pattern had been measured with the help of the local RCAF Reserve Squadron whose commanding officer had obligingly flown endless "radial" flights over the radar and out to great distances at constant height. To check on the lack of high altitude auroral echoes, Forsyth and Vawter built an upward looking antenna with a beam about 120 degrees wide in the north-south direction and about 12 degrees wide in the east-west direction. This antenna pattern could not be measured ("calibrated") using aircraft. During the Perseid meteor shower, Forsyth spent two nights lying out on the grass connected by a field telephone to Vawter, who was watching the radar screen. They established that the antenna had the intended pattern by noting the location of each meteor trail which produced an echo. Then they observed a number of auroral displays with the vertical looking antenna, but the aurora was just not producing echoes when it was high in the sky.

Peter Forsyth in about 1955. Photo by courtesy of the Communications Research Centre.

Forsyth had now become sufficiently interested in the science to take a Ph.D. in the subject, which he did in two winters at McGill and one summer in Saskatoon. At that time the University of Saskatchewan was not yet awarding Ph.D. degrees. The frequency of 106 MHz was clearly low enough to give echoes. The Gerson radar that went to Ottawa was modified by NRC to operate at 56 MHz and then was sent to Saskatoon. The NRC modification was done with great care and ingenuity. It was generally thought that the resulting radar was substantially better than the original. Since the NRC group had been primarily interested in meteor echoes, they had modified the antenna mounting so that the new 56 MHz antenna could look only at high angles. Forsyth seized the opportunity to operate the radar in this upward-looking mode and quickly established that at 56 MHz too, the aurora did not produce echoes when it was high in the sky. Then he and Vawter rebuilt the two radars into one two-frequency radar with the two antennas mounted on a single rotating tower, combined displays and single operating position. A very close match between the vertical and horizontal antenna patterns at the two frequencies was achieved. One of the more scary tasks was devising a sure-fire way of synchronizing the output of the two high power diesel driven electrical generators. This dual-frequency radar was operated over many months. A remotely controlled camera, using the lens from one of the Størmer cameras, was used to photograph the part of the sky from which the echoes were being received. There was no doubt that the echo distribution had the same configuration as the visible aurora but, of course, since the aurora was typically more than 800 km from the radar station, it was impossible to tell whether the echoes were arising from within the visible aurora or were just associated with it.

As soon as Forsyth received his Ph. D., he was hired by the DRTE (in 1951), but the Establishment left him at Saskatoon until 1952 in order to further the radio work but also because the laboratory in Shirley Bay wasn't ready. The new surge of respectability associated with the radar studies dictated that some more permanent and more comfortable arrangements should be made. Up until that time the radars had been operated from a completely noninsulated truck. For winter operations, rather than try to insulate and heat the truck, the operators wore war-surplus flying suits, heavy flying boots and three layers of gloves. Forsyth used his extra time at Saskatoon to design a proper radio building and oversaw its construction. In the best Saskatchewan (and Currie) tradition, two beginning graduate students, Morrel P. Bachynski (Bachynski later founded the highly innovative Montreal company, MPB Associates) and Ernie Kornelson (who went to the high vacuum section at the NRC), actually built the new "radio observatory" building with their own hands. The radars were installed and working within this building by the time Forsyth left for Ottawa. One of the last jobs he did was to spend a night taking star shots from the new building (as he had been taught by Bill Petrie) in order to establish sky co-ordinates for the radars and the camera tower. During the whole of his stay in Ottawa he was a frequent visitor to Saskatoon and was much involved in the research program there. Forsyth stayed at DRTE until 1958, at which time he had become Superintendent of RPL, and then he returned to Saskatoon again – that story is continued later. Forsyth was effectively replaced by Al McNamara, who had obtained an M.Sc. at Saskatchewan on betatron research, been with McKinley at the NRC; and with a new 90.7 MHz transmitter supplied by DRB built a double-Doppler radar and got the first spectra of radio returns from the aurora. The name Doppler comes from the German scientist Christian Doppler, who first observed the frequency shift of a source caused by its motion. Effectively the motion compresses or expands the wave, changing its wavelength and thus its frequency. This works equally well for sound or radio waves, so that McNamara was able to see that the single frequency of the outgoing radio wave had been converted into several nearby frequencies in the returned radio wave, indicating motions of the reflecting agent. This work led to his Ph.D. and he returned to the NRC in time to build the IGY radars that would be operated at Canadian high latitudes soon after.

The auroral radar at the University of Saskatchewan. This radar produced the first documented auroral radar echoes in Canada. Photo by courtesy of Jim Koehler

In the summer of 1951 auroral echoes were obtained in Saskatoon simultaneously on the two frequencies, along with auroral photographs, which put Saskatoon in the forefront of auroral radar research. The first dual frequency records suggested that the echoes were critical echoes of the ionosonde type, but this would imply electron densities about a hundred times greater in auroras than in the normal ionosphere. This was later found to be the wrong explanation of the echo mechanism. All other explanations of that time were also proven wrong, and the key to the explanation came only much later, from an equatorial radar in Huancayo, Peru. This radar obtained echoes from what could only be the equatorial electrojet, a band of current circling the Earth through the equatorial ionosphere. It was proposed by Donald Farley, and the theory worked out by O. Bunemann, that the electrojet current flowing in the auroral plasma causes a plasma instability that produces acoustic (sound) waves in the ion motion in the plasma. The wavelengths of these waves are just right to cause strong reflections, for radio waves of the same wavelength. It was obvious that the auroral electrojet could do the same. Many years later, Forsyth, now at the University of Western Ontario, with his student Jake Hofstee, was able to demonstrate that ion-acoustic waves do exist in the aurora, but that other scattering mechanisms were still involved.

The solution to the problem of auroral radar echoes is thus still incomplete, but it is now clear that these echoes are a much better indicator of where the auroral electrojet current is than where the bright visual aurora is. There is visual aurora associated with the electrojet, but it is the weak diffuse aurora, barely noticeable to the eye, and discovered only in later photos of Clifford Anger, another Canadian scientist, from ISIS-II, the third Canadian scientific satellite. This confirms Gerson's hunch that one could only understand the auroral ionosphere by studying the electromagnetic, the magnetic and the auroral domains in concert.

Optical Studies and Gerson's AFCRL Contract

It was into this stimulating environment that the first appointee of the new money came. Donald M. Hunten was a recent graduate of the cyclotron laboratory at McGill University. Hunten originated in London, Ontario where his father was a chemistry professor. The cyclotron laboratory, which was the first high energy physics laboratory set up after the war in Canada, was the work of J.S. Foster, a major contributor to the Canadian nuclear physics program during the war, and whose name now identifies the Foster Radiation Laboratory. This laboratory had access to the latest post-war technological developments, and Hunten's technical genius flowered in that environment. His thesis work was to measure the magnetic field strength of the proton. This was done by aligning the proton "magnets" in the cyclotron and then causing them to "flip over" using a radiowave as a stimulus. Hunten's skill at electronics was enhanced by working in a radio repair shop in the summer. He knew nothing about the aurora, but astronomy was his hobby, which meant that he had that necessary fascination for looking at the sky.

Donald M. Hunten in 1997, who kindly provided this photo.

Donald Hunten fully appreciated the significance of starting a brand new research endeavour at Saskatoon. To guide him he had the goals that Gerson had established, in consultation with Currie and Petrie. He had Petrie's experience to fall back on, but in his individual way, chose to set a different course. The existing work had mostly been done by Norwegians, and Hunten must have read all of the Norwegian pioneer Lars Vegard's work. Here was a challenge. Vegard had made a great number of observations, many of potentially great importance, but leaving uncertainties and controversies around many of them. This was partly due to the limitations of his equipment, but partly because this clever and inspired experimentalist tended to rush on to the next discovery before clarifying the last one. Hunten decided to build the most potent spectroscopic instrument that was technically feasible, and to repeat and resolve Vegard's measurements. His analysis of the situation led him to conceive the diffraction grating photoelectric spectrometer, an insrument that was to play a major role in future auroral studies, and he became a pioneer in this field.

Another essential instrument, the diffraction grater which was first developed by Joseph von Fraunhofer—the nineteenth century German optician, after whom the dark bands in the solar spectrum are named—and first used seriously by Henry Augustus Rowland at the turn of the century, is a metal plate on which fine parallel rulings are made, so close that there are typically 600 rules per millimeter. The space between the rulings is comparable to the wavelength of light, so that light beam reflected from it is diffracted through an angle, just as water waves passing through a picket fence in the water are diffracted sideways. The angle of diffraction depends on wavelength, since the geometry is controlled by the ratio of the wavelength to the distance between the rulings. These gratings were a great advance over the prism that Sir Isaac Newton had first used to split light into its colours, because the angle of separation between two given colours was greater, and at least in principle, because the grating could be made much larger. It took a genius to build the ruling engine that would take the diamond point, draw it in a perfectly straight line across the plate, lift it up, move it back, move it sideways by 1/600 of a millimeter and repeat the process. The slightest inaccuracy, such as caused by a change in the temperature of the machine, would ruin the grating, which took days to rule. The machine had to be left closed up in a temperature controlled room, and no one could enter while the ruling was in process. Needless to say, these gratings were very expensive. At about the time of the war, the Bausch and Lomb company had learned how to make "replicas" of gratings that were as good as the original. They were made like a phonograph record by pouring plastic onto the original, then stripping it off and aluminizing it, after mounting it on a glass plate. These replica gratings had just come on the market and Hunten was quick to seize the advantage they offered.

Another important post-war instrument was the photomultiplier detector. When light strikes a metal surface, the light particles—called photons—strike individual electrons in the metal forcibly enough to kick them out of the surface. It was Einstein who, in 1905, provided the correct explanation of this process. (In fact, although the Nobel Committee was well aware of Einstein's theory of special relativity in 1921 when the medal was awarded, they weren't entirely convinced about it and so pointed primarily to the photoelectric effect in their citation.) In any case these electrons can be attracted by a nearby plate with a positive voltage and caused to become an electrical current in an attached wire. These currents can be used to measure the intensity of the light but for the feeble auroral light signals Hunten was concerned that the currents would only be one billionth of a billionth of an ampere and rather difficult to measure. If, however, the positive plate has a large positive voltage, or about 100 volts, it will energize the electron as it approaches, which can kick out two or three electrons when it slams into the plate. Now add another plate, at a voltage 100 volts higher than the first, and it will accelerate all of these electrons towards it, from which will be released about 10 electrons. RCA had managed to build a device, looking like a radio tube, with nine accelerating stages in it, giving about one million electrons out for each photon that went in. This electrical gain of one million at the source was a spectacular advance in the measurement of low light levels and the 1P21 photomultiplier, as one of the best models was called, was Hunten's choice. Up to that time spectroscopists had used photographic plates as detectors of light, but it takes almost one hundred photons to produce one blackened silver grain in the film, and many grains to make a picture. Moreover it is difficult to convert the blackening of a film to a specific intensity of the light. By choosing a photomultiplier, Hunten not only made a leap in sensitivity, but would also put auroral spectroscopy on a quantitative footing.

The "diffraction grating spectrometer" thus consisted of a curved mirror to collect the light from the aurora and make it parallel, the grating which received this parallel light, and another mirror which focused the diffracted light onto a slit, behind which was placed the photomultiplier. Only one colour (wavelength) at one time got through the slit to be measured by the photomultiplier and recorded by a pen on a strip of chart paper. With a motor and cam arrangement, the grating was rotated through a small angle, causing the wavelength to change in a continuous fashion so that the pen on the chart paper traced out a record of the true "spectrum." Once each cam rotation a new spectrum was generated and this could be done in ten seconds. Imagine—a new auroral spectrum every ten seconds. This was a remarkable advance over Vegard's work. For Vegard, working with a prism spectrograph, it would take an hour or so to record a spectrum. This spectrum would only be looked at after developing the photographic plate. It would also take many hours of tedious analysis thereafter to generate an accurate intensity measurement from this spectrum. Hunten realized that he could follow changes in the aurora almost as fast as they were occurring, and promptly set to work. One problem with the photomultiplier was that it was "noisy." This was because any electrons that got loose in the tube could be accelerated down the cascade of plates and masquerade as a current. Because of the random way in which electrons let go they caused fluctuations in the recorded signal; this randomly fluctuating part of the signal was called "noise." The cure for this was to cool the photomultiplier so as to reduce the energy of the electrons in the plates, and thus inhibit their tendency to escape. So part of Hunten's invention was a technique for cooling the photomultiplier. He made a styrofoam box, with an inner metal enclosure for the photomultiplier, having a double-glazed window to let in the light but prevent frosting up, and wires delicately threaded out through the styrofoam. The box was packed with dry ice, not a commercial item in Saskatoon at that time, so Hunten produced his own from a cylinder of carbon dioxide. The procedure was simple; a cloth bag was tied over the nozzle and the valve was opened wide. There was a deafening noise, an escape of gas, with the expansive cooling filling the bag with dry ice. This ear-splitting experience was the first step in the nightly preparation of the spectrometer for a night of observation, well remembered by all of Hunten's students. The first exploratory observations were carried out with the assistance of Carl Dahlstrom, in which they studied the rapid appearance and disappearance of hydrogen emissions in the spectrum.

The second Research Associate hired with Air Force money was Alister Vallance Jones. Like Sir Ernest Rutherford, the famous father of nuclear physics, Vallance Jones had left New Zealand to go to England, to Cambridge University in to go to England on an 1851 Fellowship, and subsequently came to Canada for a post-doctoral fellowship. He spent two years in Herzberg's laboratory in Ottawa, before coming to Saskatoon. Here is the account, in his own words, of the beginning of his research activities in Canada.

"I arrived in Canada in 1949 to take up a post-doctoral fellowship in Dr. Herzberg's laboratory at NRC in Ottawa. This period was an exciting one. There were regular meetings in Herzberg's office and members of the group gave regular seminars on topics which they found of interest. One exciting aspect of the laboratory was the procession of outstanding people who came to visit and usually to give a talk on their work. One such visitor was David Bates who gave a most interesting seminar on upper atmospheric chemistry. His ability to

quote reaction rate constants and concentrations, to discuss the probability of various processes in the ionosphere, really excited my interest in the subject. Sydney Chapman was another distinguished visitor who contributed to my interest in the upper atmosphere. Perhaps most influential of all was a talk given by another PDF, Mike Feast [Note: Feast had been at the UWO Auroral Physics Conference to be described later] *who had developed an interest in the spectra of aurora. I remember this seminar well, with its discussion of the features identified in the aurora, Vegard's work on measuring temperatures from molecular band profiles, [Aden] Meinel's new work on Doppler shifted hydrogen lines in aurora, as well as the new N_2^+ band system which bears his name. About that time Herzberg became interested in the OH bands discovered in the infrared nightglow by Meinel and the possibility of finding these bands further into the infrared. However, I heard that Dr. Petrie at Saskatoon was looking for a Research Associate to explore the infrared spectrum of the aurora as part of a contract which he had received from the U S Air Force for studies in aurora and airglow. Petrie had already hired Don Hunten to work on this project so it was arranged that I would take up this position in February 1952 after returning from a honeymoon visit home to New Zealand.*

"So it was that on a frigid clear morning in February, that I alighted from the CNR 'Continental' at the CN station in Saskatoon. It was a somewhat inauspicious arrival since due to some mix-up there was no one to meet me and there I was at the crack of dawn carrying my bags up 21st St. looking for somewhere to get out of the cold. I checked in at a somewhat dilapidated hotel to await a civilized hour to telephone my new employer. Finally I got through to Dr. Harrington, the head of the Physics department. He was instantly very helpful and came to pick me up and drove me out to the university. After a somewhat alarming drive on slippery snow covered streets, we made it safely to the Physics Building.

"From this point everyone was very kind and helpful. Dr. Petrie had in fact been making arrangements to move to a position with DRB in Ottawa and by spring he departed leaving Dr. Currie to take over the USAF contract and myself and Dr. Hunten to carry on research for the project.

"On our final move west we visited Ralph Nicholls in London, Sutherland in Ann Arbor and Meinel at Yerkes before beginning the long trip across the US and Canadian prairies to Saskatoon. We were impressed by the long straight roads with only the grain elevators protruding above the level plain every 10 miles.

"So began an exciting period of about 16 years in Saskatoon, where a great deal of time was spent in teaching and doing research on airglow and aurora taking advantage of the techniques learned in Sutherland's and Herzberg's laboratories. The contributions of a number of excellent Saskatchewan graduate students was of great importance. First there was Herb Gush who did a remarkable job in building an infrared grating spectrometer equipped with a PbS detector to get the first spectra of the OH bands of the nightglow in the 1-2μm region. This work was continued by Tony Harrison, Jim Read, Don Shemansky and Dick Gattinger, who continuously improved the instrument and succeeded in obtaining better airglow spectra as well as the more elusive 1-2 μm spectrum of the aurora."

Vallance Jones' story continues in a later chapter. He was a highly trained spectroscopist, in the best classical laboratory tradition, and in Saskatoon he had the opportunity to put this knowledge to use on understanding the aurora. At Cambridge he had experience with infrared spectra, and he fully appreciated the value in studying the spectrum of the aurora in the infrared as well as in the visible region. Moreover, infrared detectors had just become available, and so he decided to build a photoelectric diffraction grating spectrometer similar to Donald Hunten's, but using a lead sulphide detector, rather than a photomultiplier. Although Hunten had his spectrometer "in full swing" when Vallance Jones arrived in 1952, he soon was in operation as well. In the spring of 1952 two new graduate students had appeared at Balfour Currie's door, and he steered Gordon Shepherd into working with Hunten, and Herbert Gush into working with Vallance Jones. Shepherd and Gush had been undergraduate Engineering Physics students together at Saskatoon (in the same class as Bachynski and Kornelson), and they continued to work together in the same institutions until much later.

For Shepherd's M.Sc. thesis Hunten proposed to measure the temperature of the upper atmosphere at different heights, using the aurora as a diagnostic. Vegard had already done that, and had shown that the blue band of the ionized nitrogen molecule, (the N_2^+ First Negative band at 391.4 nm) could be used in this way. Molecules generate bands of light in the spectrum rather than lines, but a band simply consists of a large number of closely spaced lines, each of which is associated with a particular rotation rate of the molecule. The nitrogen molecule is like a little dumbbell, rotating about a line perpendicular to its axis. The "rotation" rate is governed by the temperature so that the hotter the gas, the faster the molecules rotate. Each rotation rate is associated with a particular line in the band, and so it turns out that by accurately measuring the band profile, one can deduce the temperature of the gas emitting the light. Thus by using Hunten's spectrometer on the ground, one could analyse the temperature of the atmosphere where the aurora was. The spectrometer was superbly suited to this task. Vegard had produced a puzzling result – that all auroras gave the same temperature regardless of their type or apparent altitude. This value was minus 150 degrees Celsius, a rather low value, but since it seemed impossible that the upper atmosphere had only one temperature value, it also meant that the method was suspect. The theory seemed all right but the missing element was the auroral height. For this, Hunten decided to employ the parallactic height measurement that Carl Størmer had invented in Norway and Balfour Currie and Frank Davies had used at Chesterfield Inlet. Størmer and Vegard had never put the two techniques together, perhaps because of personality, but more because to get an auroral height one had to freeze the aurora in an exposure of a few seconds, while Vegard's spectrograph would take an hour to get one temperature. Hunten's remarkable spectrometer could take an "auroral temperature" in ten seconds, just about the same time as required to get a photograph. Hunten mounted one of Currie's Chesterfield (actually from Størmer) cameras on the front of the spectrometer, which itself was on a gun-mount that could be pointed to any direction in the sky. For the second site he chose Langham, a small town to the north, where there was a high school, and therefore a high school teacher, who might be persuaded to spend his evenings taking pictures of the aurora. This turned out to be true. For communications two Air Force two-way radios were commandeered from the surplus. The way in which the work unfolded was striking for Shepherd, who found himself grinding the crystals by hand to get the right frequencies, and cutting the antennas to the right wavelength, which were also installed by hand. This do-it-yourself style of Hunten's was a characteristic of his work, and made his work so creative that is enthusiasm infected all of his students.

Donald Hunten's photoelectric diffraction grating spectrometer with a Størmer auroral camera mounted on the bracket at the upper left. The shutter is just a flap, raised and lowered by hand.

It was a three-man operation, and at the first sight of aurora all hands were to give up whatever they were doing and head for their posts. The radios were turned on, the dry ice loaded and when all three were in place, the work would commence. The spectrometer was mounted on the cubicle attached to the sloping roof of the physics building, once occupied by the Forsyth's first auroral radar, which could be reached by a hatch from inside. There was still no guard rail. Shepherd stood on the small cubicle roof and pointed the spectrometer, while maintaining communications with Rudy Penner, the high school teacher in Langham; they operated their camera shutters together. After a winter of operations and the writing of the thesis the result was clear: different auroras at different heights did give different temperatures—the results could even be used to generate a temperature profile. At 100 km, near the base of the aurora the temperature was near room temperature, about 30 Celsius, and it increased by six degrees for each kilometer of elevation above that.

This was only the first of Hunten's achievements in auroral spectroscopy with his photoelectric spectrometer. With succeeding students Lyle Broadfoot, later at the University of Arizona and an eminent spectrometric investigator on NASA's Mariner and Voyager missions to the outer planets missions, Ed Lytle who later went to CARDE, Eric Rawson, Don Shemansky (now at the Jet Propulsion Laboratory of the California Institute of Technology and an investigator on the Cassini-Huygens mission to Saturn) and others, he proceeded to investigate the chains of cascade processes in the molecular nitrogen systems that are so prominent in the aurora – the First Negative Bands, the Second Positive, the First Positive, the Vegard-Kaplan bands and the Meinel bands. The details are rather complex but suffice it to say that he put the studies of molecular nitrogen excitation processes on a quantitative footing – he made it a scientific field of study and not simply observations of phenomena.

Even at Saskatoon, auroras were not always present. In order to make the most use of the spectrometer, it would be helpful to have an object of study that was present every night. Such a source is the airglow, an emission of light from the atmosphere that has close relationships with the aurora. It occurs in a similar height range, and some of the spectral lines, such as the auroral green 557.7 nm line appear in both phenomena. The difference is that the auroral light is produced by particles impacting into the atmosphere from outside, while the airglow results from chemical processes that go on as atoms and ions formed during the daytime under the action of sunlight, reform during the night and emit light. Hunten considered the range of airglow possibilities and selected one that does not occur in aurora. It was sodium, the same sodium that is placed in streetlights and produces the same yellow glow from the high atmosphere that appears on the streets (and is a constituent of table salt).

As described earlier, Balfour Currie was probably the first person to record the spectrum of airglow sodium on a photographic plate during his year at Chesterfield Inlet. However the quality of the spectrum in Chesterfield Inlet did not permit an identification, so the discovery is credited to R. Bernard, in 1939. The yellow sodium lines (there are two, at 589.0 and 589.6 nm) are said to be resonant, meaning that ordinary sodium can absorb a solar photon that excites the atom in this transition, whereupon after about one millionth of a second the atom emits an identical photon and returns to the unexcited state. Thus the upper atmospheric sodium that is illuminated by sunlight is fluorescent and radiates the sunlight at this narrow wavelength toward the ground. During twilight, when the Earth is dark, but the atmosphere high up is still illuminated, this fluorescence can be measured with the spectrometer. Hunten was intrigued by a further possibility. As twilight progresses the Earth's shadow moves upward through the layer of sodium atoms, so that the illuminated region becomes thinner and thinner. The resulting decrease of sodium light intensity can be measured with the spectrometer and in fact the data can be used to deduce the height of the layer, its thickness, and even the concentration of sodium atoms in it. The mathematical details of this procedure had never been worked out before and this was a further inspiration for Hunten. He was one of those physicists who could approach a completely new problem, analyse it in detail, and lay out a line of development that was insightful, accurate and yet simple in expression. His approach to the sodium problem was imitated by others who followed his original procedures, and there was no advance in the method until the development of the laser.

The first spectrometer measurements worked out much as expected, and did give both a height and concentration for the sodium. The emission was very bright and easy to see in one-minute scans of the spectrometer. During the summer Hunten returned to his parent's home in London, Ontario for an extended period, leaving Shepherd in charge of the observations. During the summer in Saskatoon, twilight occurs at the most enjoyable part of the evening, but on every clear night he had to leave whatever he was doing and bicycle up to the physics building for a "twilight run." Morning twilights were also desirable, but it is exceedingly difficult to determine whether it is clear or not during the night except by getting up and going outside and becoming sufficiently dark adapted to see stars. Very few morning twilights were obtained. As the summer wore on, Shepherd noticed that the sodium signals were deteriorating. He checked and readjusted the instrument but to no avail, the sodium signals got weaker and weaker until they were barely detectable. Naturally he was worried about what his supervisor would think of his summer's efforts on his return. Hunten didn't seem particularly perturbed, but went away with the data and returned with a plot of what was now nearly one year of data, which showed very dramatically that the sodium concentration had a seasonal variation; it was high in the spring and low in the summer. Shepherd had made one of the most common mistakes that students make—assuming that he was making a mistake rather than nature was telling him something.

Nate Gerson and the Legacy of John Arthur (Jack) Jacobs

Saskatoon was not the only Canadian site that Gerson visited. He wanted also to establish geomagnetic research, and hung this hope on J. Tuzo Wilson, then head of the Geophysics Group at the University of Toronto, later Principal of Erindale College, later still, following his retirement as Director-General of the Ontario Science Centre. Wilson has already been introduced in Chapter 1 as witnessing the formation of COSPAR and leading the International Union of Geodesy and Geophysics through the 1957-58 International

Geophysical Year. Tuzo Wilson was a highly creative and innovative physicist, who applied his ideas to what was once known as Continental Drift and thus helped establish the modern science of plate tectonics. He had little experience in Earth magnetism, let alone auroral magnetic variations, but was not one to pass opportunity by, (Gerson would support this with a contract) and was always willing to be helpful. In the Applied Mathematics Department at the University there was a highly skilled theoretical geophysicist, Jack Jacobs, who would be ideal if he could be persuaded to turn his attention from the temperature structure of the Earth's interior to its geomagnetism. But Jacobs would need an able graduate student. The best geomagnetism was being done in Germany and so Gerson asked the great Julius Bartels of the University of Göttingen if he could recommend an able student who would come to Toronto. After a year or so, one emerged, by the name of Schmidt, but at the last minute he changed his mind. Gerson then turned to the renowned T. Nagata of Japan. After some time, including a dinner meeting during the IGY meeting in Rome, and further letters, Nagata produced a student, Tatsuzo Obayashi. Obayashi was from a simple rural background. In 1950, Japanese culture was much more removed from North American culture than it is today and Japan had recently been an enemy power—Obayashi would be one of the first Japanese students in Canada following the war. He phoned Jacobs from Vancouver to say he was in Canada but how was he to get to Toronto as he had no money? Japan had paid for his travel to Vancouver, but no more. Wilson provided the needed money from the contract.

From left to right, Jack Jacobs, Tuzo Wilson and Ron Farquhar, examining an IGY gravimeter. Taken from Derek York's history of Tuzo Wilson in GSA Today, photo by MacLeod-Gilbert A. Milne and Co. (Derek York died of cancer on August 9, 2007).

They were an unlikely combination. Jacobs was a Cambridge graduate, with a capacity to speak at a rate about three times as fast as any ordinary person could comprehend – though likely still much slower than his mind functioned. Obayashi was a careful, solid thinker who had to consolidate at every step, and his grasp of English was still poor. Nevertheless, he had a quick mind and an extremely industrious nature and with his intense concern for excellence, he succeeded marvellously. His enthusiasm carried Jacobs "kicking and screaming into the world of geomagnetic micropulsations" (Gordon Rostoker's words).

Obayashi was later to become a driving force in the development of space research in Japan. With Nagata's encouragement, Minoru Oda, Kunio Hirao and Obayashi established an Institute of Space and Aeronautical Sciences (ISAS) at the University of Tokyo that not only built its own satellites, but its own rockets to launch them as well. This university group made Japan the third nation to develop its independent launch capability, after the USSR and the USA. It was only in 1980 that the combined efforts of the major European powers, in ESA, the European Space Agency, achieved the same with its Ariane rocket. While Obayashi never held the top leadership position, he made major contributions to ISAS and Japanese space science. All of this can be said to have flowed from Gerson's desire to establish a school of geomagnetism and Jacobs' inspiration and leadership. Eventually the budget of ISAS exceeded that of the University and it became an independent institution outside the university.

In 1957, Jacobs left Toronto to take up a professorship at the University of British Columbia (UBC), and Obayashi went with him. Together they carried out pioneering work in the study of micropulsations in the Earth's magnetic field. When Obayashi graduated, he returned to Japan and was replaced by a new Japanese graduate student, Atsuhiro Nishida. Like Obayashi, Nishida returned to Japan after completing his doctorate under the supervision of Jacobs and he, too, was to have a powerful influence in the development of space research in Japan. It is worth pointing out that, before he started to exert his influence on the field of space research in Canada, Jacobs had already planted some seeds from which the strong Japanese development of space science and technology ultimately evolved.

Jacobs was born on April 13, 1916 and graduated from University College London with an M.A. degree in mathematics and hydrodynamics. On the outbreak of the war he joined the Royal Navy in the training division, achieving the rank of Lieutenant Commander. After the war he became lecturer at Royal Holloway College, where he received a Ph.D. in 1949. In 1951 he moved to the University of Toronto. The remainder of this story and that of Sir Charles Wright is told by Gordon Rostoker.

"At UBC, Jacobs began his efforts to develop infrastructure for research in the earth and space sciences that were to leave his unique mark on science and technology in Canada. He was instrumental in establishing the Institute of Earth Sciences at UBC at the beginning of the 1960s and succeeded in bringing in top flight scientists to serve on its Faculty. [This Institute subsequently evolved to become the Institute of Earth Sciences and Astronomy and eventually merged with ocean sciences to become the Department of Earth and Ocean Sciences at UBC.] Among the first individuals to join the Institute was a young scientist from Tohoku University in Japan, Tomiya Watanabe. Watanabe was a first-rate theoretician who carried out studies in non-linear space plasma physics at a time when this now burgeoning field was in its infancy. Jacobs, whose interest was always in the origin of the earth's magnetic field, arranged for Watanabe to supervise graduate students who were interested in the upper atmospheric component of geomagnetic research. With Jacobs providing the resources and constant encouragement, a number of young space researchers graduated from UBC. One of them was Gerry Atkinson, who initially went to the Jet Propulsion Laboratory of the California Institute of Technology but later returned to Canada, initially to the Communications Research Center in Ottawa, subsequently to the National Research Council of Canada, and ultimately to the Canadian Space Agency, where he served as Chief Scientist.

A second graduate was Gordon Rostoker, who initially was an overseas post-doctoral fellow of the National Research Council based at the Royal Institute of Technology in Stockholm, Sweden and spent the balance of his career at the University of Alberta.

"While he was at UBC, Jacobs established close scientific ties with the Pacific Naval Laboratory (PNL) [later known as Defense Research Establishment Pacific] *just outside Victoria, which was involved in research related to the detection of submarines. The idea was to use magnetometers to detect the metallic hulls of submarines, but in order to do this the effects of the earth's magnetic field and its fluctuations had to be understood. A department led by Erwin Lokken was charged with carrying out this research, and PNL resources helped to support a group at the University of Victoria which also addressed these problems, led by Harry Dosso and John Weaver. As well, research went on at Royal Roads Military College not far from PNL, led by John Duffus and at the federal Department of Energy, Mines and Resources based in Victoria, led by Bernard Caner. Jacobs interacted with all these groups by having members of his Institute in Vancouver carry out joint research activities with researchers in Victoria and environs.*

"In the 1960s, Jacobs established a close working relationship with Sir Charles Wright, a senior researcher at PNL. Charles Seymour Wright was born in Canada in 1887, and is best known as the radio operator on the British Antarctic Expedition (1910-1913) led by Robert Scott. Later in 1946, Wright was knighted for his work with the British Admiralty during the war years and became chief of the Royal Naval Scientific Service. After his retirement from that organization, he eventually ended up joining the staff at PNL and spent the rest of his life studying his last great passion, pulsations in the earth's magnetic field. Sir Charles established a close working relationship with Robert Helliwell of Stanford University, studying the conjugate properties of geomagnetic pulsations through the setting up of instruments at Byrd Station in Antarctica and at Great Whale River (now known as Poste-de-la-Baleine) on the west coast of Hudson Bay. The working relationship between Helliwell, Wright and Jacobs was a significant driver of experimental space science in Canada in the 1960s and 1970s and made the Institute of Earth Sciences a major player in the international effort to study geomagnetic disturbances at auroral latitudes in general, and magnetic pulsations in particular. (Over the period March 25-26, 1965 UBC hosted a Conjugate Point Studies Meeting in Vancouver which brought together most of the key players in geomagnetic and auroral studies in North America to plan future research activities involving international collaborations.)

"In 1967, Jacobs left the University of British Columbia to join the Faculty of the University of Alberta as its first Killam Memorial Professor of Science. One of his first actions was to recruit Gordon Rostoker (who was at that time at the Royal Institute of Technology in Stockholm) who joined the Faculty at University of Alberta in 1968. Rostoker later developed arrays of ground-based magnetometers that culminated in the large-scale CANOPUS project and its present day successor, CARISMA. As he left UBC, Jacobs also convinced a young graduate student, John Samson, to come to Alberta as a doctoral candidate supervised by Rostoker. Samson was to develop into a key member of the space science community and ultimately a force in the growth of the space physics group at the University of Alberta. Jacobs further exercised his ability to develop infrastructure in the Canadian scene by spearheading an effort to obtain a Negotiated Development Grant from the National Research Council of Canada. This effort was successful, and permitted the formation of the

Institute of Earth and Planetary Physics at the University of Alberta at the beginning of the 1970s. [This Institute was later renamed the Institute for Geophysics and Meteorology, and is now known as the Institute for Geophysical Research.]

"In 1975, Jacobs left the University of Alberta to return to his native England where he obtained the prestigious position of head of Geophysics at Cambridge University. After reaching the age of retirement, Jacobs moved to the University of Wales in Aberystwyth where he spent his last days researching and writing before his death in 2003."

Sir Charles Seymour Wright

"Wright's close relationship with Jack Jacobs has just been mentioned. But Sir Charles is such an eminent Canadian, who contributed so much to Antarctic science and the understanding of the Earth's magnetosphere that he deserves a separate brief section.

Sir Charles Seymour Wright, KCB, OBE, MC, MA, was born in Canada in 1887. He was educated at Upper Canada College and the University of Toronto. He won a scholarship to Gonville and Caius College, Cambridge, England, undertaking research in cosmic rays at the Cavendish Laboratory. It was while studying in Cambridge that he met Douglas Mawson, who had been part of the British Antarctic Expedition, 1907-1908 under Ernest Henry Shackleton. Wright applied to join the forthcoming British Antarctic Expedition, 1910 - 1913 of Robert Falcon Scott. He was accepted as physicist, and along with five other scientists spent the first winter at Cape Evans studying glacier ice, snow and sea ice. Magnetism, gravity and aurora were added to these studies the subsequent winter. Scott appointed Wright to be a member of the first supporting party on the polar journey with Edward Leicester Atkinson, Apsley Cherry-Garrard and Patrick Keohane. Wright was later part of the search party of eight men and seven mules who searched for Scott and the pole party. On 11 November 1912 he discovered the party's tent on the Ross Ice Shelf.

On returning to England, he lectured in cartography and surveying while also writing up his scientific work. In 1914, he joined the Royal Engineers as a second lieutenant and served in France. He rose to the position of General Staff Officer in wireless intelligence and was awarded the MC and OBE. Wright joined the Admiralty Research Department in 1919, becoming superintendent at Teddington ten years later. Between 1934 and 1936 he was director of scientific research at the Admiralty. He played an important part in the early development of radar and detection of magnetic mines and torpedoes. He received the KCB in 1946 and took the post of chief of the Royal Naval Scientific Service. He took up several positions in subsequent years, firstly as scientific advisor to the Admiral at the British Joint Services Mission, Washington DC, then in 1951, director of the Marine Physical Laboratory of the Scripps Institute of Oceanography at La Jolla, California. He joined the staff at the Pacific Naval Laboratory at Esquimault, Canada in 1955.

Wright revisited Antarctica in 1960 and 1965. In 1967, he joined the Institute of Earth Sciences, University of British Columbia and Royal Roads Military College, Victoria, British Columbia. In 1969, he retired to Saltspring Island near Victoria in British Columbia. He died on 1 November 1975."

Gerson finds the University of Western Ontario

Gerson was an assiduous listener and one Saturday morning in the fall of 1950 he found himself to be one of the few remaining people in the audience for one of the last papers of the American Institute of Physics meeting in Cleveland. Ralph Nicholls, a young assistant professor from the University of Western Ontario, newly imported from Imperial College, London, England, was giving his paper on laboratory excitation of atmospheric spectra to a very small audience. At the end, Gerson came forward and asked if Nicholls would like a USAF contract to extend this work. Thus he founded a school of laboratory studies in atmospheric spectra that would help in the interpretation of auroral spectra produced in Saskatoon. Nicholls, too, suffered from an initial negative reaction from NRC, but later with Harold Schiff from McGill, went on to establish the Centre for Research in Experimental Space Science (CRESS) at the new York University in Toronto (now the Centre for Research in Earth and Space Science, but still CRESS).

The University of Western Ontario had earlier linkages to space research, particularly radio science, going back to the 1930s, with Gar Woonton, who led the work on radio development there during the war. In 1948 Woonton moved to the Eaton Electronics Research Laboratory at McGill where he made important contributions.

Nicholls not only established this field in Canada, he assisted Gerson in another important way. In order to lay out a master plan for this new study of auroral physics, Gerson wanted to organize an auroral physics conference, to bring together the best and most experienced minds from Europe, and put them alongside the new North American workers. But where to hold it? In principle, Saskatoon would have been ideal, but logistically it would be too difficult. Currie was concerned about the onerous trip to Western Canada (by CN rail or Trans Canada Airline's North Star propeller aircraft), and many visitors were elderly and frail. So London, Ontario was chosen instead, and the young Nicholls thereby became its organizer. The elder giants – Chapman, Størmer, Vegard – all came from Europe must have been a great inspiration to Gerson's proteges, Forsyth, Hunten, Nicholls, and Meinel (from Yerkes Observatory). Meinel reported on new observations of hydrogen emissions in aurora, and his interpretation of incoming protons, spiraling down magnetic field lines. The date was 1951 and it marked the launch of Canadian energies into the auroral field. Forsyth reported the first results from the new radars and Millman reported on results from the NRC radar that Gerson had provided. Forsyth remembers the event as a turning point for the radio community. He showed pictures of the long range radar echoes and of the aurora near the horizon which seemed to have the same configuration. After he had finished, the grand old man of aurora, Carl Størmer, stood up and said that although he, himself, was too old to get involved, he was excited to see the development of this new method of looking at aurora and felt that it was going to produce new insights. From that time on, what had been regarded as a passing fad suddenly became respectable. Radio studies of aurora were finally accepted as being part of the mainstream of auroral studies.

At this time only a little was known about the Earth's atmosphere. There was a rough idea of how its composition varied with altitude, but only a poor idea of its temperature structure. Knowledge of the chemical reactions going on were rudimentary and based on guesswork. Two young Britishers, Harrie Massey, later to found the British Space program, and David Bates, later to establish the world's foremost school of atomic and molecular process theorists at Queen's University, Belfast, were both present. They were later Sir Harrie and Sir David. Both gave the best interpretations of optical auroral observations possible at the conference and established future directions of this work. Another was the great Sidney Chapman, who made major contributions to atomic oxygen and ozone chemistry, to geomagnetism and the origin of the aurora.

CONFERENCE ON AURORAL PHYSICS
UNIVERSITY OF WESTERN ONTARIO, LONDON, ONTARIO
23–26 JULY 1951

Front row, left to right: H. E. Moses, J. F. Carlson, A. T. Vassy, C. Störmer, M. E. Warga, N. J. Oliver, R. M. Chapman, B. T. Darling, L. Herman.
Second row, left to right: A. B. Meinel, T. Y. Wu, S. Borowitz, L. Katz, N. C. Gerson, Chairman, D. Barbier.
Third row, left to right: R. W. Nicholls, D. M. Hunten, S. Chapman, H. S. W. Massey, C. W. Gartlein, J. Vandertuin, E. Vassy, W. Petrie, R. G. Turner.
Fourth row, left to right: D. Schulte, A. L. Aden, H. Alfvén, A. D. Misener, D. R. Bates, R. W. B. Pearse, S. Altschuler, C. E. Montgomery, O. Oldenberg.
Back row, left to right: M. W. Feast, J. H. Blackwell.

The attendees at the University of Western Ontario Conference on Auroral Physics, held in July, 1951. Most of the intellectual giants in the field were present.
Photo by courtesy of Ralph W. Nicholls.

Nate Gerson's impressions of these giants gives us insight into the dynamics of the conference: "Barbier, imposing and dignified; Chapman, imperious and haughty; Alfven, cheerful and imperturbable; Bates, witty and sharp; Massey, quiet and reserved; Størmer, venerable and warm; Ta Ya Wu, formal and correct; E. Vassy, professorial and effective; Herman, careful and weighty; Herzberg, impeccable and precise. ….Chapman always called me Major Gerson (he never realized that after six years in the National Guard I had only attained the rank of corporal)."

Colin Hines at Cambridge University

There was another university at which something important to Canadian space science was happening in the early fifties and it was, yet again, Cambridge University in England. Colin Hines was born in Toronto, attended Jarvis Street Collegiate, and the University of Toronto. In the summers following his third, fourth and M.Sc. years, he was employed in the DRB Radio Propagation Laboratory, and on receipt of his B.Sc. they proposed to support him in going to Cambridge to work with the prominent scientist Jack Ratcliffe, where he arrived in 1951.

As Hines describes it, *"That was the Cambridge of Hermann Bondi, Tom Gold and Fred Hoyle, with their exhilarating and widely promulgated cosmological theory of continuous creation; the Cambridge of the graduate students Francis Crick and James Watson, as yet unknown to the world at large or even to much of the university community, with their nascent unraveling of the double helix."*

It was also the Cambridge of Jack Ratcliffe and his famous ionospheric group. Hines was then interested primarily in fundamental electromagnetic theory, so initially he began work with the renowned Hermann Bondi. This led to a dead end, so he returned to Ratcliffe's group, that was at that time addressing the problems of moving irregularities in the ionosphere, which they called "drifts." Various radio techniques had revealed the irregular radio signals attributed to the motion of perturbations in ionization, but their origin was completely unknown. There seemed then to be three alternatives: 1) electrostatic forcing such asby hydromagnetic influence which would imply an "outside" origin such as the magnetosphere; 2) turbulence carried by winds and 3) atmospheric waves of some sort. As Hines had been studying hydromagnetic waves (waves in the magnetic field) prior to coming to England, this challenge attracted him back to Ratcliffe. However, his knowledge of this phenomenon soon led him to conclude these waves would be too fast to explain the observations and so he turned to atmospheric waves, in density and pressure, as a possible cause, and added atmospheric pressure gradients and gravity to his hydromagnetic formulation. For a very slow atmospheric wave, gravity is important. This progress was sufficient for a Ph.D. and a published paper, and he was back at the Radio Propagation Laboratory by the summer of 1954. It was not the same RPL, as it had been relocated to Shirley Bay by this time, renamed the Radio Physics Laboratory and combined with the Communications Laboratory and the Electronics Laboratory into DRTE.

Here Hines joined a team led by Peter Forsyth on communication using meteor trails. Radio scientists had recently been astonished to find that 50 MHz signals could be received from a transmitter over the horizon. This is too high a frequency to be reflected from the ionosphere, so there must be some other scatterer present. Forsyth, with Eric Vogan, were able to show that the signals came in short bursts and ultimately identified the origin as meteor trails. There was a need in Canada for long range (i.e. over the horizon) communications and the RPL group had shown that this could be done in short bursts by having the outgoing wave reflected from the ionization of meteor trails. The JANET system, with the name adapted from Janus, who looked both ways, was shown to be capable of transmission over 1000 km or so. This offered theoretical questions that were a challenge to Hines. It will help to appreciate the situation to know that there was a dispute going on in the literature between two groups about the nature of the meteor ionization irregularities: were they turbulence or waves? This related back to Hines' thesis, although his interest in that was fading by this time. It was revived by an invitation to write a review paper on "Motions in the Ionosphere" as part of a special issue of the Proceedings of the Institute of Radio Engineers to commemorate the International Geophysical Year, just completed. This was a turning point, but the impact occurred after the IGY. This story is continued in Chapter 5.

Prince Albert Radar Laboratory (PARL)

In 1957 a joint venture of DRB and the United States Air Force (USAF) was agreed upon to investigate the problems associated with the detection of intercontinental ballistic missiles that might be launched over the North Pole destined for North America. The specific problem was the presence of the aurora borealis that might screen the missiles from radar detection. The agencies assigned to implement this project were DRTE in Canada and the Lincoln Laboratory of the Massachusetts Institute of Technology (MIT) in the USA. The USAF provided the high power radar, a sister unit to the one operated by MIT at Millstone Hill near Boston, and a prototype of the radars subsequently installed on the Ballistic Missile Early Warning Line. It operated at 448 MHz with a peak power of 2.5 MWatt and average power of 100 KWatt. The 25 meter antenna was capable of rapid tracking. Del Hansen of DRTE was in charge from 1957 to 1963 and the information here is taken from his notes.

One can't think of Prince Albert in Northern Saskatchewan without thinking of John Diefenbaker, who was Prime Minister at the time. Clearly a northern site was desired and it was also thought important that it be close to a University, particularly one already engaged in auroral research, as the Institute of Upper Atmospheric Physics was established at the University of Saskatchewan in this same year, 1957.

PARL was officially opened on June 6, 1959, just at the end of a Canadian Association of Physicists Congress at the University of Saskatchewan, and many of Canada's leading physicists had made the trip 140 km north of Saskatoon. The feature of the opening was a congratulatory message from President Eisenhower to Prime Minister Diefenbaker, to be delivered from Millstone Hill to PARL via a reflection off the moon is believed to be the first time this had been done.

This spectacular facility made many remarkable demonstrations and observations, beginning with a similar communications relay using the 60 meter reflective orbiting balloon called Echo 1 instead of the moon. After that, satellite tracking became a major activity. Glen Lockwood observed Sputnik IV separate into seven objects including a cabin that would contain a future astronaut. They also observed the decay of several satellites into the atmosphere, Sputnik III which decayed on April 6, 1960; the rocket body of Sputnik IV on July 17, 1960 and the rocket body of Sputnik V, which decayed on September 23, 1960. Later it observed Alouette I on its first pass, traveling south over Alaska following its launch southward from Lompoc, California. It also tracked ARCAS rockets launched from Cold Lake, Alberta and numerous Black Brant rockets launched from Churchill.

A remarkable scientific development was taking place in parallel with this. In 1958 Bill Gordon gave a seminar at Cornell University in which he described the scattering of radio waves from the ionosphere by a mechanism called "incoherent scatter." The ionosondes used by Henderson and Rose to observe the 1932 solar eclipse depended on "coherent scatter" in which the ionospheric electrons oscillate together as a body to generate the relatively strong returned radio signal. In incoherent scatter each electron acts independently and it is the randomly superposed reflections from each that form the returned signal, predicted to be extremely weak. It was expected to be weakened further, since the returned frequency would be broadened by the thermal velocity of the electrons, spreading out the signal. However, when Ken Bowles, a Cornell graduate student made the first observations using equipment at the University of Illinois in 1959, he found the returned frequencies to be contained in a narrow line, characteristic of the slower moving ions. This was completely unexpected and it took some time for the theory to be worked out, as plasma physics was still in its infancy. What happens is that the electrons do scatter, but even though they are "free" electrons, separate from their atoms, their motions are still controlled by the slower ions, to which they are tied by electrical forces.

This gave the University of Saskatchewan and DRTE scientists a remarkable window of opportunity, as they had by coincidence a radar powerful enough for these observations, right on their doorstep, and this research began in 1960. However, because of an interruption in the observations caused by a fire, impacting heavily on some student theses, coupled with Forsyth's departure and the closing of the facility in 1967, the potential scientific benefits were never achieved. PARL is still being used today, not as a radar but as a receiving station for satellite downloads of data by the Canadian Space Agency.

A facility for incoherent scatter is now scheduled to return to Canada, by a strange route. The National Science Foundation (NSF) of the US had operated incoherent scatter facilities in a number of locations including Arecibo, Puerto Rico (the largest in the world), Huancayo, Peru; Chatanika, Alaska; and Sonde Stromfjord in Greenland, but wanted to locate one near the magnetic pole, in Resolute Bay, Canada (74° N). When they put this $25 M proposal forward to the Senate Appropriations Committee, the Chair at that time was Senator Ted Stevens of Alaska. This was the time of the dispute between Canada and the US (Alaska) over salmon on the Pacific Coast, the so-called "salmon wars." The response was that the NSF could have their radar, and put it anywhere they liked, so long as it was not in Canada. Since then, the NSF regrouped, and designed a highly advanced system that is mobile. This mobile $44 M system, called AMISR (Advanced Modular Incoherent Scatter Radar) has established its first element at the University of Alaska in Fairbanks and the second is currently being set up in Resolute Bay, Canada. The salmon wars over, in 2007 incoherent scatter came back to Canada.

Postscript

Space science took root in Canada because of individuals with experience in studying space from the ground, and these individuals held key positions in university and government. However, while the individuals were ready, the institutions were not. This onset of activity was triggered from outside the country, by grants and gifts from the US Air Force.The NRC, the DRB and the universities responded very quickly to the challenge, however, and the enterprise soon took shape as a Canadian activity with US collaboration. The military rationale for space science was based on the need for northern communications, but this was never a problem on university campuses, despite the sensitivity to war that many students felt strongly at this time. Balfour Currie made clear from the beginning that classified research would not take place on "his" campus, and that was sufficient.

References:

Friends of the CRC website: http://friendsofcrc.ca/
Gerson, N.C., Collaboration in geophysics – Canada and the United States 1948-1955, Physics in Canada, Volume 40, No. 1, pages 3-8, 1984.
Hines, Colin O., Earlier days of gravity waves revisited, PAGEOPH, 130, 1989.
Nicholls, R.W., Nathaniel C. Gerson, Physics in Canada, March/April, 2002.
York, Derek, J. Tuzo Wilson, Rock Stars in GSA (Geological Society of America) Today, September, pages 24,25, 2001.

Chapter 4

The IGY and its New Moons

How the IGY began

The IGY was a grand idea that could have become mired in its own bureaucracy, but it worked extremely well because of the enthusiasm of the participating scientists that were allowed to organize it in the ways they knew best. How does one organize 30,000 scientists working at 4,000 stations from 67 different nations? In the period following World War II, each country had an agenda in relation to its security. The Arctic was of particular concern because it was where the next war might have been fought – the knowledge of the Arctic would then have been of paramount importance. It was in the best interests of each country to know just much scientific knowledge its potential adversary had. In the US this was handled by Vannevar Bush, then president of the Carnegie Institution of Washington (this is well described by Korsmo, 2007). Bush headed the US Research and Development Board, formed in 1947. The Board had a civilian chair and two representatives from each of the army, navy and air force, and it reported to the secretary of defence. It was divided into committees, panels and working groups that covered all of the physical, medical, biological and geophysical sciences. Out of this the responsibilities were divided among the different authorities. Thus, for example, sea ice was the responsibility of the navy, and land-based snow that of the army. The Executive Secretary for the Board was Lloyd Berkner. Berkner also worked for the US Department of State where he advanced the idea of solving global problems through scientific knowledge and technological advances.

The originating event of the IGY is said to be a dinner party held on April 5, 1950 at the home of James Van Allen in Silver Spring, Maryland, where the attendees were Lloyd Berkner, J. Wallace Joyce, Ernest Vestine, S. Fred Singer and a special guest from England, Sydney Chapman, on his way to an upper atmosphere meeting at Caltech (sponsored by the armed services). During the evening the idea of a third international polar year arose in a conversation between Berkner and Chapman. The latter strongly endorsed the year 1957-58, because it would be a maximum of solar activity. The idea went to the Caltech meeting as a group of twenty, now joined by Marcel Nicolet of Belgium, and then in July to a meeting at Penn State University. From there it went through three scientific unions, the IAU (International Union of Astronomy), URSI (Union of Radio Science International – translation from the French) and IUGG the International Union of Geodesy and Geophysics to the senior worldwide body, ICSU (International Council of Scientific Unions). ICSU approved the proposal and set up a special committee to organize it – CSAGI (Comité Spécial de l'Année Géophysique Internationale) and invitations went out to all nations and unions worldwide in the spring of 1953. There was an enthusiastic response, but also a suggestion that the scope be broadened from a polar study to a worldwide study and Chapman suggested the name International Geophysical Year (IGY). The USSR was not an adherent of ICSU at that time, but got a special invitation later, and responded to it some 18 months afterwards. At the CSAGI meeting in Rome, in the fall of 1954, the USSR delegates sat in silence as the US plans to launch a scientific satellite were discussed. Finally, at the Brussels meeting in 1955, the USSR plan was put on the table. They offered to provide 15 of the 48 ships required for oceanography. They would establish three new permanent seismic stations in the Arctic, and would implement a comprehensive program on sea ice, permafrost and hydrology in the Arctic. This stimulated the US to enhance its own plans for the Arctic and Antarctic. The Working Group on Rockets and Satellites was meeting in Washington, D.C. when Sputnik 1 was launched; its member were stunned.

Nate Gerson was Recording Secretary for the US National Committee for the IGY and Hugh Odishaw was Executive Secretary, later Executive Director. Joseph Kaplan of the University of California (Los Angeles) was the chairman and Alan Shapley Vice-Chairman. (As an aside, Kaplan and the Norwegian scientist Vegard shared the name of one of the molecular nitrogen bands that Hunten had studied – the Vegard-Kaplan band). During the CSAGI meeting in Rome in 1954, mentioned above, the delegates had an audience with Pope Pius XII (Eugenio Pacelli), a large-group audience held specifically that morning for all the delegates, and the wives who were present. It was held in the Sistene Chapel and the Pope repeated his speech in at least four different languages. It was on the benefits worldwide of scientists collaborating on geophysical research. This story was kindly provided by one of the attendees, Nate Gerson's wife Sareen Gerson.

The point is, however, that all countries, whatever their political or security interests, underlying their generosity in funding the IGY, had turned this activity over to their scientists. This was no less true of the USSR, whose military had agreed to launch scientific satellites using military capability. The CSAGI had to invent many new concepts to make the activity work, including the World Data Centers and an Antarctic Treaty declaring that "Antarctica shall be used for peaceful purposes only", putting to an end some previous posturing about territorial claims to this region. In fact, the formal agreement did not come about until well after the IGY. On May 2, 1958, President Dwight D. Eisenhower issued identical notes to the relevant governments proposing that a treaty be concluded to ensure a lasting free and peaceful status for the continent. Talks by 12 governments began in June 1958 in Washington, D.C. and continued for more than a year. The Antarctic Treaty was signed on Dec. 1, 1959 and enacted on June 23, 1961. Fifty years later, however, nations are talking about claiming their continental shelves in the Arctic. In many ways the IGY continued long afterwards; COSPAR may be thought of as one extension of it. The measurements of CO_2 that began then, fortunately became permanent and much later established the basis for the global warming alert.

Canadian response to Sputnik

The following story of the Sputnik 1 launch was written by Peter Forsyth and is taken from the "Friends of CRC" website.

"And the IGY brought what was for me my last truly exciting few weeks in RPL. We had for several months been expecting the launching of the American satellite but its orbit was to be at low latitudes so that we would not be able to make any use of it. Suddenly in October 1957, Sputnik was launched. Not only was it passing regularly over Canada but it carried a radio beacon that was intended to assist in the tracking process. Since the launch was unexpected there were few, if any laboratories outside the USSR set up for the determination of the satellite orbit, but the determination of the orbit at many places in the world was important and the earlier the better, because this would give new information about the earth's gravitational field and about its atmosphere.

"Because the launch was unexpected all the radio observatories and laboratories in the Western World were starting even in a light-hearted 'race' to see who could first determine and describe the satellite orbit.

"I remember hearing the first announcement of the satellite and its radio beacons on the CBC News. Within minutes Clare Collins had agreed to meet me at the lab and within hours we had picked up the Sputnik signal and were devising methods to determine the precise location of the satellite each time it approached Ottawa. As they showed up to work on Monday morning, others, including Colin Hines, were recruited. As word spread of our initial progress offers of assistance came in from the National Research Council's Radio and Electrical Engineering Division and from the Department of Transport's monitoring station. All such offers were gratefully accepted, because we were learning as we went. Most of us hadn't thought about orbital motion since undergraduate days but we relearned what was needed in a few days, or at least what we thought was needed. Later, nature (and the Russians) trapped us neatly. Sputnik 2 was launched into an orbit for which many of the approximations that we had made for Sputnik I proved to be invalid. Fortunately, since we were still doing all our calculations by hand we realized something was wrong. Some of the other groups who were using computers kept churning out quite ridiculous orbital parameters for some time. But for Sputnik I all went smoothly. After three nearly sleepless days and nights of observations and calculations we had narrowed the possible orbits down effectively to two and here the NRC people were able to give us a single observation that eliminated the ambiguity. We had the orbit and happily sent it off by telegram to the World Data Center in Washington. Later it was confirmed that this was the first valid orbital determination made and reported, at least in the Western Hemisphere and probably one of the first, if not the first in the Western World. An accomplishment of no great lasting import because we all learned quickly to use much more sophisticated techniques for tracking satellites. But it did represent one more of those occasions which seemed to come often at RPL when a group of scientists could share the high excitement, the unique comradeship and the rare sense of fulfillment that comes from tackling together a challenging and demanding physical puzzle."

What lay behind the IGY?

If the Second IPY of 1932-33 (IPY-2) was a major effort for Canada, the IGY of 1957-58 was a mammoth undertaking, involving some 78 observing stations across the country and its territories. By 1957, the world, as well as science, had greatly changed. IPY-2 was carried out during the depression, with very limited technology, and with few individuals capable of conducting the field measurements. In 1957, with the rooting of space science in Canada following the war, as described in the previous Chapter, all of the technologies evolving from their development during World War II, and the crop of new "space scientists" emerging from Canadian universities, it was feasible, given the organization and the funding, to construct a multi-dimensional picture of Canada. As it turned out, both were available. The dimensions were broad, from geomagnetism, aurora, solar activity, meteor studies and cosmic rays to those then seemingly remote from space science: glaciology, oceanography, seismology and the Earth's gravity. Later all of these disciplines would be encompassed by the space age, but at that time glaciology meant trekking across Canada's glaciers. Surprisingly, the reader has already met the glaciologists involved: the glaciology project (Project IX in the Canadian IGY program) was under the direction of J. Tuzo Wilson, and the expedition to the Salmon Glacier in British Columbia was headed by Jack Jacobs.

The stimulus of the IGY was that of exploring a country that was still scientifically unknown in many of its aspects. The US had a similar desire, but on a global scale; for them the Antarctic was the vast unknown and they invested $12 million in the IGY, then an enormous figure for science. Similarly, Russia wanted to know more about its broad regions. All of these national desires resonated in a vast spirit of collaboration, coming at the end of one war, and at a time when the Cold War was looming.

A large Canadian IGY Advisory Committee was set up under the Chairmanship of Frank Davies (Assistant Chief Scientist, DRB). It has to be remembered that this activity wasn't considered "space" – it was much broader, so this committee was constituted as a subcommittee of the NRC's Associate Committee on Geodesy and Geophysics, the national committee relating to the International Union of Geodesy and Geophysics (IUGG), of which Tuzo Wilson happened then to be President. The other members of the Canadian committee were C.S. Beals (Dominion Astronomer), B.W. Currie (University of Saskatchewan), K. Davies (affiliation unknown), P.A. Forsyth (DRTE), H.B. Hachey (Joint Committee on Oceanography, New Brunswick), J.A. Jacobs (University of Toronto), R.F. Leggget, (Division of Building Research, NRC), J.E. Lilly (affiliation unknown), J.L. Locke (Dominion Observatory), R.G. Madill (Dominion Observatory), Don McKinley (REED, NRC, A.G. McNamara (REED, NRC), P.M. Millman (REED, NRC), W. Petrie (DRB), G.W. Rowley (Northern Affairs and National Resources), J.C.W. Scott (DRTE), and A. Thomson (Meteorological Branch). Frank Davies resigned as Chairman in 1957 and was succeeded by Don Rose, who continued until 1960. However, progress was slow until Rose established a smaller Coordinating Committee, consisting of Beals, Davies, McKinley, Currie and Mahoney (Secretary to both Committees). From that point on, planning moved rapidly in an organized fashion. The participating groups were as shown in the list below.

Participating Group	Name	Affiliation
Meteorology	W. Godson	Meteorological Branch
Geomagnetism	R.G. Madill	Dominion Observatory
	B.W. Currie	University of Saskatchewan
	G.D. Garland	University of Alberta
	F.H. Sanders	Pacific Naval Laboratory
Aurora	P.M. Millman	REED, NRC
	A.G. McNamara	REED, NRC
	F.R. Park	REED, NRC
	B.W. Currie	University of Saskatchewan
	D.M. Hunten	University of Saskatchewan
	A.V. Jones	University of Saskatchewan
Ionosphere	J.S. Belrose	DRTE, DRB
	B.W. Currie	University of Saskatchewan
	W.B. Smith	Department of Transport
Solar Activity	A.E. Covington	REED, NRC
	D.A. MacRae	University of Toronto
	T.R. Hartz	DRTE, DRB
	V. Gaizauskas	Dominion Observatory

Cosmic Rays	D.C. Rose	Pure Physics Division, NRC
	J. Katzman	Pure Physics Division, NRC
	H. Carmichael	Atomic Energy of Canada
	J.F. Steljes	Atomic Energy of Canada
Latitudes and Longitudes	M.M. Thomson	Dominion Observatory
Glaciology	J.T. Wilson	University of Toronto
	S. Orvig	McGill University
	T.A. Harwood	Physical Research, DRB
	G.F. Hattersley-Smith	Physical Research, DRB
	R.F. Legget	Building Research, NRC
	L.W. Gold	Building Research, NRC
Oceanography	H.B. Hachey	Fisheries Research Board
	N.G. Gray	Mines & Technical Surveys
	W.L. Farquharsen	Mines & Technical Surveys
Rockets and Satellites	D.C. Rose	Pure Physics Division, NRC
	J.H. Chapman	DRTE, DRB
	J.W. Cox	Physical Research, DRB
	D.W.R. McKinley	REED, NRC
Seismology	J.H. Hodgson	Dominion Observatory
	P.L. Willmore	Dominion Observatory
Gravity	M.J.S. Innes	Dominion Observatory
Nuclear Radiation	C. Garrett	Applied Physics, NRC
	J.L. Wolfson	Applied Physics, NRC
	P.M. Bird	National Health & Welfare
Meteor Studies	P.M. Millman	REED, NRC
	I. Halliday	Dominion Observatory

The makeup of these groups was the one existing at the end of the IGY, satellites did not yet exist in the minds of the Canadian planners. The scope of the effort was striking, with a major institutional role for NRC and lesser ones for others such as National Health and Welfare, and involved even disciplines as apparently unrelated as Building Research. Warren Godson, an extremely able atmospheric scientist was from the Meteorological Branch of the Department of Transport, since at that time the weather was of prime concern to transport, particularly aircraft; the Department of Transport was also involved in the ionosphere because of communications. These activities later became part of Environment Canada when that Department was created. The Dominion Observatory included geomagnetism, solar activity, longitudes and latitudes, seismology, gravity and meteor studies. Later these activities were divided between the Herzberg Institute of Astrophysics of NRC and Natural Resources Canada. The Pacific Naval Laboratory of DRB in Esquimalt B.C. was also involved – in geo-magnetism. And in cosmic rays, Atomic Energy of Canada, Ltd. As already noted, many of these areas have since been recognized as space activity, although this was not obvious at the time. Longitudes and latitudes, for example, fell under stellar astronomy then (in the Dominion Observatory) but it now is very much identified with the Global Positioning System (GPS), the well-known US system of satellites, which the public now widely use to find their locations while hiking or driving. Globally this technique is known as GNSS (Global Navigation Satellite System), including the European Galileo satellites and the Russian GLONASS (Global Navigation Satellite System). However, the discussions here are limited to those areas recognized as space-related at the time.

The observations were of two types, continuous recording, or following a set schedule. The International committee established a system of "world days" on which observations would be concentrated. These began on June 8, a few days before the IGY official starting date. During the summer of 1957 the days were July 4, 26, 27 (new moon), August 12, 25 (new moon), 26, September 1, 23 (new moon), 24 and 30. The new moon days were always included, with an adjacent day for additional observations.

However, the early days of planning were rather rough. On January 6, 1955 Jack Meek, then at the University of Saskatchewan, sent a letter to R.G. Madill of the Magnetic Division of the Dominion Observatory, outlining the cost of operating eight magnetometers in addition to the existing ones at Saskatoon and Meanook; these would cover the province of Saskatchewan, from Val Marie at 49° N latitude in the south to Uranium City at 59° in the north. The cost of the equipment, one year of operations and compilation of data was estimated at $52,500. He wanted to add an auroral intensity recorder, or panoramic camera. In a hand-written undated document, Don Hunten estimates the cost of the intensity recorders as $1,500-$2,000 and patrol spectrographs as discussed later at $1,000 - $1,500. Meek also asked for four years of salaries. On February 9, Davies wrote to Currie to report on the conclusions of a small group meeting on the IGY. While endorsing the strong science, Meek's proposal was criticized as being a "local" experiment not coordinated to the IGY global plan; the budget was also thought to be underestimated by a factor of two – there was also concern about whether Meek would be in Saskatoon to run the program.

Davies also chaired the meeting of the full committee on February 14, 1955 with Forsyth acting as Secretary, in place of Bill Petrie, who was now in DRB headquarters. They reviewed the proposed Canadian program in light of recommendations made by CSAGI. Geomagnetism would fall under the Dominion Observatory, the Ionosphere under the Radio Physics Laboratory of DRTE while aurora was mainly assigned to the NRC, with recommendations for Meinel type auroral all-sky cameras at Meanook, Resolute Bay, Baker Lake, Churchill, Saskatoon, Victoria, Yellowknife and Norman Wells. It was also recommended that auroral intensity recorders be operated at the same sites, but also in Winnipeg, Eskimo Point, Gillam, and The Pas. If patrol spectrographs became available commercially they should be operated at Meanook, Resolute Bay, Baker Lake, Churchill and Saskatoon. It further recommended that the NRC be responsible for developing a Meinel-type camera, a prototype auroral intensity recorder and prototype auroral radar equipment. Concerning cosmic rays, a search was on for a mountain site in Western Canada above 2100 meters altitude, accessible by road and with 15 KVA of electrical power.

By March 28 Rose had taken over and had written a letter to Currie about how to get the auroral work started. In the letter Rose says that "The Physics Division at N.R.C. has no background in auroral work and could not do much to help with the instrument. I think, therefore, that the details of the Aurora program have to be reconsidered. You will have seen by now that I am leading up to saying that your knowledge and experience in the auroral field are so much greater than anyone else's that we would have to rely on your advice to such an extent that we may as well rely on you for a large part in the organization of the auroral work." Currie responded quickly with an affirmation that he was willing to take on this work and an estimate for an "all up" Saskatoon station at $47,500 per year for two years, and $17,000 the year following. Then he described a "small network" of three stations for which the equipment would cost $15,600. Meek's proposal was estimated at $32,000. By June they were discussing the details of who should make the intensity recorder, and getting Currie an advance of $2,000, mostly for his travel. Currie then wrote letters on June 20, 1955 to Christian Elvey of the University of Alaska about the all-sky cameras. Although Meek already had one operating, Currie wanted the Canadian and US cameras to be compatible. He wrote to Norman Oliver of the Air Force Cambridge Research Center on June 20, asking whether there was a source where all-sky cameras and patrol spectrographs could be purchased. He also received a letter from Franklin Roach, a US pioneer in airglow studies, encouraging him to attend the upcoming Technical Panel on Aurora and Airglow (of the IGY) in Boulder. Roach wanted to establish a line of stations connected with Canada that ran down through Montana, Sacramento Peak (near Alamogordo, New Mexico), Mexico City, Huancayo (in Peru) and San Juan (Argentina). Currie also received a hand-written note from Don Rose, written on the weekend, urging him to attend and confirming that he would be reimbursed. On July 5, in a letter to Carl Gartlein of Cornell University, New York, head of the visual auroral observations activity, Currie said that it would be difficult for him to attend, for family reasons, but on the same day he wrote to Roach, saying that he had requested airline reservations and arrived in Denver on July 17.

On July 20, Currie sent a letter report to Don Rose on the meeting. He noted that the US Congress had allocated $12 million for the IGY, of which aurora and airglow would receive receive $950,000, about one-third of that to be spent in Alaska, and one-half of that would go to radar observations. The all-sky cameras had first priority, but Meinel's version would not be used, for it was the version developed by Elvey that was favoured, which was very similar to Meek's design. Currie somewhat reluctantly admitted that the Canadian units would have to be built in Canada, at Saskatoon if necessary, but he would prefer to have this done by a "small firm." At the meeting Currie committed Canada to installing cameras at Meanook, Resolute Bay, Baker Lake, Churchill, Saskatoon, Yellowknife and Victoria. The Panel wanted more: Moosonee, the Rockies, somewhere north of the Gulf of St. Lawrence and Chimo. Elvey wanted to install a radar at Aklavik, and would likely add an all-sky camera as well, but Canada could fill an important gap at Uranium City, or Lac La Ronge. Currie was skeptical about Gartlein's plan for large numbers of hourly visual observations, even if the results were to be put on punched cards. The patrol spectrographs also had high priority and the panel had already spent $11,000 getting the design work done by the well-known optical company, PerkinElmer; on the basis of which they would go out for bids – the estimated cost was $7,000, much higher than Hunten's estimate (for a simpler instrument developed by Vallance Jones) of $1,000 to $1,500. The problem with Vallance Jones' instrument was that an operator was required. Currie hoped that Canada could buy five of these along with the US purchase of 15 units.

Currie reported on the Canadian radars, based on the design by McNamara. The US requested four of the scanning spectrometers developed by Hunten; and Currie agreed to send them drawings. There was interest in Hunten's auroral intensity recorder, but no decisions made. One important aspect for the IGY as a whole was the handling and exchange of data. It was agreed that any country contributing data would be entitled to receive all the data to be distributed. This was in fact one of the great concepts and successes of the IGY. There was agreement (by CSAGI) that there should be a World Data Center for auroral, magnetic, ionospheric, cosmic and solar data, storing all of the IGY data. The US was one obvious location. But CSAGI was aware that the Soviet Republics, who also had a major stake in the IGY would want to host it. The other problem with the US was the problem for many scientists to get visas to go there to look at the data. In the end, the centres were located in different countries.

Meteorology

The Canadian meteorological observations were ambitious, with aerological balloon measurements, either radiosondes or rawinsondes of pressure, temperature, humidity, and wind direction and speed. These were performed twice daily, at 33 stations. Larger balloons were also used at seven sites in order to reach into the stratosphere. At Resolute Bay (74° N) the number of flights was doubled to four per day. The total amount of ozone in the atmosphere was measured with Dobson spectrophotometers, but the only three IGY sites were Alert (82° N), Edmonton, and Resolute Bay. As described in more detail later, Dobson spectrophotometers work by measuring the absorption of solar ultraviolet light in the ozone spectral bands, but at high latitudes during the long winter the sun is not available, and the moon has to be used instead. The monitoring of ozone in the stratosphere that began during the IGY continued on afterwards and reaped enormous benefit in the recognition of the "ozone hole" in the Antarctic when it was first observed. This long length of data record helped scientists and policy-makers worldwide in establishing the Montreal protocol for limiting the release of chlorofluorocarbons into the atmosphere. Although "global warming" was not in the vocabulary then either, they did measure the incoming solar radiation at 17 sites, as well as the light reflected by the Earth's surface (Moosonee was an important site). Resolute Bay was a major station with many additional measurements including the ozone concentration at the Earth's surface; profiles of temperature, humidity and wind using an instrumented 30-meter tower; and heat flow, through the Earth's surface and cloud heights.

Geomagnetism

Measurements of the Earth's magnetic field using photographic recording were made at seven stations, Agincourt (near Toronto), Meanook, Victoria, Baker Lake, Churchill, Resolute Bay and Yellowknife. Magnetic variations were measured with electronic magnetometers at the same stations, but additionally at Eskimo Point, Flin Flon, Gillam, Ottawa, Saskatoon, Swift Current and Winnipeg. Earth currents, currents flowing through the Earth induced by magnetic variations were measured at Meanook. About half of these sites were operated by the Dominion Observatory and the other half by the University of Saskatchewan.

Aurora

Seven types of instruments were deployed for auroral observations: 1) all-sky cameras; 2) auroral intensity recorders; 3) auroral radar; 4) patrol spectrograph; 5) scanning spectrometer; 6) visual observations and 7) auroral height determinations. This program was under the joint direction of the University of Saskatchewan and REED at the NRC. The U of S took the prime responsibility for the organization of the field observations while the latter acted as the collecting and data distributing headquarters.

The auroral all-sky camera was invented by Aden Meinel, at Yerkes Observatory of the University of Chicago. However, as described earlier, a simple version was independently conceived by Jack Meek of DRB, during the time he was at the University of Saskatchewan. It consisted simply of a 16-mm movie camera, mounted on legs 115 cm above a convex parabolic mirror, achieving a 160° view of the overhead sky. Pins on the legs indicated the elevation above the horizon and needed because the system had significant distortion. Exposures of 15 to 30 seconds were made, once each minute and then the film was advanced. A more professional system was developed by REED at NRC, using 35 mm film. The latter were deployed at nine stations: Alert, Baker Lake, Resolute Bay, Victoria, Yellowknife, Churchill, Meanook, Ottawa and Saskatoon. These were supplemented by the 16-mm cameras at Flin Flon, Regina, Saskatoon and Uranium City. Ottawa obtained the largest amount of data, 162 rolls of 30 meter film, with Baker Lake the least, at 26 rolls.

Jack Meek's All Sky Camera, on the roof of the Physics Building at the University of Saskatchewan. By courtesy of the Department of Physics and Engineering Physics, University of Saskatchewan.

Supplemented by similar systems in Alaska and Northern Russia, this assembly of cameras provided a near-total view of the aurora over the polar regions, shared by auroral scientists worldwide. Syun-Ichi Akasofu, a US scientist of Japanese origin at the University of Alaska was one of the most dedicated analysts, in this laborious task of piecing together thousands of images to reveal how the global distribution of the polar aurora changed with time. In so doing he discovered the concept of the "auroral sub-storm." From a single location, the auroral behavior appears almost random, with no "rhyme or reason." Akasofu's discovery was, that when viewed with a network of cameras, the aurora normally existed globally as thin arcs, roughly extending along the lines of constant magnetic latitude (latitude based on the magnetic pole) and this condition could persist indefinitely, sometimes for days. Since Churchill was close to this "home" position of these arcs, it was possible to see these practically every night. For Saskatoon, for example, this quiet-time aurora would be a little too far away, over the horizon. Akasofu identified disturbances, or events, that would begin by a brightening of the most equatorward arc, then an equatorward expansion of the whole system, followed by a rush northwards, with the

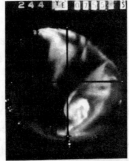

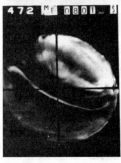

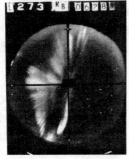

All sky auroral images taken in Canada during the IGY, taken from the Canadian IGY report.

arcs folding back on themselves in the evening sector. This he called a "substorm", a small storm. After roughly an hour, the activity would die away, leaving patches of aurora, particularly in the morning, where the aurora became patchy and pulsating, with the patches appearing and disappearing every ten seconds or so. Then it would eventually settle down to the quiescent state, whereupon a substorm could re-occur later the same evening, or perhaps not for several days.

Across the Arctic Ocean, Yasha Feldstein was piecing together the USSR observations. Since the magnetic pole is south of the actual north pole and in Canada, one has to go to much higher geographic latitudes in Europe to achieve the same magnetic latitude as in Canada. Thus the quiet-time aurora is viewed at higher geographic latitudes, which have more hours of darkness in winter, and Feldstein had access to all-sky images from stations that had darkness for 24 hours per day in mid-winter; these were well inside the Arctic Circle. Feldstein was able to follow Akasofu's night-side auroral arcs around to the dayside, tracing out a complete ring of aurora. Feldstein found that on the Earth's dayside the aurora was much closer to the magnetic pole, at about 78° magnetic latitude, compared to the night-side quiet-time location of about 67°. In other words, the ring pattern was offset from the magnetic pole, leading Feldstein to call it the "auroral oval." Thus Feldstein was able to describe the complete spatial pattern of the quiet-time aurora, superimposed on which Akasofu described its dynamical behavior. To return to Saskatoon, at 60° magnetic latitude, this picture makes sense of what an observer sees. In the early evening, the quiet-time aurora is over the horizon, but as midnight is approached the gradual shift of the auroral oval equatorward causes it to appear over the northern Saskatoon horizon. In fact it is more accurate to think of the auroral oval as fixed to the sun, with the Earth rotating under it, carrying Saskatoon beneath the oval. If the night is quiet, the arc hangs in the northern sky until midnight and then retreats northwards. But if a substorm occurs, an auroral "display" is seen, with aurora all over the Saskatoon sky. What is actually seen depends on whether the substorm occurs before Saskatoon's midnight, or after. If it is before midnight, the folded arcs (called bands) become evident, rushing westward across the sky, but if it is after midnight, the observer sees the appearance of diffuse and pulsating aurora.

Balfour Currie was personally very interested in the all-sky camera results, with his long association with the behavior pattern of the aurora. He assigned a graduate student, John Black to analyze the Canadian data. Black concentrated on Saskatoon and the lower latitude stations and found that by cranking the 30 m rolls of film rapidly, by hand, through his viewer, he could identify auroral motions as in a movie that were difficult to detect any other way. In this way he identified one feature not reported before, an extended (east-west) sheet of weak aurora, located equatorward of the bright features (arc and bands) that have been described. This placed it considerably closer to Saskatoon, where it could be seen very regularly, although it took care, and a dark location because of its lower intensity. Black reported on this feature very precisely in his Ph.D. thesis but as it didn't relate to anything anyone had reported for the aurora previously, he didn't have an explanation for it; he was stumped, and so were his examiners, in spite of their more extensive experience and knowledge. This unusual aurora wasn't named until after the global aurora was observed from a satellite, where its features could be easily seen. The first such sighting was from the Canadian ISIS II satellite by Clifford Anger, who called it the "diffuse aurora." This led to a distinction between the "discrete" more poleward auroras and the "diffuse" aurora. Unfortunately, Black, who took up a faculty position at Brock University, never received recognition for his finding, but it was a significant Canadian discovery and it is appropriate to record it here.

The auroral intensity recorder was an early example of technology transfer from university to industry sector. Based on an earlier design by Bill Penn and Balfour Currie, and at the instigation of Peter Forsyth, such an instrument, also called a meridian scanning photometer, had been constructed and operated on the roof of the Physics Building at the University of Saskatchewan. It had a 45° mirror, rotating about a horizontal axis so that it scanned a spot on the sky from the northern horizon, up to overhead and down to the southern horizon (along the meridian), with the incoming light reflected in the mirror sent off in a horizontal direction to a photomultiplier detector. This would give less information than an all-sky camera, but the intensities would be more accurate, and more sensitive, because the detection was done electronically. However, one has to distinguish between auroral light and moonlight or local lights scattered off clouds, but that can be done using a spectral filter, since the auroral light occurs at specific wavelengths, in spectral lines, and the well-known auroral green line, from atomic oxygen at 557.7 nanometers wavelength, is one of the strongest lines in the aurora. Hunten had developed a facility for making interference filters in the laboratory and developed the concept of wedge interference filters, where the transmitted wavelength varied across the filter. This was made the basis of a new instrument, but to manufacture the required twelve instruments, to be installed from Alert to Winnipeg, would be too much for the Physics Department machine shop, so the job was contracted out to industry. The effort required to transfer such knowledge and technology to industry was underestimated, particularly when these instruments were intended to operate untended, in harsh environments, with operators having no detailed knowledge of the instrument. Not only that, but the recording was to be done on coded paper tape, to be processed on an early computer, an LGP-30, for which the capability and experience were not mature. In the end, only the Saskatoon and Churchill instruments, which had experienced operators, were able to provide data. But it was an interesting and educational experience with industry.

Donald Hunten (left) and Alister Vallance Jones inspecting the IGY auroral intensity recorder.
Photo by Courtesy of the Saskatchewan Archives Board StarPhoenix Collection, call number S-SP-B-2393-2"

The auroral radars were built at the NRC, based on Al McNamara's earlier experience at the University of Saskatchewan. These operated at 6 m wavelength, with the results recorded photographically and on paper tape. They were located at Baker Lake, Ottawa, Resolute Bay and Saskatoon. Additionally the earlier radars at Saskatoon were operated intermittently at 56, 106 and 196 MHz. At Saskatoon, it had been found that the relationship between radio aurora and optical aurora was very complex, involving the angle between the look direction and the magnetic field. At each of these stations, the relationship to the magnetic field was different, and each had to be understood individually. This work produced a wealth of data that has yet to be fully understood.

IGY laboratory buildings at Resolute Bay. The optical instruments are mounted in the tower at the right-hand side, with the all-sky camera sitting on top. Photo by courtesy of A.G. McNamara.

The patrol spectrographs were ingenious instruments in which the spectra were recorded along a strip of sky from north, to overhead, to the south horizon (like the intensity recorder). The photographic recording would display the position on the sky in one direction and wavelength in the perpendicular direction. This was important as auroral light at different wavelengths had different characteristics that would vary independently across the sky. One such spectral line, for example, was from hydrogen atoms, produced from auroral protons coming into the atmosphere. Although incoming electrons are responsible for producing most of the auroral light that observers see, the protons provide important information as their magnetospheric origin is different. They follow trajectories different from those of the electrons; these can be distinguished in the recorded images. It was planned that five of these be operated during the IGY, two by the NRC (Baker Lake and Resolute Bay), one by DRB (Churchill), one at Meanook by the Dominion Observatory and one at Saskatoon by the University there. In fact, only the unit at Saskatoon obtained results, from January 1 to December 31, 1959.

The scanning spectrometer was the one that had been built by Hunten; it was used for specific studies, but not used on a routine basis.

The visual auroral observations, even with the advantage of many observers, and a modernization through recording their observations on IBM computer cards, could not compete with the more high technology observations.

The auroral height finding was based on earlier work at Saskatoon (and before that in Chesterfield Inlet and in Norway) in which parallactic photographs were taken from two stations, with the heights determined used to associate with the rotational temperature measurements obtained with a spectrometer. This was undertaken at Churchill—which brings to the fore the Defence Research Northern Laboratory (DRNL), established in 1947 and operated until 1965—before it was handed over to another agency. At the time there was concern about possible land invasion of the Arctic and it became imperative to determine the equipment, clothing, feeding and tactical deployment, navigation, re-supply and a host of associated problems for Canadian defence forces. The early days of the laboratory involved working in very primitive conditions. Clothing and living quarters (tents) were of primary concern. In summer, mosquitoes were a problem, and research was done on that too, with as many as 250 bites per minute on one arm between the wrist and elbow being recorded in tests. There were also tests on petroleum products and the use of cold weather lubricants. But in 1956 with the IGY looming, the main focus of DRNL shifted to geophysics. Ray Montalbetti was one of the first two Ph.D. graduates of the University of Saskatchewan (in 1952), but in nuclear physics. He was hired by DRB and worked for a time in Ottawa at DRTE before arriving in Churchill. Harry Lutz was involved in selecting sites for the IGY measurements to be made there. For the auroral height finding they needed two sites, a suitable distance apart. One place they visited (by air, much easier than for Currie and Davies) was Eskimo Point, where they were hosted by Father Ducharme, the same priest who had been at Chesterfield Inlet during the Second International Polar Year, throughout the auroral height finding measurements there. While his experience there and his ability as a radio operator made him an excellent choice, he wasn't able to offer adequate electrical power. They settled on two sites on the railway line south of Churchill, Belcher and O'Day. They also obtained the support of the RCAF for the installation of an auroral intensity recorder, a magnetometer and an all-sky camera at Bird which was a Mid-Canada Distant Early Warning (DEW) station. It was arranged that Don McEwen from the University of Saskatchewan be employed to work under this project. Montalbetti and McEwen carried out a very detailed set of measurements and learned that the altitude of the aurora depends on its intensity. McEwen's reflections on his experiences are given at the end of this chapter.

Ionospheric Physics

A very extensive set of ionospheric physics measurements was proposed. Radio wave absorption measurements were to be carried out at ionospheric stations, where the amplitude of the echo is measured in order to determine how much of the radio wave is absorbed during its travels; this was proposed to be done at five stations: Baker Lake, Churchill, Ottawa, Resolute Bay and Winnipeg, but was done only at Resolute Bay and Churchill.

Whistlers are gliding tones of radio signals that occur at audio frequencies and are described in the next section, interpreted by Owen Storey. The gliding tones are caused by the fact that the signals of different frequencies travel at different velocities, causing the frequencies to be dispersed. This was a very new field of endeavour, and DRTE proposed to make measurements at Flin Flon, Halifax and London. Dartmouth College in the US operated two more closely coordinated sites. There is no mention of these measurements in the final report.

The transmission path length between Winnipeg and Resolute Bay was measured for each ionospheric layer as a function of frequency, using oblique radio sounders. Receivers were also located at Alert, Baker Lake, and Churchill.

Scintillations and cosmic noise were measured by using a radio source in the constellation Cassiopeia and continuously recording the signal at 50 MHz in Saskatoon and Ottawa. The ionosphere causes the signal to vary, or scintillate. A zenithal receiver at 30 MHz records the ionospheric absorption at this frequency.

Forward scatter is the technique used to study ionospheric irregularities by monitoring the transmission between two transmitters and a ring of receivers at a distance of about 1000 km. It was proposed that the transmitter be located at Yellowknife and the receivers were to be located at Baker Lake, Churchill, Saskatoon, Sulphur Mountain and The Pas. In fact, Ottawa was used to receive signals from Coral Harbour, from Greenwood (Nova Scotia), Cape Henrietta Maria and Churchill.

Normal vertical incidence ionosondes were proposed to be operated at Alert, Baker Lake, Meanook, Victoria, Winnipeg, Yellowknife, Churchill, Ottawa, Resolute Bay, and St. John's, but in actual fact only the latter four were operated during the IGY.

Winds in the atmosphere were to be studied by the movement of ionization irregularities. It is presumed that at the low altitudes to be used (frequency of 2 MHz) that the collisions between the neutral atmospheric species and the ions would be sufficient to make the velocity the same for both. This technique was later developed to a high level of expertise in Saskatoon.

Solar Activity

In Ottawa the Dominion Observatory fitted a telescope with a filter transmitting atomic hydrogen emission to image solar flares on the sun. These are enormous outbursts of energy emitted sporadically in small regions, lasting an hour or so. Some 75 flares were recorded during the first half of the IGY; there was no report on the latter half at the time of reporting.

The technique developed by Covington at the NRC to detect radio noise at 10.7 cm wavelength to monitor solar activity that was described earlier was in operation during the IGY. Fortunately those measurements are still being continued today, at the Dominion Radio Astrophysical Observatory in Penticton and provide one of the best monitors of solar activity. Accordingly the data were forwarded to the Regional Warning Agency at Fort Belvoir in the US, and are still being distributed today, by the NGDC (National Geophysical Data Center) in Boulder, Colorado.

Cosmic Rays

Cosmic rays are extremely energetic particles, with the energies corresponding to an energization through an electrical voltage of billions of volts, coming from outer space. They are interesting for several reasons: one is to understand their cosmic origin; the second is that when these particles strike the constituents of the atmosphere they shatter their nuclei, producing fragments that are of fundamental interest to high energy particle physicists. This was how such scientists studied high energy particles before high energy accelerators were available. For space physicists there is another reason. The cosmic rays are deflected by the Earth's magnetic field and when this field is stronger, the deflection is greater and the cosmic ray intensity at the Earth is less. Thus cosmic rays provide a way of monitoring the Earth's magnetic field that can be extended backwards in time. The basis of this is that the cosmic rays produce a radioactive isotope of carbon in the atmosphere, carbon 14 (^{14}C), that is taken up as carbon dioxide by vegetation, trees for example. The ratio of ^{14}C to normal carbon, ^{12}C, can be used to determine the age at which the carbon dioxide was taken up (because of the radioactive decay subsequent to that) and the ^{14}C abundance gives information about the cosmic ray flux at that time, and thus the value of the magnetic field then. One current topic of hot interest is whether there is a connection to climate; cosmic rays penetrate deeply into the atmosphere, even reaching the surface, and in so doing they create ionization, which in turn can act as condensation nuclei for cloud particles, thus influencing the Earth's radiation budget. Thus there may be a connection between the Earth's magnetic field, which is changing all the time, and climate. However the processes are all much more complicated than suggested here, and no firm conclusions can be drawn at the present time.

Don Rose was successful at establishing five cosmic ray observatories in Canada for the IGY. There is an interest in having a range of latitudes, because the ability of the cosmic rays to penetrate to the Earth's surface increases with increasing latitude, a result of the particles (they are not rays, although the name has stuck) moving at smaller angles with the magnetic field. There is also an interest in having a range of altitudes, to determine the depths to which the cosmic rays penetrate – thus mountain tops are of interest. The sites operated during the IGY were Ottawa, Deep River, Ontario (the location of AECL), Churchill, Resolute Bay and Sulphur Mountain (Banff). The IGY final report lists a number of published papers, including Brian Wilson and Doraswamy Venkatesan of the University of Calgary as co-authors. This arose from Rose arranging for their involvement in the Sulphur Mountain observatory.

Rockets and Satellites

Churchill has been mentioned many times in the preceding pages; it was now to become a centre-piece of Canadian space science. Archaeology in the Churchill area shows evidence of human presence dating back 4,000 years as aboriginal groups came into this area to hunt during large mammal migrations. Its written history goes back to 1619, nine years after Henry Hudson discovered what is now called Hudson Bay. At that time Jens Munck, under commission from King Christian of Denmark, sailed into what is now Churchill harbour and wintered over, during which he and all but two of his crew of 64 persons perished. It was under the Hudson's Bay Company that the name Churchill was first applied to this region, in honour of Lord John Churchill, later to become the Duke of Marlborough. Construction of the company post began in 1689. In 1730 the construction of a mammoth stone fort, Prince of Wales Fort, was authorized on the side of the Churchill River opposite to the present town; it took forty years to complete and is now a national historic site. The town of Churchill came into being in its present location in 1931 with the completion of the railroad. Fort Churchill, so named to distinguish the camp from the town of Churchill, was originally established as a military base in 1942 as part of an air route from the UK to the US. The camp was maintained by the US until taken over by the Canadian Government in 1944. During 1944-1954, the main use of Fort Churchill was as an experimental and Arctic training centre.

The Canadian committee was fully informed of the US plans for building and operating a rocket range at Fort Churchill during the IGY. The proposal was for the launch of 41 rockets—37 Aerobees and 4 Nike-Deacons. The schedule for this is listed in the planning document. Seven types of measurements were proposed: 1) The density of electrons and ions in the ionosphere, measured by transmitting a radio signal from the rocket; 2) pressure, temperature and density of the atmosphere and 3) temperatures and winds in the atmosphere, using grenades, geophones and DOVAP (Doppler Velocity And Position). The atmospheric characteristics can be measured by tracking the path of a sound wave emitted by a grenade on the rocket, to the ground; 4) measurements of incoming auroral particles and of magnetic fields; 5) chemical and ion composition of the atmosphere using mass spectrometers; 6) atmospheric density and temperature using falling spheres; these are measured from the rate of fall of the ejected spheres, using tracking radar and 7) the vertical profile of airglow emissions, using filter and photomultiplier detectors. In return for the use of the site, NASA made available to Canada two rockets which they could instrument as they wished.

The US made an enormous investment in this global program and identified Churchill, Manitoba as a location from which rockets could be launched directly into the aurora borealis, to determine their true nature. The US Army Engineers, supported by the Aerojet General Corporation, began construction of the basic facilities in the summer of 1956. An Aerobee launch tower was specially designed for the liquid fuel rockets Aerobee 150 and 350; it had a tilt capability of 10° to allow for the effects of surface winds. A Nike-Cajun launcher was also built, as well as payload and rocket preparation buildings, telemetry trailers, a blockhouse, a generator building, and tunnels connecting the various buildings. Support was provided with tracking radars, a meteorological station, ballistic wind balloon facilities and complete communications and transport systems. Six rockets were fired between October 20 and November 20, 1956 to test the range and its operations, and it opened for IGY officially on July 1, 1957. During the 18 months of the IGY (July 1, 1957 – December 31, 1958) about 200 Aerobee and Nike-Cajun rockets probed the high atmosphere above Churchill.

The responsibility for the gift to Canada of two Nike-Cajun rockets to be launched containing their own instrumentation was given to CARDE. Two identical payloads were constructed, with the goal of measuring the altitudes of two airglow layers, sodium (Na) and hydroxyl (OH). The sodium emission has already been mentioned with respect to Balfour Currie at Chesterfield Inlet and Donald Hunten at the University of Saskatchewan. The existence of sodium in the upper atmosphere was a complete mystery; one possible origin for it was from meteorites and the other was salt from the oceans. From the ground, one could only determine that the emission was coming from the atmosphere high above; a rocket experiment was needed to determine the altitude at which the light was being created. The hydroxyl emission is now known to originate from $H + O_3 > OH^* + O_2$, involving a mesospheric ozone layer high above the dominant layer of ozone in the stratosphere. OH^* is an excited state of the hydroxyl radical that emits light in the so-called Meinel bands, discovered by the same Meinel who designed one model of all-sky camera. These rocket flights were pioneering experiments in the investigation of an as yet little studied region of the atmosphere, the mesosphere, lying between about 50 and 85 km. Four infrared radiometers were placed inside the nosecones, using lead sulphide detectors for the infrared OH emission. Two visible wavelength photometers for the Na emission at 589 nm used photomultiplier detectors. Although the first rocket functioned well, no useful data could be obtained from it. The second rocket, reaching 143 km, obtained good results. The altitude of the hydroxyl layer could not be precisely determined because of the low signals, and low sensitivity of the rather primitive infrared detectors; but it was somewhere between 75 and 90 km, which is consistent with the above reaction. The sodium emission was found to come from a thin layer centred at 92 km, which eliminated some tentative explanations that suggested it should extend as high as 140 km.

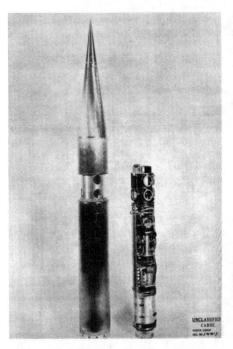

The CARDE payload for the measurement of OH and sodium emission in the upper atmosphere. The instrumented section is shown on the right, with the circular windows of the two photometers at the top. This was inserted inside the payload cylindrical section shown on the left. After leaving the atmosphere the nose cone was moved forward to expose the photometer windows as shown.

Following the end of the IGY the Churchill range was inactive for a period, but then launchings took place during the summer of 1959. Six rockets with Canadian experiments were launched in September, 1959. Two flights were with Aerobee-Hi rockets, carrying payloads that had been instrumented under the coordination of Walter Heikkila at DRTE for ionospheric measurements, but including experiments from Rose's group for the detection of auroral particles. The first flight was successful, but on the second flight the rocket failed. The other four rockets were of a Canadian design, the so-called PTV (Propellant Test Vehicle) developed by CARDE in Valcartier, Quebec (the CARDE story is told later). Their success led to the later development of a scientific sounding rocket vehicle by Bristol Aerospace in Winnipeg, as is also described later.

Photograph of one of the DRTE payloads being transported. From left to right the personnel are Don Rose, Bud Budzinski and Ian McDiarmid of NRC, then Don Awry and Peter Forsyth of DRTE. Photo by courtesy of the Communications Research Centre.

A CARDE rocket ready for flight. The name CARDE is plainly visible on the rocket body, but the actual flight is not identified.

The IGY final report stated that plans had been approved for Canadian experiments to be carried in a US satellite to be launched in 1961. The proposed experiment was a swept-frequency sounder, called a "topside sounder." As is described in the next Chapter, this small passenger experiment soon turned into a complete satellite, Alouette I and entirely built in Canada.

What was learned from the IGY

This is an appropriate place to summarize what was learned from the IGY and at the same time to briefly describe the Earth's environment as seen in the context of space science.

The Sun

Although one tends to take the sun for granted it is the source of all life and controls almost everything that we do. Its energy is maintained through nuclear reactions in its interior, keeping the visible surface that we see, the photosphere, at a temperature of around 5500 °C. It is the heat from this surface that keeps us warm, but the sun emits many other wavelengths as well, from very short wavelength X-rays to long wavelength radio waves. The smooth disc that we see is in fact very highly structured. Galileo observed with his telescope in 1610 that there were small dark regions on the sun's surface, called "sunspots", that rotate with the sun. Beginning in 1825, Heinreich Schwabe, a German pharmacist, observed the sun every day and kept a record of the sunspots. After 42 years of observations he had confirmed that the number of sunspots rises and falls with a period of 11 years – there is a sunspot cycle. The total amount of energy emitted by the sun also varies with the solar cycle by a fraction of a percent, but, according to atmospheric computer models, this is not enough to change the Earth's temperature. Yet a solar cycle influence on climate is observed. An example of this is the Little Ice Age of the mid-17th century in Europe when the sunspots virtually disappeared. At this time glaciers in the Swiss Alps advanced, gradually engulfing farms and crushing entire villages. The River Thames and the canals and rivers of the Netherlands often froze over during the winter, and people skated and even held "frost fairs" on the ice. The first Thames frost fair was in 1607, the last in 1814. This means that the mechanism causing the change is still not understood.

During a solar eclipse the corona can be observed, wisps of light extending outward from the sun. Here the temperature reaches millions of degrees. Along with this there is an outflow of particles, electrons and protons, traveling at about 600 km per second, called the solar wind. When the solar wind encounters the Earth's magnetic field, the magnetosphere is formed. One motivation for the timing of the IGY in 1957 was that it would occur at a maximum in the solar cycle, when the geomagnetic perturbations would be greatest, the aurora the most frequent and the ionosphere the most disturbed. This turned out to be a wise decision. Not all solar maxima are equal, and the 1957 maximum turned out to be the granddaddy of them all, with many spectacular geomagnetic storms and auroras. Much was learned about the sun during the IGY.

The Atmosphere

By this time the general structure of the atmosphere was known. As one ascends in the atmosphere, as in any aircraft flight (or in the ascent of a mountain), it gets colder very rapidly. Heat from the sun is absorbed at the Earth's surface, which is thereby kept warm, but as the air rises above this it cools with the decreasing pressure; the temperature falls by about 8°C every km up until roughly 10 km, when it suddenly begins to rise again. The region of falling temperature is called the troposphere, its upper limit is called the tropopause and the region of rising temperature is the stratosphere, the region of the ozone layer. Here the absorption of ultraviolet light by the ozone, created by the absorption of ultraviolet light of a different shorter wavelength, heats the atmosphere, causing the temperature to increase with increasing altitude, so that at 50 km altitude the atmosphere has a temperature about as warm as on the Earth's surface; this is the stratopause.

At higher altitudes, the temperature begins to fall again as the ozone concentration falls – this region of falling temperatures is called the mesosphere. At 85 km, in summer the temperature can reach as low as -125°C, the minimum is called the mesopause—the coldest place on the planet. At these low temperatures ice crystals can and do form, creating high-altitude clouds of ice that can be seen in late sunset in northern Canada during July and the first part of August, when the lower atmosphere is dark but the high-altitude clouds are still in sunlight. These clouds were first seen in Scandinavia in 1885, and were given the name noctilucent clouds. Later they were observed from satellites where they can be viewed in full daylight; for a reason that is not entirely clear, they were given a different name (perhaps in case they turned out to be a different phenomenon), polar mesospheric clouds (PMC). In fact they are the same, just viewed by different methods. That they were not recorded before 1885 is taken by some as an indication of climate change; others argue that they were seen but not recognized for what they were and this question is not resolved.

Of course water freezes at 0 °C, but that is for the air pressure at the Earth's surface. At the mesopause the amount of water is extremely small, only a few parts per million of the whole atmosphere, but at this mesopause temperature and pressure the conditions are right for condensation into ice. The connection to climate change is that the water at these altitudes is produced by the photodissociation of methane, a greenhouse gas. Given Canada's huge Arctic region, it is surprising that noctilucent clouds never became a major field of research in Canada – that field still belongs to the Scandinavians (and Germans, with facilities in Scandinavia). NASA currently has a satellite in orbit specifically for the study of these clouds called AIM (Aeronomy of Ice in the Mesosphere); it was launched April 25, 2007. In fact Balfour Currie did hire an operator to photograph these clouds during the summers, but this never developed into a scientific program. However, Canadian space instruments have observed these fascinating and beautiful clouds. Because of the strong dynamics at 82 km where they are observed, they take up remarkable patterns, as are shown on a number of web-sites devoted to these.

A final question is why the mesopause is so cold in summer, much colder than in the winter. This is because there is a large-scale motion of the atmosphere at these altitudes from the summer to the winter hemisphere. This motion is associated with rising air in the summer, causing cooling, while the associated downward motion in the winter hemisphere causes warming – one of many peculiarities of the high atmosphere.

Above the mesopause the temperature rises again because another still shorter wave-length of solar ultraviolet radiation is absorbed here, where the temperature can rise to any-where from 500 °C to 800 °C, depending on the state of the eleven-year solar sunspot cycle. This enormous temperature change is a result of the large change in output of extremely short wavelength solar ultraviolet radiation during the solar cycle; this region is called the thermos-phere.

Many things were learned about the atmosphere from ground-based measurements and rocket measurements during the IGY, but one of the most difficult is the measurement of the density of the atmosphere, often expressed as the number density, how many molecules there are in one cubic meter of atmosphere. Sputnik 1 provided this by simply burning up in orbit on January 4, 1958, following its launch on October 4, 1957. What happened was that friction from the atmosphere, thin as it was, caused the satellite to lose energy and its orbit to move closer and closer to the Earth, increasing the "atmospheric drag" even more. At a crit-ical point the drag became so great and the temperature rise so high that the satellite burned up. This information was used to calculate the density of the atmosphere at different heights and satellite drag became a highly valuable technique for determining the densities in the ther-mosphere, where satellites fly.

The Van Allen belts

One of the scientific discoveries most specifically associated with the IGY was the "trapped radiation belts", often called the "Van Allen belts" after their discoverer. After the initial US Vanguard satellite launch failures, Wernher von Braun and his Redstone and Jupiter rockets came to the fore. The first of these, with Explorer 1 aboard, carried Geiger counters provided by James Van Allen of the University of Iowa. Van Allen had been studying cosmic rays using balloons, and small rockets launched from balloons, so he had suitable equipment at hand. Once in orbit they were stunned to find that the particle counts were very low, almost zero. It looked like the instrument was faulty, but in experimenting with a duplicate instrument in the lab they discovered that if the radiation levels were very high the count rate would saturate, giving very low values, like those observed in orbit. For the next launch they modified the experiment to accept very high count rates, and confirmed that there was a region around the Earth, some 4-6 Earth radii away that contained this highly dangerous radiation.

It was a completely new discovery, except that the Norwegian scientist Carl Størmer (the inventor of the Størmer auroral camera), in the early 1900s had laboriously calculated the trajectories of particles approaching the Earth's magnetic field; charged particles are deflected by magnetic fields (as are cosmic rays, as described earlier). He discovered that there was a region around the Earth into which these particles could never reach; he called them "forbidden regions." Turning the argument around, if by some means one could get a particle into a forbidden region it would never get out – it would be trapped. Thus it turned out that there are mechanisms that allow particles to slowly enter this region where the population builds up and they finally become the trapped radiation belts. This was an important element in the understanding of the magnetosphere.

The magnetosphere

A second and related great concept that came out of the IGY was the existence of the magnetosphere itself. The term was coined by Thomas Gold, an astronomer of Austrian origin who studied at Cambridge University in England before moving to Cornell University in the US in 1956. It has already been mentioned that there is a "solar wind" flowing out from the sun, and embedded in this "plasma" of electrons and protons is a piece of the sun's magnetic field. Thus this extended solar magnetic field, called the "interplanetary magnetic field" (IMF) reaches to the Earth and far beyond. The Earth also has a magnetic field and the two do not mix, so when the IMF touches the Earth's field there is a kind of territorial competition, forming a sharp boundary between the two, called the magnetopause, with the region inside called the magnetosphere. However, the interaction of the two cause a startling deformation of the Earth's otherwise dipole type field, as around a bar magnet. The Earth's field is pushed much closer to the Earth on the side facing the sun, and stretched far out into space on the side away from the sun, extending even beyond the moon. Deep inside the magnetosphere are the radiation belts, already described. The particles that cause the aurora must somehow find their way through this system to find their way onto magnetic field lines that lead them down to the high-latitude atmosphere where they produce light through collisions with the atmospheric constituents. There is a Canadian involvement here too, that is described later.

What causes the aurora?

This question is thousands of years old. The oldest written description is considered by some to be that found in the Old Testament book of Ezekiel, written around 593 B.C.E.

"And I looked, and behold, a whirlwind came out of the north, a great cloud, and a fire infolding itself, and a brightness was about it, and out of the midst thereof as the colour of amber, out of the midst of the fire.....................I saw as it were the appearance of fire, and it had brightness round about. As the appearance of the bow that is in the cloud in the day of rain, so was the appearance of the brightness round about."

Chinese records exist covering the period 100 C.E. to 1000 C.E., including drawings, some of which indeed resemble the aurora. The Greeks developed theories about the different lights in the sky, including meteors. Aristotle believed that as vapours rose from the earth or sea they became dry and could therefore ignite. Large amounts of vapour would produce lights of longer duration, and this category included the aurora. Much later, Suno Arnelius in 1708 in Uppsala, Sweden postulated also that the origin was in vapours from the Earth. However, he believed that these formed ice crystals in the cold high latitudes and that the aurora was the reflection of the sun's rays from these crystals. Scientists with a better background in fundamental science got involved in that century, including the Danish scientist Ole Rømer, the first person to measure the velocity of light, and Sir Edmund Halley, the famous British astronomer.

But it was the French scientist Jean Jacques de Mairan (1678-1771) who wrote the first serious scientific work, *Traité Physique et Historique de l'Aurore Boréale,* inspired by a display over Paris in 1726. He related the aurora to the zodiacal light, which is sunlight scattered by interplanetary particles, raising the aurora from the near-Earth to space. The Russian M.V. Lomonosov (1711-1765) was the first to propose that the aurora was a phenomenon of electrical discharges, but still in the atmosphere. The Swedish physicist Anders Celsius (1701-1744), of the Celsius's temperature scale, produced a major scientific work that laid out the scale of the problem, stating that many simultaneous observations from different locations would be required to acquire true knowledge of the aurora. He also kept the phenomenon rooted in the atmosphere, but his brother-in-law Olof Peter Hiorter (1697-1750), following Celsius suggestion, followed up a possible connection with the compass needle. Hiorter took hourly observations of the position of a compass needle for one entire year, interrupted only by a holiday trip, and Christmas; his observations established that there was a connection between the variations he observed and the movement of the aurora (something Halley had suggested earlier). The English clockmaker, George Graham, in collaboration with Celsius, established that these magnetic variations occurred simultaneously in England and Sweden.

Although the compass was well known, the pattern of the Earth's magnetic field was not, and it was Carl Friedrich Gauss who proposed setting up eight magnetic stations around the globe to determine it. Lacking the logistical means to do this he approached the British navy, which certainly did. This network was in operation by 1841 and one of these stations was established on what is now the University of Toronto campus; the plaque can be found near the Sandford Fleming Building. As an aside, Sir Sandford Fleming himself was a remarkable Canadian, who emigrated from Scotland. He advocated the system of Universal Time that scientists use today, based on the local time at the Greenwich meridian; it was accepted in 1884. It took longer for the world to accept his concept of time zones, but it eventually did. He designed Canada's first adhesive postage stamp, the "Threepenny Beaver", in 1851, and he surveyed the route for the first trans-Canada railway, the CPR. But to get back to the magnetic data, Sir Edward Sabine found a connection between the Earth's magnetic field variations, as measured in Toronto, and sunspots; both had the same 11-year cycle. Thus the linkage was made between sunspots, the Earth's magnetic field and the frequency of occurrence of aurora. The evidence was fast accumulating that the aurora was driven by the sun and that it was an electromagnetic phenomenon. It is important to recall that understanding electromagnetic phenomena at that time was impeded by the fact that the electron was discovered only in 1895, by J.J. Thompson at Cambridge, a discovery that allowed this field to advance.

Kristian Birkeland (1867-1917) of Norway developed the idea that electrons streamed out of sunspots at such great velocity that, guided by the Earth's magnetic field, they could reach the atmosphere, giving up their energy in collisions with atoms and molecules there, this energy being converted to the light that is seen as the aurora. To confirm this theory he set up a simulation in his laboratory, with a magnetized Earth (a magnetized sphere with a phosphorescent surface) and an electron gun with which he could direct streams of electrons towards his Earth. In a convincing demonstration in 1896, Birkeland did indeed generate rings of light around the poles of his simulated Earth, a convincing demonstration. But this was not a proof that the same thing happened in the solar terrestrial system. This demonstration was thrown into question when lines of atomic hydrogen were discovered in the auroral spectrum. Much later, in 1951, Aden Meinel of the Yerkes Observatory in Chicago went further by showing that these lines were broadened, something he attributed to the Doppler shift of the hydrogen atoms streaming into the atmosphere. These would originally be bare hydrogen nuclei, single protons, but in the atmosphere protons can grab electrons and become atoms, allowing them to emit the light of the hydrogen spectrum. But this did not prove that protons produced all of the auroral light that is seen.

It was only at Fort Churchill, during the IGY that Carl McIlwain, of Van Allen's group at the University of Iowa flew rockets carrying proton and electron detectors into the overhead aurora. He found that it was indeed electrons that produced the auroral light, not protons. The protons do stream into the atmosphere, but in a different location. They produce what is called hydrogen aurora, and it can be distinguished from normal aurora only with optical instruments.

As mentioned earlier, the individuals promoting the IGY particularly favoured 1957-58, in part because it was 25 years after the 1932 polar year study, but also because it would be at the peak of a solar cycle. This turned out to be a wise choice and the famous red aurora of February 10-11, 1958 was a spectacular global event that was well-studied, in particular at the University of Saskatchewan where unique spectra of this event were captured.

A personal reflection of the IGY

The facts and figures of the IGY still hold interest, but a personal reflection of what it was like to participate in such a major adventure at one of Canada's most active locations, Fort Churchill is more revealing of what was involved in doing science fifty years ago. Don McEwen was a student at the University of Saskatchewan when Balfour Currie recruited him to spend his IGY year at Fort Churchill. Following that he was employed at DRTE for a time, but took time out to go to the University of Western Ontario to obtain a Ph.D. with Professor Ralph Nicholls. Later he moved to the University of Saskatchewan where he had a distinguished career in auroral studies. In his retirement years he began experiments at South Pole, which he still visits. Here are his reflections.

"My enduring memories of the International Geophysical year (IGY) at Fort Churchill, Manitoba are of the spectacular auroras that I saw so frequently there in the winter night sky, and the many interactions with all the scientists and engineers who had come north to explore the upper atmosphere with rocket instruments. These adventures actually continued over a whole four years up to 1961—for the duration of the IGY, the International Cooperation Year which followed, and then on the research staff of the Defence Research Northern Laboratory (DRNL). It was especially gratifying to work with Ray Montalbetti who was the officer in charge there during most of that time. My auroral studies which began then have continued to this day!

"I was very lucky, as a physics student at the University of Saskatchewan in 1956, to be hired by B.W. Currie to go to northern Manitoba to undertake auroral measurements in support of rocket firings at the Fort Churchill Rocket Range during the upcoming International Geophysical Year. One of the persuasive arguments that he used to convince me to accept was that he had got his start this way in the North at Chesterfield Inlet during the Second Polar Year, in 1932-33! Little did I know then that this would start me on a career so different from the one I had planned.

"I boarded a train in Saskatoon for the 2-day passage to Churchill, Manitoba in late June, 1957, after a year of studying and instrument preparations. The latter half of the trip, from The Pas northward to Churchill, was very slow as the tracks were laid on a base of permafrost. With thawing in the summer, the rails were somewhat unstable and the train was limited to a speed of only 25 mph. It also had to pull off to sidings occasionally as empty southbound freight trains (coming from the Churchill grain port) had priority. We finally reached our destination to find a real hum of activity at the newly established rocket range. The crew there had already had a few test firings and was about to herald the opening of the IGY with a number of rockets designed to explore the aurora and other features of the upper atmosphere.

"DRNL had agreed to provide logistical support for such auroral studies and had already established two sites, Belcher and O'day, on the railway line some 100 km south of the rocket range, for the auroral parallactic photography. I found a keen group of technical people assembled there under the able direction of Harry Lutz.

"What an exciting prospect—an adventure in the North, to experience the Northern Lights AND to be involved in rocket explorations of the upper atmosphere! There had been a few test firings at Ft. Churchill already, but the program got underway officially on July 1, 1957. The Churchill rocket range was operated by a competent US army group from the White Sands Proving Ground in New Mexico—quite experienced from earlier flights there but rather new to remote Arctic conditions—and what a contrast it was between the Arctic permafrost and the desert-like terrain of their alternate base in the south.

"Probing the secrets of the upper atmosphere with rockets was a new endeavour. Two rockets had been developed, the Aerobee which was liquid fueled, and the Nike-Cajun, which had solid fuel. Both had large accelerations on lift-off, so instruments on board had to withstand large g-forces. Flights with Black Brant rockets, designed and built in Canada, had a longer burning period which resulted in less stress on the packages aboard. They later became the rockets of choice for research purposes.

"One of the first projects at the Fort Churchill Range was to determine upper atmospheric temperatures. Two techniques were tried—one, firing a series of grenades during the up-leg flight of the rocket and picking up the sound from spaced ground microphones—the other, releasing a sphere from apogee and measuring its rate of descent to the ground.

"The combined techniques allowed extraction of the atmospheric temperature profile up to approximately 100 km. There were two series of these launches, the first in July, 1957, the second in January, 1958. The latter gave the surprising result that the Arctic winter upper atmospheric temperatures were higher than those at low latitudes such as over White Sands!

"Each rocket launch, from the Range some 15 km from the military base at Ft. Churchill, attracted much local interest—the payload preparations, the lengthy 'countdown' prior to firing, the tracking during flight, and of course the detailed 'post-flight conference' the next day to report results and any problems encountered. One major challenge was always the tracking of the rocket. The standard tracking means was by radar, then in rather early days, and that required skill in manually pointing the radar dish antenna at the rapidly moving rocket. The target was usually lost before it reached 100 km in ascent, and sometimes it was the rocket booster motor instead that was tracked, in error. Skills improved as the IGY year progressed. Radar beacons were later added to payloads to allow more reliable tracking, and a new DOVAP technique was also developed to give more extended tracking.

"My major responsibility was auroral height measurements during the many auroral rockets scheduled. They were mostly during the winter months, so the autumn was spent readying the two field-stations down-range from the launch site. DRNL had already installed two Wannegans for living, and large semi-cylindrical US army Jamesway tents were later assembled to house generators and supplies. (Identical tents are still in use at the South Pole 50 years later, in 2007—each sleeping 16 persons). Antenna towers were also assembled, to allow HF radio communications with the Rocket Range. These stations were on the CNR railway line connecting Churchill with Winnipeg, so provided us with transportation to the sites. The return trip was more uncertain however, as we had to wave a lantern to flag down the passenger train about two in the morning, with the engineer not expecting passengers on the winter tundra!

"With the full support and assistance of DRNL personnel, we occupied these remote sites most of the 1957-58 winter, making nightly auroral measurements. We had installed a scanning spectrometer (similar to the U of S one designed by Don Hunten), an all-sky camera and a magnetometer there, in addition to the height-finding cameras, so we were all set as winter approached! But near-continuous cloudy skies prevented much success until December, when the nearby coastal Hudson Bay finally froze.

"Noteworthy of the auroral rockets scheduled that winter were 2 Nike Cajuns instrumented by Carl McIlwain of the University of Iowa, then a graduate student of James Van Allen. His assigned task was to determine what caused the aurora. As we were all awaiting suitable auroras for his flights, there was an amazing spectacle one evening - an aurora in the twilight sky which was completely red! This turned out to be the onset of the worldwide Great Red Aurora of February 10-11, 1958. It moved southward during the evening and was later seen as far south as New Mexico. It was so intense that many measuring instruments went off-scale. McIlwain launched his two rockets later that month successfully through auro-

ras and his thesis finding was that 'the luminosity of the two auroras investigated was primarily due to the incidence of energetic electrons upon the atmosphere.' His direct measurement of those particles was a first and a real breakthrough in understanding of auroral phenomena. Other IGY scientists were looking for such answers with similar studies. They included groups led by Leo Davis and Les Meredith (NRL) using Aerobee rockets and Kinsey Anderson (University of Minnesota) using large balloons.

"Somewhat later, Bill Fastie arrived at the rocket range with a spectrometer to investigate the far ultraviolet spectrum of the aurora. This was a novel instrument that he had designed and which later became very widely used worldwide (the Ebert-Fastie spectrophotometer). It filled most of the nose cone of an Aerobee rocket. One problem he hadn't anticipated was that the helium used to pressurize the rocket fuel also poisoned his photomultiplier detectors, and they became noisy. He obtained a uv auroral spectrum in a flight up to an apogee of 125 km, the first such measurement. There had been much speculation about just what emissions there might be from the aurora at wavelengths below the visible region viewable from the ground. He identified major nitrogen band emissions, but did not detect the hydrogen Lyman line at 121.5 nanometers.

"Very many of the rocket flights during the IGY produced results which only the project scientist fully understood, I'm sure, but which greatly added to our knowledge of space. Rocket science was new though, and there were failures—some motor malfunctions, some mechanical or electronic problems, some due to wind changes. During lift-off and early flight the rockets were very vulnerable to wind gusts and shears which could put them off course. Range safety was always a prime concern—any rocket in danger of landing outside the reserved impact area had to be quickly destroyed. This put great stress on the radar-tracking crews. One rocket was needlessly destroyed when a faulty radar track seemed to indicate it was off-course. The project scientist was furious, needless to say! During the course of the IGY there were many such emotional times, but fortunately most of them were of excitement with successes.

"Churchill was a harsh environment which challenged even the hardy. The winters were long and bitter. With no trees or natural shelter, the winds seemed to always blow! The windchill often exceeded 2500 (corresponding to about -40 °C with a 20mph wind) and then everyone was confined to quarters; frequent blizzards blocked roads for days. Summers were not much better. Black flies and mosquitoes were so thick during midsummer that headgear and netting were required. Fortunately, military stores were well stocked with all environmental gear for loan. The autumn was perhaps the only consistently pleasant season. Outings during the remainder of the year had to be planned with care and an eye on the weather. That situation really dictated mainly indoor recreational activities, which most residents seemed to adapt to quite well.

"The assembly of several military branches, both Canadian and US, as well as civilian agencies at Fort Churchill during the IGY made for great social times but resulted in some stresses due to differences in cultures and habits. July 1st and 4th were celebrated on a common day, and the first event was a challenge US vs Canada softball game. A sudden snow storm forced its cancellation however, perhaps fortunately, as the only playing field there was just stones and gravel!"

Postscript

The timing of the IGY could not have been better. The scientists who became involved in the IGY had the technologies they needed to embark on this vast examination of the Earth and its environment. The Earth also needed the IGY, as this global study was long overdue. In the larger framework policy-makers learned the value of establishing centralized data bases, open to scientists worldwide. They also learned that in addition to taking the pulse of the Earth periodically it was necessary to monitor certain characteristics on a continuous basis. The space age, the age of COSPAR, has made that possible, and now the state of the Earth is being measured from space in greater and greater detail. At the same time, the atmospheric models that go hand-in-hand with the observations are greatly improved. All of this capability will be needed to face the environmental challenges that lie ahead.

References:

Auroral image website: http://www.spaceweather.com/aurora/gallery_20mar01.html

Brekke, Asgeir and Alv Egeland, The Northern Lights – Their Heritage and Science, Grondahl Dreyer, Oslo, 1994.

Korsmo, Fau L., The genesis of the International Geophysical Year, Physics Today, July, p. 38, 2007.

Noctilucent cloud website: http://www.spaceweather.com/nlcs/gallery2005_page1.htm

Space Research Facilities Branch, Space and Upper Atmosphere Research in Canada: Balloons, Rockets and Satellites, National Research Council of Canada, SRFB 024, January, 1969.

Sullivan, Walter, Assault on the Unknown: The International Geophysical Year, McGraw-Hill, New York, 1961.

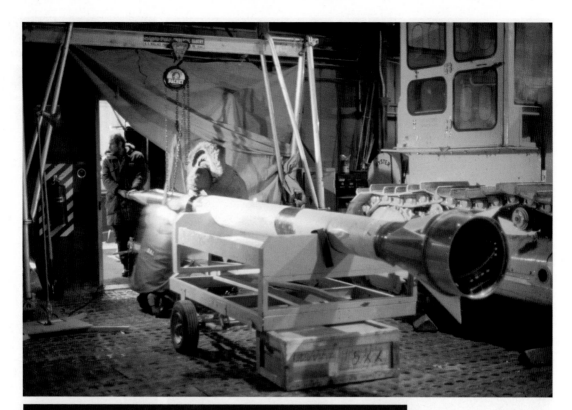

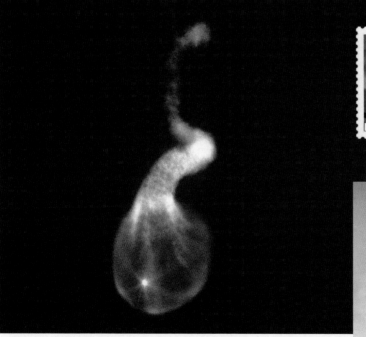

(Top) The second stage of Black Brant rocket AAF-IVB-32 (essentially a BB III) with payload attached is moved into the hanger prior to mating with the first stage at Cape Parry, December, 1974. (Above) The vapour trail from the twilight launch of Canada's ISIS II satellite on April 1, 1971. As the rocket passed from darkness into the sunlight the trail began to fluoresce. The vehicle is the bright spot close to the centre of the large bulb of gas. "Down" is "up" in this photo - it is a matter of perspective. (Right) Preparation for the launch of ISIS II, on a Delta E-1 vehicle, from Vandenberg Air Force Base, near Lompoc, California.

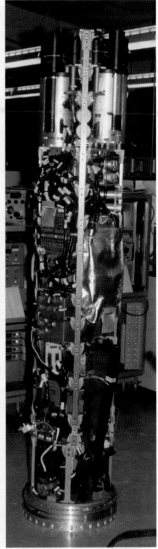

(Above) Gordon Shepherd with the payload for AMF-VB-41, a Black Brant VB rocket launched from Cape Parry on the Arctic coastline, December, 1974. (Above, right) Payload for the York University Black Brant AMD-VB-34, launched from the Churchill Research Range on January 23, 1974. The aluminum cylinders with black tubes at the top are auroral photometers, the black booms of the atomic oxygen absorption experiment are at the bottom, folded up against the payload.

(Right) An aerial view of the Churchill Research Range showing the Auroral Launcher in the extreme upper left, and to its right the Universal Launcher; with the assembly building immediately below. The right-hand of two linear features extending downwards to the left is the service shaft terminating at the blockhouse. To the left of the left linear feature is the Black Brant outdoor launcher. Moving to the centre right of the photo is the Aerobee launcher with the visible tower, and to the left the Nike launcher. At the bottom left is the Operations Building.

(Above) Rocket AMD-VB-34 on the rail in the Auroral Launcher at the Churchill Research Range, January, 1974. (Middle) One of four Black Brant III rockets being mounted on its launcher at East Quoddy, Nova Scotia, for the July 10, 1972, solar eclipse campaign. (Bottom) Lift-off of AMD-VB-34 into the aurora (not visible in the photo), on January 23, 1974, from the Auroral Launcher at the Churchill Research Range.

The payload for AMD-VB-34 in the Operations Building at the Churchill Research Range, all "buttoned up", and awaiting transport to the vehicle assembly area. This is an exceptionally long payload, with the nose-cone tip just a few cm from the ceiling. The York University team is on the right, from left to right Mike Walton, Ashley Deans and Jacques Pieau. (Above) — The Black Brant X rocket developed by Bristol Aerospace for the Cape Parry "cusp" launch by David Winningham of Southwest Research Institute of San Antonio, on the rail in the tent, December, 1981. The BB X had a Terrier MOD 1 as the first stage, a BB VC as second stage, and a newly developed Nihka rocket as third stage. (Below)

(Above) Nike-Apache ready for launch from Fort Churchill Manitoba Photo by courtesy of NASA

(Right) The exterior of the Auroral Launcher, a Canadian design.

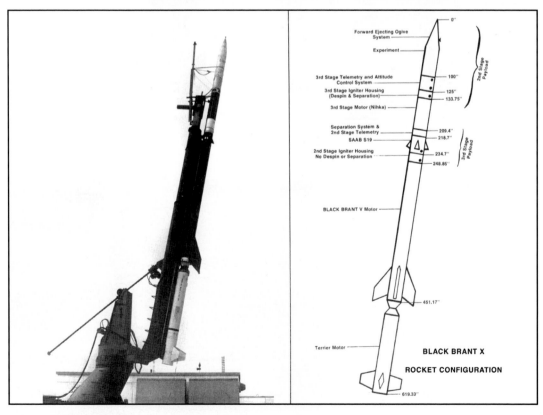

Forward Ejecting Ogive System — 0"

Experiment — 2nd Stage Payload

3rd Stage Telemetry and Attitude Control System — 100"
3rd Stage Igniter Housing (Despin & Separation) — 125"
— 133.75"
3rd Stage Motor (Nihka) —

Separation System & 2nd Stage Telemetry — 209.4"
SAAB S19 — 218.7"
2nd Stage Igniter Housing No Despin or Separation — 234.7"
— 248.85"
3rd Stage Payload

BLACK BRANT V Motor —

— 451.17"

Terrier Motor —

BLACK BRANT X

ROCKET CONFIGURATION

— 619.33"

(Above, left) The Black Brant X test rocket on the rail at the NASA Wallops Flight Facility on Virginia's eastern short; the flight was successful. (Above , right) Drawing of the Black Brant X, showing the three stages.
Photo by courtesy of Bristol Aerospace

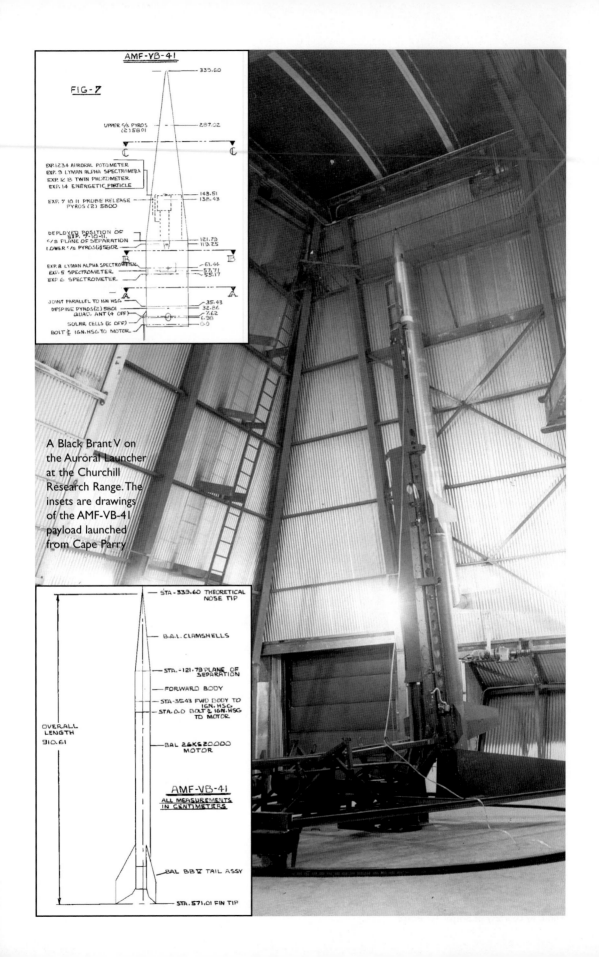

A Black Brant V on the Auroral Launcher at the Churchill Research Range. The insets are drawings of the AMF-VB-41 payload launched from Cape Parry

CHURCHILL RESEARCH RANGE
ROCKETS

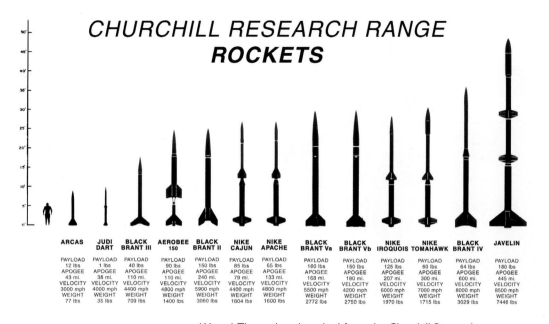

	ARCAS	JUDI DART	BLACK BRANT III	AEROBEE 150	BLACK BRANT II	NIKE CAJUN	NIKE APACHE	BLACK BRANT Va	BLACK BRANT Vb	NIKE IROQUOIS	NIKE TOMAHAWK	BLACK BRANT IV	JAVELIN
PAYLOAD	12 lbs	1 lbs	40 lbs	90 lbs	150 lbs	85 lbs	65 lbs	160 lbs	150 lbs	125 lbs	60 lbs	64 lbs	180 lbs
APOGEE	43 mi.	38 mi.	110 mi.	110 mi.	240 mi.	79 mi.	133 mi.	168 mi.	180 mi.	207 mi.	300 mi.	600 mi.	445 mi.
VELOCITY	3000 mph	4000 mph	4400 mph	4800 mph	5900 mph	4400 mph	4800 mph	5500 mph	4200 mph	6000 mph	7000 mph	8000 mph	8500 mph
WEIGHT	77 lbs	35 lbs	709 lbs	1400 lbs	3060 lbs	1604 lbs	1600 lbs	2772 lbs	2750 lbs	1970 lbs	1715 lbs	3029 lbs	7446 lbs

(Above) The rockets launched from the Churchill Research Range. (Below) The Bristol family of rockets, as of 1980.

Bristol Rocket Series

X   VIIIC   VA   IIIA   VB IIIB   VC   IVA   IVB

Note on rocket notation. In the rocket name AAF-IVB-32, for example, the first A identifies the funding agency; A is NRC. The second character, also A here identifies the Principal Investigator institution as NRC. In AMD-VB-34, M stands for York University. The third character, F, identifies the payload contractor as Bristol Aerospace while D identifies SED Systems. Following that is the name of the rocket, here a Black Brant IVB, while the number, 32 for example, means that it was the 32nd IVB rocket launched.

Photo by courtesy of Bristol Aerospace

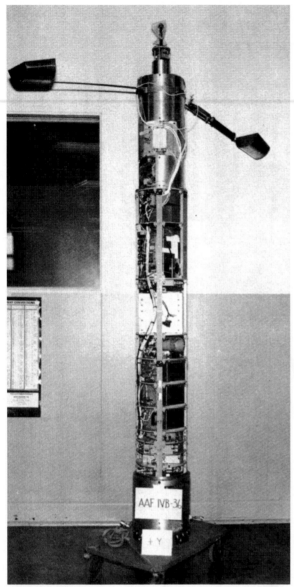

AAF IVB-36

+ Y

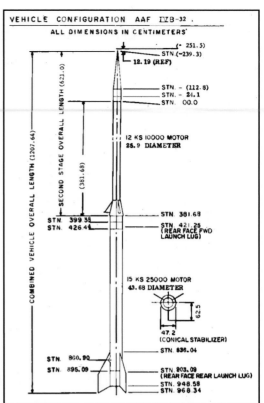

VEHICLE CONFIGURATION AAF ⅣB-32.

ALL DIMENSIONS IN CENTIMETERS

(- 251.5)
STN. (~239.3)
12.19 (REF)

STN. - (112.8)
STN. - 24.1
STN. 00.0

12 KS 10000 MOTOR
25.9 DIAMETER

STN. 381.68
STN. 399.38
STN. 426.44
STN. 421.25
(REAR FACE FWD
LAUNCH LUG)

15 KS 25000 MOTOR
43.68 DIAMETER

62.5

47.2
(CONICAL STABILIZER)
STN. 836.04

STN. 860.90
STN. 895.09
STN. 903.09
(REAR FACE REAR LAUNCH LUG)
STN. 948.58
STN. 968.34

COMBINED VEHICLE OVERALL LENGTH (1207.64)
SECOND STAGE OVERALL LENGTH (621.0)
(381.68)

(Left) An unusually long Black Brant IV payload, AAF-IVB-36, fabricated by Bristol, with antennae extended; it was launched on January 26, 1981. (Above) Drawing for the Black Brant AAF-IVB-32 led by NRC Space Physics and launched at Cape Parry in December, 1974. Photos by courtesy of Bristol Aerospace

(Below) ARCAS being prepared for launch (Right) A Black Brant VI rocket ready to launch at White Sands Missile Range in New Mexico. Unfortunately this rocket never came into regular use. Photo by courtesy of NASA

Chapter 5

The Sixties – A Decade of Exponential Growth

The Stimuli

Canadian space science, having taken root quickly just after the Second World war, was ready to catch the crest of the wave that was the IGY. The investment of effort and infrastructure along with the impact of artificial Earth satellites led to a remarkable growth in space science during the 1960s. This growth was nothing short of phenomenal. A figure from the Chapman report, a milestone in Canadian space history, shows the exponential growth of funding for university space programs and the growth in numbers of personnel. From 1961 to 1971 the funding increased from about $0.7 million to almost $4 million a year. The number of professionals increased from about 30 to 300 and the student population grew from about 70 to 600.

What drove this incredible growth? Fundamentally it was a response to the emergence of space exploration in the rest of the world, primarily the USSR and the USA, but global in scope as expressed by the formation and activity of COSPAR. Behind all this was an awareness of the importance of this new capability to explore the environment above the Earth's surface. Canada's pre-existing knowledge of space and its well-developed experience in ground-based observations allowed it to fully grasp the opportunity presented to this large thinly populated country that was fortunate to have the Earth's north magnetic pole on its territory. This became the COSPAR decade. But to create space activity required support—government support. In Canada this arose initially from two directions, the Defence Research Board (DRB) and the US Air Force. As the sixties progressed, this support waned and was replaced, as far as the universities were concerned, by the more sustained support of NRC. It should be remembered that at this time NRC had two separate responsibilities, the support of its own laboratories and the support of universities. The latter responsibility was later transferred to a new body, the Natural Sciences and Engineering Research Council (NSERC) on May 1, 1978. But as will be seen, the NRC was concerned not just with its own laboratories, but with leading and coordinating national initiatives such as the Churchill Research Range and later the Astronaut Program. The NRC grew in parallel with this during the sixties.

The importance of the Defence Research Board (DRB)

When DRB was first formed in 1946, the Radio Propagation Laboratory became an integral part of it. Omond Solandt, DRB's first Chairman, remained at its head until 1956 and guided it through its formative years. Born in Winnipeg, Manitoba, he graduated in medicine from the University of Toronto. After his internship and graduate work he accepted a permanent position on the staff of the Department of Physiology at Cambridge University, England. In January 1941, he was appointed Director of the Medical Research Council Physiological Laboratory, became Deputy Superintendant of the Army Operational Research Group in 1943 and Superintendent in May 1944.

Solandt joined the Canadian Army in February 1944 with the rank of Colonel and continued as Director of the Army Operational Research Group until 1945 when he was appointed Director of the Operational Research Division, South-East Asia Command (SEAC), and scientific advisor to Lord Mountbatten, then Commander-in-Chief of SEAC. Returning to England in June 1945, he was soon appointed to the War Office as a member of the joint Military Mission sent to Japan to evaluate the effects of the atomic bomb.

The Canadian government appointed him Director General of Defence Research on December 28, 1945, so that he could help plan postwar military research. In 1947 he became the founding chairman of the Defence Research Board and served as such through 1956. He was an ardent canoeist and paddled many of Canada's rivers. After his official retirement he was, from 1956-63, Vice President for research and development at Canadian National Railways and then from 1963-66 Vice President for research and development at De Havilland Aircraft. From 1965 to 1971 he became Chancellor of the University of Toronto. He was the founding Chairman of the Science Council of Canada. Later still from 1966 to 1972 he acted as chairman of the Council and was thus one of the most influential voices in the science policy debate of that period.

Solandt realized the enormous scope of defence research and the monumental level of activity that would go on in the USA and the UK. He chose therefore to concentrate on those projects in which Canadians were particularly qualified and which had direct bearing on aspects of Canada, such as geographical location and climate. Because implementation of a new technology is extremely expensive, he chose to emphasize the quality of the research and not to demand immediate benefits. By focusing on a few distinctive projects he hoped to find for Canada a place on the international defence scene that would allow it to participate fully with other countries. The capability within the DRB laboratories could be multiplied by funding research at universities, and he favoured institutes as a way of achieving the critical size required. Communications was clearly one project area that met all of the above criteria. But there was another.

This other important group that came under the new DRB was the Canadian Armament Research and Development Establishment (CARDE) in Valcartier, Quebec. As described earlier, Don Rose was its first Superintendent, from 1945 to 1948, after which he returned to the NRC. From there he drew on CARDE to create a Canadian rocket program as will be described later. CARDE housed an unusual scientist, Gerald Vincent (Gerry) Bull, the youngest Ph.D. graduate of the University of Toronto. At CARDE he was, at 23 years of age, given the task of building a supersonic wind tunnel to test missiles, and then realized that it would be a lot easier to make a supersonic missile. He developed a model with a wooden casing, called a "sabot" which was essentially shot from a gun, with the sabot falling away. It worked extremely well. By age 31 he was head of the Aerophysics Division of CARDE. Advances in Soviet aircraft technology led to the termination of his program, so he left CARDE. Donald Mordell, Dean of Engineering at McGill University, believed that Bull's technology was capable of placing a satellite in orbit and invited him to McGill. A persistent belief in his technology was in fact a driving force behind Bull's career. He gained the support of both the Canadian and US military establishments, but lost it later when the desired product did not emerge with the proposed budgets. His story is told in more detail by Chris Gainor (2007).

He turned to commercial missile production, and selling guns to South Africa, but this at the time of an embargo landed him in jail. It was an understandably bitter man that emerged, who linked up with Saddam Hussain, apparently to build a "big gun." To Bull this would have been a dramatic demonstration of his techniques, but others noted that its projectiles could reach Israel from Iraqi soil. This made him a marked man. It was perhaps ironic that it was a small gun that killed Gerald Bull on March 22, 1990, while unlocking his apartment door in Brussels, a 7.65 millimetre pistol in the hand of an assassin. He still died a popular individual and 600 people attended his funeral in Montreal. Soon after, the big gun program was stopped and the parts needed for its completion were embargoed. This tragic story is perhaps an unsettling way to describe the remarkably broad range of university programs that existed at this time, but for most it was a period of great enthusiasm and optimism.

The Theoretical Studies Group at DRTE

The IGY was a trigger for many things. In the case of Colin Hines, a request to prepare a review paper on "Motions in the Ionosphere" was critical. David Martyn, who had originally accepted the task, and whose paper had stimulated Hines' first thoughts about waves, later withdrew and had he not, the story would have ended here. Instead, the writing of this paper carried Hines forward to the point of saying that current observations of deformations in meteor trails could indeed be explained as waves.

Shortly after, the ICSU took steps to resolve the issue of turbulence and ionospheric motions by asking URSI (International Scientific Radio Union), IUTAM (International Union of Theoretical and Applied Mechanics) and the IUGG to organize an International Symposium on Fluid Mechanics in the Ionosphere, to be held at Cornell University, Ithaca, New York, July 9-15, 1959.

Hines declined his original invitation, but his subsequent rapid progress on what one now calls "gravity waves" caused him to get re-invited, but too late to be included in the official program. He was given ten unscheduled minutes to speak, stretched it to twenty, and made his case before this distinguished audience of ionospheric, atmospheric and fluid dynamics speakers. It was a convincing story. Small perturbations, near ground level, such as the flow of winds over mountain tops, or from severe storms, launch waves that travel into the upper atmosphere, like ripples emanating from a pebble dropped into a pond. As a wave rises, it grows because with the decreasing density of the atmosphere the amplitude of the wave motion has to increase in order to carry the same amount of energy—energy that is carried upwards into the atmosphere. A peculiarity of these gravity waves is that although the energy propagates upwards, the individual wave fronts (called the phase) appear to be moving downwards, an appearance resulting from their horizontal motion. Such downward phase motions had been observed in the F-region of the ionosphere.

His presentation was well received by the ionospheric scientists, as it explained their data. But the meteorologists present were slower to convince. They felt their current understanding of the atmosphere was adequate; that they didn't need gravity waves in "their" atmosphere. Later, when atmospheric computer models called Global Circulation Models developed, it was found that gravity waves were needed to transport energy and momentum in the models. Now they are a completely accepted feature of atmospheric science and observations of them are routinely made by radio and optical methods. Hines' famous paper on these waves was published in the Canadian Journal of Physics in 1960.

Colin Hines had returned to RPL in the summer of 1954 and was assigned to work under Peter Forsyth on meteor studies. Soon after, Owen Storey, a student of Ratcliffe's arrived at RPL with the intent of making whistler observations. As we have already seen, "whistlers" are the ephemeral gliding tones one used to hear on old technology AM radio, electromagnetic sounds generated by lightning flashes, following the Earth's magnetic field to distant locations and then converted to sound by the radio. Storey invited Hines to work on the theory and when they worked together they discovered that the magnetic field guidance would exist up to only some limiting frequency, called the hybrid frequency, allowing ion (plasma) densities to be determined along the field line. About this time, Eugene Parker introduced the concept of the "solar wind", a name given to the continuous outflow of plasma from the sun and he took note of Storey's work that indicated high plasma concentrations on field lines rather distant from the Earth. This implied that the plasma environment of the Earth was not limited to the ionosphere, but extended much farther into space. This ionspheric bulge of low-energy plasma, extending some four times the Earth's radius out into space, became known as the plasmasphere. The plasmasphere was discovered independently by a Russian scientist, Konstantin Gringauz, and a US scientist, Don Carpenter from Stanford University. Gringauz made his discovery from a 1959 spacecraft, Lunik 1, bound for the moon, when he noted a sharp drop in plasma density a large distance from the Earth. Carpenter was following in the footsteps of Owen Storey, and found from whistlers a kink in the proton density curve that became known as "Carpenter's knee." While neither measurement alone would be fully convincing, the two taken together proved the existence of the plasmasphere. Before this, the first US satellites were launched and James Van Allen discovered the existence of high energy plasma there as well. Shortly afterwards, Tommy Gold coined the name "magnetosphere" for the magnetic region which belonged to the Earth, but extended far into space—but how far?

Just at this time Frank Davies was assigned to DRB Headquarters with the military administrative personnel; Jim Scott was named to replace him as head of DRTE. Forsyth, who had been Superintendent of RPL, was about to leave for Saskatoon, so Scott named Hines to replace him. Hines immediately offered Storey his full support, only to find he had already decided to move to the US National Bureau of Standards in Boulder, Colorado. Hines then managed to recruit someone recommended very highly by Jack Ratcliffe, Jules Fejer from South Africa. Hines made "the disturbed ionosphere" the focus of RPL and appointed T.R. (Ted) Hartz leader of the observations group, supported by George Reid and Jack Belrose. R.E. (Ron) Barrington took over whistler observations – and Fejer fitted well into this group. Walter Heikkila, who had been studying the scattering of radio waves in the troposphere was persuaded that rocket techniques for auroral flights to be made from Churchill would provide a better future. The final group was called synoptic studies, their responsibility was to meld the studies from the other groups, along with others reported in the literature and so achieve a comprehensive integration. Irvine Paghis led this group, supported by Doris Jelly, Clare Collins and Luise Herzberg, the wife of Gerhard Herzberg, a very capable scientist in her own right, who had just joined RPL.

The invitation to prepare a review paper on "Motions in the Ionosphere" with its consequences has already been mentioned, but there was another. This came at a time when, in 1959, Davies had returned from Headquarters, in 1959, impressed with the need to make DRTE research more relevant to the military. He objected to the new RPL focus of "the disturbed ionosphere" and this led to a near-impasse with Hines, but a new structure was worked out with the support of A. Hartley Zimmerman, Chairman of DRB. Hines would retire from RPL in 1960 to be replaced by Irvine Paghis and be made head of an independent theoretical group, reporting directly to the Chairman, who had a vision of something like the Institute of

Advanced Studies in Princeton. Six positions would be assigned to the group. During the summer of re-adjustment Hines attended a symposium directed at disturbed ionospheric and auroral conditions in Kiruna, Sweden. Here he heard a paper about the work of Neil Davis of the University of Alaska, who had discovered from auroral imaging that auroral forms moved westward before midnight and eastward after, as well as equatorward in the midnight sector. This led Hines to draw a picture of a circulation motion in the magnetosphere in his program book. Later during his summer travels he passed through Cambridge, where he met a young officer of the New Zealand Air Force, W. Ian Axford who was about to be seconded to RPL. Axford had completed a Ph.D. degree on shock waves in interstellar plasmas and was then sent to Ratcliffe's group to get some exposure to the ionosphere. Hines was sufficiently impressed that he took steps to ensure that he ended up in the Theoretical Studies Group and not RPL. Paghis reluctantly agreed, provided Hines would "raid" RPL no more.

Even before taking up his new position, Axford began to investigate the implications of Hines' plasma motions in the magnetosphere. Together in Ottawa the Kiruna drawing evolved into something that satisfied both of them, as well as the other information available to them and this they called a "convection pattern." The results were presented to the American Geophysical Union the following spring, and the International Conference on Cosmic Rays and the Earth Storm in Kyoto, Japan in that summer. It was published in the Canadian Journal of Physics in 1961 under the name Axford and Hines, in 1961, but by this time it had already been widely accepted by the scientific community. Thus Canadian scientits had created an early viable description of the Earth's magnetic environment, the magnetosphere and the fundamental plasma motions within it. Other models appeared about the same time; each seemed to explain certain aspects of the magnetosphere but none explained all – the magnetosphere today is still not fully understood. But the truths of the Axford and Hines model and its magnetospheric convection remain valid today.

Despite these successes for Hines personally, his Theoretical Studies group was not developing well. Axford returned to New Zealand to complete his Air Force duties and later to many other things. He became a professor at Cornell University and then the University of California at San Diego, Director of the Max Planck Institute for Aeronomy in Germany (now the Max Planck Institute for Solar System Research), the President of COSPAR, and Vice-Chancellor of the Victoria University of Wellington, New Zealand, finally becoming Sir Ian Axford. Jules Fejer left for a warmer climate in California. However, Hines had two spectacular applicants, Ivor Robinson and Roger Penrose, and so set up a meeting with the Management Committee of DRB to propose a Canadian Institute of Advanced Studies. This may have been a mistake; Hines had strong support from the Chairman, Hartley Zimmerman, but the other members of the Committee had little understanding of what he was about. He made one further proposal, in which the NRC and Atomic Energy of Canada Limited (AECL) would become partners with DRB in this institute, but this was also rejected by the committee. This was the final attempt; Hines had several offers from which to choose, and selected the University of Chicago. After a number of productive years there he returned to his home country and the University of Toronto, in 1967.

The University of Saskatchewan

Formation of the Institute of Space and Atmospheric Studies

The University of Saskatchewan establishment of optical and radar studies of the aurora has already been described. Saskatoon is at a relatively high latitude and the aurora can be observed on its northern horizon several nights a week. This made it an ideal place to study the auroral effects on the ionosphere that disturb communications and it was extremely well equipped for both radar and optical auroral observations. Balfour Currie had taken personal responsibility, as far as the University was concerned, for all this equipment and the large grants and contracts. He wanted to give it some institutional structure and Frank Davies suggested a research institute. Thus the University in 1957 created the Institute of Upper Atmospheric Physics (usually referred to as UAP), under an enabling grant from DRB. Its name was later changed to the Institute of Space and Atmospheric Studies (ISAS). Balfour Currie was its first Director.

By the time Forsyth arrived in 1958, Hunten and Vallance Jones had become university faculty members as Associate Professors. Forsyth was appointed Professor with limited teaching duties and paid out of Institute funds; this caused some resentment from the Physics Department faculty members with heavy teaching loads. This would be like a Canada Research Chair today, but that was fifty years ago and the system wasn't really ready for this kind of arrangement. Shepherd had come directly from his Ph.D. at the University of Toronto to become Assistant Professor in 1957 (on university funds). There was another Assistant Professor, Alex Kavadas, who arrived in 1956 and who was funded from Institute funds. Kavadas had just obtained his Ph.D. at the McLennan Laboratory of the University of Toronto. He worked there with Professor Anderson in gaseous electronics, one floor below Herbert Gush and Gordon Shepherd. Shepherd was impressed with Kavadas' theoretical skills and thought they could be put to good use on auroral radar studies in Saskatoon. He suggested this to Currie and a meeting was arranged for Ottawa during Currie's next visit there. It took place in the Lord Elgin Hotel Coffee Shop and after some conversation Currie said, "What is it you want anyway, Alex, would you like a job"? Kavadas replied, "Yes I would, thank you" and thus was concluded an event of great subsequent importance to Saskatoon.

The situation was in fact more complex. Kavadas had fought in the Greek underground during the war, and however patriotic an act this was at the time, its communist leanings were not appreciated by the US authorities when he arrived in New York as an immigrant. The rise of McCarthyism led to his deportation, but he managed to gain entry to Canada as a student at McGill, and then the University of Toronto. However the Canadian immigration authorities seemed to be much influenced by the US interpretation of Kavadas' validity as a citizen, and he was repeatedly threatened with deportation. That he managed to complete his Ph.D. studies, support a family, and pay his lawyer was a remarkable achievement. Only after his meeting with Balfour Currie was he left in peace, to eventually become a Canadian citizen and make the significant contributions to the country that he did.

The development of rocket instruments began at Saskatoon in 1960 under the stimulus of Peter Forsyth, with launches first conducted in 1962. In about the same year, Currie sent Kavadas and Shepherd up to Churchill to witness the launch of a CARDE payload, designed to explode a spherical canister of nitric oxide in the high atmosphere, an atmospheric chemical modification experiment. After many nights of repeated attempts Kavadas expressed his frustration by saying that "this is a ridiculous way to do space research." Shepherd, equally frustrated by Alex's criticism, then asked, "Well, then what approach would you recommend?" After some thought, Kavadas replied, "A very tall stepladder." (Today, space futurists are talking not about a space stepladder, but a space elevator.) His opinion was soon to change, however. In 1961 Dennis Johnson was appointed to the professional staff of the Institute to support the design and development of rocket experiments for individual projects. Two senior technicians, Frank Hector and Dave Glass had also been added. One rocket-borne experiment on electron densities and another on electric fields were developed during 1961-62 and launched in 1963. By 1963-64 ten flights had been made. Discussions took place with the College of Engineering in order to expand the scope of the work and to provide thesis topics for Engineering students.

The creation of SED Systems

In 1965, the Space Electronics Section of the Radio and Electrical Engineering Division (REED) of the NRC decided it would no longer do the payload engineering work it had been doing to support the rocket program. Kavadas recognized an opportunity for the University and proposed that a branch of the Institute of Upper Atmospheric Physics, in collaboration with the College of Engineering, take over this work. Under this agreement the University accepted a contract from NRC to construct rocket payloads for research groups across Canada; the new group formed to undertake this work was called the Space Engineering Division (SED) of the Institute and Kavadas was named as the Director.

Three years later the division had evolved into three groups, the payloads implementation group, the research group and a third concerned with the design and development of new experiments or instruments for industrial concerns. At the time seven graduate students were working on topics related to the SED work, while six more had already received their degrees: four in Engineering, one in Physics and one in Commerce. SED was then employing 37 permanent professionals and technicians. Kavadas placed great emphasis on the training of students in this practical project-oriented activity. They were already looking ahead to the development of small satellites, for surveying, water resources, pollution, weather, plant diseases and "general changes in the behavior of the plant world over very large areas" (this some forty years before global climate change).

SED, under the management of Dennis Johnson, grew rapidly, and soon was too large for the available quarters on campus. They then located space near the airport, on what had been an Air Force base during the war, in a building that had been the officer's mess. This was soon converted to a rocket payload fabrication facility. As their capability expanded they began to look at other applications. For example, a rocket radio transmitter could be used to telemeter measurements from locations other than on rockets, difficult locations such as high voltage power lines. They developed a device to measure flexure in power lines, with the results telemetered to ground, and providing a warning prior to power line failure, just one example of many applications.

The evolving nature of SED combined with its rapid growth caused certain stresses and strains. The primary income was from the NRC contract, to build rocket payloads, but the needs of the graduate students also had to be met. This was the responsibility of the University, but as graduate students preferred to work at the SED building at the airport, their needs were often overlooked, including furniture, telephones, secretarial time, drawing office, machine shop and so on. There were also university researchers across Canada contracting to SED to build their experiments. In a very carefully crafted letter prepared on June 9, 1969, Kavadas made this argument and asked for a grant from the University to cover the indirect costs associated with the students. SED was also paying off a $100,000 debt to the university incurred by the renovation of the building.

The 1968-69 SED report shows five sections: 1) Management – F.B. Perrot; 2) Operations – Mike Hodson; 3) Telemetry and Data-Processing – C.F. Pensom; 4) Development – Dennis Johnson and 5) Research. Sections 3 and 5 were independent of the others that constituted the original mandate of the SED. Development activities at that time included a rocket auroral scanner for the University of Calgary, a probe ejection system for the University of Western Ontario and a rocket spin-test facility. The research section included Bob Besant and K. Williams from the College of Engineering, Jim Koehler from the Physics Department, as well as Kavadas. Contracts from researchers totaled $48,500 which together with the NRC contract resulted in a total budget of $371,397. The staff included 17 professionals, 20 technical and 8 support staff.

On January 7, 1969 Kavadas sent a memo to Leon Katz, Head of the Physics Department, informing him of a potential disaster. It was a response to a letter to the University from Dick Rettie, Chief of the NRC Space Research Facilities Branch (SRFB) written on December 10, 1968. As described later this group was formed by the NRC in 1965 to coordinate the contract work and the Churchill Research Range. Rettie was warning the university of its intention that SED be phased out. The main reason given is that the evolved structure of SED, with its five sections as outlined above, is not what the NRC intended to support and indeed the government funds allocated (from vote 10) were allowed only for operations and not for support. They suggested that SED had become a quasi-commercial operation. There was also a claim that SED costs were 20% higher than would be charged by industry. Overall, the NRC was becoming uncomfortable in supporting this kind of operation inside a university, despite all the advantages this offered, and after 3.5 years of operation, wanted a commercial solution.. In a parallel letter to Currie, Kavadas urged that a solution be found in the context of the national space program, noting the recent Chapman report and its recommendation of the establishment of a Canadian Space Agency. The NRC offered to help out by saying that if a purchaser bought SED that all the NRC-owned tools, equipment and facilities would be left in place on a contract loan basis. The other alternative offered was that SED scale down to the fabrication of payloads originating at the University of Saskatchewan only.

In many ways this commercialization made sense, as, with increasing success, the linkages between the two groups separated by the South Saskatchewan River had weakened, and so the University accepted the decision of the NRC and SED became an incorporated body—SED Systems. It was good for the space industry, but removed a large segment of what had been considered university capability. As President of SED Systems, Kavadas developed an unusual horizontal management style. Any employee could come to his office at any time, if there was a complaint, and there were many. This was a heavy burden on the President, but SED flourished. Dealing with such problems finally led to Kavadas' retirement (to an island between the BC Mainland and Vancouver Island). SED Systems then became a more conventional company, and eventually moved back across the river to Innovation Place, the research park just north of the University. The company, now a communications firm primarily, has 250 employees and an annual sales volume of $50 million. Writing in 1986 Alex Curran, then President of SED wrote that "It has happened in Saskatoon where SED Systems was the nucleus for an industrial community which is now home to over 100 high-tech companies employing close to 3000 people." (Kirton, 1986). It can be added that if Currie had not his abiding faith in human nature and the initiative to intervene with the Canadian immigration authorities on Kavadas's behalf, it might never have happened.

Dennis Johnson on the left and Alex Kavadas, founder and first President of SED Systems on the right, standing in front of a rocket console. Photo by courtesy of SED Systems, with the assistance of Brian Whalen.

The story of O_2 singlet delta

Meanwhile, Vallance Jones had also been busy with his students. Herbert Gush, ultimately at the University of British Columbia (now Professor Emeritus, retired to a farm in Sicily) had built the first infrared spectrometer, and obtained the first spectra of the hydroxyl bands in the 1-2 μm (micrometer) region in the nightglow. Hydroxyl, with the chemical notation OH, is called a radical, an incomplete molecule, which exists in the upper atmosphere because of the chemical reactions there. It emits across a very wide region of the spectrum in what are called the Meinel bands, after the discoverer, Aden Meinel. Meinel was an astronomer, then located at the historical Yerkes Observatory of the University of Chicago, located at Williams Lake, Wisconsin. The OH radical was a target of the first Canadian rocket payload by CARDE as described in the last chapter. After getting his Ph.D. in Toronto, Gush took a post-doctoral fellowship at the Laboratoire Aimé Cotton in France, where he implemented the technique of Fourier Transform Spectroscopy to obtain remarkable hydroxyl spectra, with many consequences as are described later.

Following Herbert Gush's departure, Tony Harrison, Jim Read, Don Shemansky and

Dick Gattinger continued this work. Working in the infrared is rather different from working in the visible region. The atmosphere is not uniformly transparent here, because the small amounts of water vapour and carbon dioxide in the lower atmosphere are effective absorbers of infrared light at certain wavelengths. Only in the "windows" between these absorbing bands could the upper atmosphere be "seen" at infrared wavelengths. At longer wavelengths the atmosphere itself emits light. A heated red-hot poker emits red light, but a cooler "black" poker emits longer wavelength light—infrared light. The atmosphere does the same. Since Vallance Jones and his students were all pioneering, they had a difficult job to sort out these new effects and this required real detective work. Vallance Jones was an extremely thoughtful and careful scientist, with impressive background knowledge and his patience was well suited to this kind of investigation. Quite by accident Tony Harrison found that during evening twilight, one band of infrared radiation persisted after the light had vanished in the other regions. They soon identified the source of this light as coming from a long-lived excited state of the oxygen molecule. Such long-lived excited species are called metastables. A metastable oxygen molecule is still an oxygen molecule, but it has distinguishing characteristics and participates in different chemical reactions. It's a little like an actor, who assumes different characteristics for a time, but he eventually reverts to his true self. Metastables do not appear readily in laboratory equipment because they are destroyed (by revision to their ordinary selves) by collisions with the walls of the apparatus. It was precisely this effect which made it so difficult for Gordon Shrum earlier in the century to produce the atomic oxygen green line, which is also produced by a metastable form of oxygen, but of the atomic variety. In the upper atmosphere there are no walls, and metastable species abound; the lack of laboratory information makes their study in the atmosphere both more difficult and more important. In any case, Harrison and Vallance Jones had discovered atmospheric emission called the Infrared Atmospheric Band coming from a metastable that was identified in Herzberg's book as "singlet delta", written $^1\Delta$, but very little else was known about it. The 1.58 μm transition that Harrison had observed was from the zeroth vibrational level of the $O_2(^1\Delta)$ state to the first level of the "ground" state of ordinary molecular oxygen. Their immediate questions were: "what is the altitude of the emission, what are the chemical reactions that produce it, and what are the ones that destroy it?" Dick Gattinger was able to show that this emission was produced in the daytime by photodissociation, the splitting of ozone, O_3, by ultraviolet sunlight, into the metastable $O_2(^1\Delta)$ with an atom of oxygen left over.

The success of this work brought new funding to extend the observation of the spectrum to longer wavelengths and in this they were aided by a new Research Associate, John Noxon, fresh from his Ph.D. at Harvard University. John Noxon was attracted to Saskatoon by the strong impression that Hunten had made upon him during a visit to Boston, and it attested to the rise of Saskatoon's international stature. Noxon and Vallana Jones worked together; Noxon's New England accent was even slower than Vallance Jones' New Zealand one. Their quiet thoughtful personalities matched well, and Noxon's stay was beneficial to both the University and to him. Noxon had some interesting talents; he was an accomplished organist and a mountain climber. As a member of the Harvard Mountaineering Society he had been twice to the Himalayas. He was also a good story teller. A career bachelor, or so it seemed then, he identified with the students more than the faculty, and was the focus of much of the auroral and airglow aura that characterized the nighttime activity in the attic and on the roof of the physics building. A burst of aurora resulted in a flurry of activity with students running across the roof in different directions, from instrument to recorder, rarely colliding in the darkness. Noxon wanted to look at the much stronger emission to the zeroth vibrational level of the ground state at 1.27 μm, but this is absorbed by all the normal molecular oxygen between where the emission is produced in the high atmosphere and the ground-based observer. To get above this absorption a higher altitude observing platform was needed. They built a balloon-borne spectrometer which was flown by the CARDE group with the assistance of Bob Lowe and John Hampson. As a result, Noxon was able to establish that the much stronger 1.27 μm band could be easily observed in the airglow and the aurora from balloon and even aircraft heights.

The US Air Force built a copy of the infrared spectrometer and mounted it in a CS137 aircraft which flew out of Hanscom Field near Boston to altitudes near 13 km. When Noxon returned to Harvard, his experience made him a natural operator, and he often flew on these missions. Many were inside the Canadian Arctic, so Noxon returned many times to Canada for observations of singlet delta. One visit was disastrous. During a stopover in Goose Bay, Labrador, a snow-clearing bulldozer ran into their parked aircraft and damaged it. It had to be abandoned because it could not be repaired there, and it could not be flown out. After that Noxon did not fly again for a long time.

W.F.J. (Wayne) Evans was a "sixties student", though perhaps not a typical one. He did his M.Sc. thesis with Vallance Jones on pulsating aurora. Early in his observations he detected huge pulsations that always began at the same time in the evening. After more investigation he realized that this was the time that the drive-in movie ended in Sutherland, just outside Saskatoon. As the cars rounded the corner on the highway back to Saskatoon, their headlights swung across his photometer. Happily this was never published, and he did observe, in the morning hours, the auroral pulsations that became part of his thesis. However, the "Evans effect" became part of the Saskatoon folklore.

For his Ph.D. thesis Evans wanted to fly balloons to measure the 1.27μm molecular oxygen emission, but with a much smaller instrument than the spectrometer flown by CARDE. Together with Edward J. (Ted) Llewellyn, who had arrived from Wales as another post-doctoral fellow, but was immediately appointed as a professor, Evans and Vallance Jones developed a miniature photometer that would fly on a small balloon. These were launched from the Saskatoon airport. About a dozen people were needed to support the long train of the balloon, the tether and the payload while the balloon was being partially filled with helium. It is partially filled because as it rises in the atmosphere the pressure falls and the balloon expands. When sufficient helium has been added, the balloon is released at the "top" end and it begins to rise, to be carried by the wind, which must be low. Steve Peteherych normally got the job of holding the payload at the opposite end and his task was to run with the wind in order to keep the package underneath the rapidly rising and drifting balloon. Finally the balloon line became vertical and taut, and he released the package. With large balloons this is done with a truck, but these hand-held launches are more exciting. This was also always done at night, but the balloon could usually be seen in the morning twilight, like a bright star, and good results were obtained.

But to determine the vertical distribution of this emission the balloon was not enough, a rocket would be needed. This was arranged by Hunten, who was now at the Kitt Peak National Observatory in Tucson, Arizona, so Evans, along with Llewellyn, Vallance Jones and Hunten measured the vertical profile of the emission by flying a rocket through it. There was a strong peak near the top of the stratosphere near 50 km, which was expected from the known ozone distribution, but they found another small peak just above 80 km that some others (including Hunten) were initially reluctant to believe. This was eventually shown to result from recombining atomic oxygen. Evans then went off to France on a post-doctoral fellowship, to the Service d'Aéronomie, in Verrières-le-Buisson, on the southern outskirts of Paris. He came home two years later, expecting to find a university faculty position. But the sixties had ended, and along with it the spectacular growth of space science in the universities. From now on, university positions would be hard to find, but Evans was attracted to the Atmospheric Environment Service of Environment Canada, in an elegant new building on Dufferin Street in north Toronto. Here he used his balloon experience to attack the stratosphere with a series of large balloons he called STRATOPROBE. That is described in the next Chapter.

Vallance Jones got into an additional line of auroral work with the sudden resignation of Bill Petrie. Petrie was a quiet, gentle and private person and no one ever claimed to know the reason. There seemed to be an emotional aspect since he declined to accept the briefcase which the students, at the Annual Physics Club banquet, offered as a parting gift (they then raffled it off, with its initials, W.P.). Petrie went to Ottawa to take a position at the DRB, where he was able to establish another auroral laboratory in the Defence Research Northern Laboratory (DRNL) at Churchill. But it left Vallance Jones with Petrie's spectrograph, and also a very new and elegant one that arrived soon after. It had been designed by Meinel at Yerkes Observatory and built by the USAF for use in Saskatoon. With this instrument, Vallance Jones accumulated the best atlas of the auroral spectrum achieved to date, and made a number of important observations. The most important was the great sunlit aurora of February, 1958, during the IGY, a storm that reached down to Arizona. This was one of the greatest storms in decades, and it was fortunate, and typical that the Saskatoon group were ready and alert enough to catch it, since there was no warning. Hans Koenig, a technician, gets much of the credit for that success.

Hunten and alkali metals in the upper atmosphere

The seasonal concentration changes of sodium were the first important result to come from Saskatoon, but it did not answer one of the major questions that Hunten wanted to resolve, which was: Where does the sodium come from? Is it carried up from the sea by wave action and diffusion, or is it carried in from outside by meteorites? To resolve this next step the solution seemed to be to look for a relative of sodium, namely lithium. The relative concentrations of sodium and lithium are different in the sea than from meteorites. For the weaker lithium emission something more sensitive than the grating spectrometer would be needed. Hunten took advantage of a new instrument, called the birefringent photometer. A photometer is the name given to a photoelectric device which measures the light in a fixed selected wavelength band, rather than in many bands, as does a spectrometer. Birefringence is the property of certain crystals such that different directions of polarized light travel at different speeds in the crystal. If such a crystal were illuminated by light passing through polaroid sunglasses, and the sunglasses were rotated a quarter turn, the travel time of the light would be changed. It turns out that by setting up such crystals in a certain way, a rotating polaroid filter will "chop" the intensity of a narrow band of wavelength (or colour) on and off, but will leave white broadband light unchanged. Such a device would be ideal for looking for a weak narrow lithium emission against a white twilight background. Mica is a birefringent substance but to use it one needs a sheet of precisely selected thickness. Hunten became very adept at splitting sheets of mica to the desired thickness, by inserting a pin at the end of the thicker sheet at just the right place. To work well one also needs to moderately restrict the initial band of wavelengths, and for this an interference filter was the best candidate. Interference of light in thin films is what causes brilliant colours in soap films or in oil slicks on water. By using multiple layers, of precise thickness, one can generate a sandwich of layers that will, by interference, transmit a narrow band of wavelength and reflect the rest. To make permanent layers one uses a solid material like cryolite, melts it in a miniature electric furnace resembling the filament of a light bulb. The material then evaporates onto a glass plate held in front and forms a solid thin layer. This must be done inside a vacuum chamber, and there must be two materials, for which the speed of light is much different, and alternate layers of the two materials, each precisely one quarter of a wavelength thick, must be evaporated. Hunten set up a filter-making facility in the basement of the physics building, in the same room occupied by Herzberg's spectrograph which was no longer used. It took about twenty-one layers to make a good filter and this required a highly developed skill. Steven Peteherych (along with Wayne Evans, later at the Atmospheric Environment Service in north Toronto but now deceased), made some of the first filters and Bill Ryhorchuck later became a full time filter technician.As usual, Hunten was ahead of his time, since there were no other physicists making their own interference filters. Nowadays one would not think of making one's own filter since they can be purchased from commercial firms, albeit at steeply rising prices because of the technical skill required.

Hunten was an innovator but he had no time for patents or business ventures. He was too busy applying his new technology to new scientific applications. Howard Rundle (later a professor at the University of Saskatchewan, now deceased) built the first sodium birefringent photometer for his Ph.D. thesis and the measurements were continued by Ralph Bullock, later to take a major role in the Black Brant rocket developments at Bristol Aerospace. Harry Sullivan (later a professor at the University of Victoria) built the first lithium photometer for his Ph.D. thesis. Lithium brought some surprises. One was that the lithium intensity, normally very low, jumped up following well-known meteor showers at specific dates of the year. So some lithium was carried in by meteors but this wasn't necessarily true of the baseline level, and anyway sodium showed no response to these showers. But Lithium is also a component in the hydrogen bomb and in 1962 high altitude thermonuclear explosions were conducted in the atmosphere. The birefringent photometer was able to see this lithium, at about 85 km, drifting over Saskatoon a few days after explosions in the South Pacific and in the USSR. It was in part the recognition that these explosions could be detected, but also the physical evidence for environmental modification, that fortunately brought these tests to a halt soon after. Another relative of sodium is potassium, and Bill Gault (who later made major contributions to the WINDII project at York University) constructed for his Ph.D. thesis a potassium device in an effort to bring further understanding of the origin of these metals. But even this did not bring the final answer. Hunten had Joe Scrimger (later at the Pacific Naval Laboratory in Victoria) set up Herzberg's spectrograph so it could measure the sun's spectrum. Sunlight passing through the sodium layer at precisely the sodium wavelength would be absorbed by the sodium, and so there would be a "bite" out of the solar spectrum at the sodium wavelength. In fact there is sodium in the sun's atmosphere, which causes a large bite, but Scrimger and Hunten found at the bottom of the sun's broader bite a small narrow bite due to the earth's sodium. In this way they could measure the daytime concentration of sodium. Even this did not yield the final answer. After moving to the US, Hunten eventually gave up the sodium work, going on to study other planets. But many years later, sodium re-appeared to him, in a most unexpected way, in the atmosphere of Jupiter's moon Io. Earth's sodium is still studied in detail by other workers using laser radars (lidars) to pinpoint the layer height accurately. This turns out to be an accurate way to measure atmospheric temperatures at these altitudes, but sodium still refuses to give up completely the mystery of its origin.

Currie relinquished his Directorship of ISAS in 1966, which after nine years had brought $1,437,500 in support to the Institute with the largest annual amount being $260,000 in 1966-67. During the nine-year period 29 Master students and 17 Ph.D. students had received their degrees, a considerable achievement. In 1959-60 the Institute employed 8 technicians, 1 observer (instrument operator) and one clerk-stenographer. After laying what seemed a firm foundation for space science, Currie was concerned about future support, with the termination of the US Air Force funds and perhaps those from DRB as well. Hunten had left for the Kitt Peak Observatory in Tucson, Arizona in 1964, and Vallance Jones left for the NRC in 1967. These were replaced by Howard Rundle, a student of Hunten's and Ted Llewellyn, as mentioned earlier.

Currie steps down

Currie was succeeded by John Gregory, a recently appointed Professor from New Zealand. Gregory had pioneered in using medium frequency radio waves for tracking ionospheric motions from which the atmospheric winds could be determined – this introduced atmospheric dynamics to ISAS. He accepted the challenge of putting the Institute on a firm financial footing but in fact this would not be easy. Since SED had already physically separated from the University, its commercialization didn't have a drastic impact – its services were still available for contract and good relations continued. But it certainly removed a large capability from formal university activity. Then on November 8, 1967 Frank Davies wrote to Balfour Currie telling him that cuts in the budget of the Department of Defence had impacted heavily on DRTE and that the contract to ISAS would have to be terminated soon. The support from the USAF was also winding down.

However, during this time ISAS achieved stabilization. Gregory established a site west of Saskatoon for a MF radar facility for wind measurement, called the Park site and began a regular program of wind measurements. After his retirement these were continued by a new faculty member, Alan Manson, who continued them over two solar cycles, providing a definitive climatology of winds for the mesosphere and lower thermosphere. In this he was ably assisted by Chris Meek, son of Jack Meek.

Another blow was the closing of the Prince Albert Radar Laboratory. Gregory wrote to John Spinks, President of the University of Saskatchewan on March 1, 1967 informing him about this, and that the NRC was unable to take it over. He had already written to Rennie Whitehead, Deputy Director of the Science Secretariat, pleading for a new assessment of the role of PARL in the space community. While this facility was lost to the university community as a radar, PARL survives to this day as a telemetry ground station for the CSA.

Other universities

Aeronautics had been taught at the University of Toronto since 1938, but space science in the form of re-entry physics was initiated only in 1946. This introduced the shock tube for high-speed shock waves, developed as a supersonic wind tunnel. With support from DRB, the University of Toronto Institute for Aerospace Studies was formed in 1949 with a modern laboratory built with financing by DRB in 1959. At about the same time, rocket research was begun and a plasma dynamics laboratory established. This had clear connections to the rocket and missile development work at CARDE, but extended it into new areas.

McGill University, following World War II, expanded its interests to include shock-tube research, aerodynamics, plasma physics and upper atmospheric chemistry. But all this was dwarfed by the High Altitude Research Project (HARP) introduced earlier, initiated in 1962 under a contract from the US Army Ballistic Research Laboratories. The goal was to probe the upper atmosphere with gun-launched projectiles, eventually leading to an orbital launch capability. DRB would not provide support, but support was forthcoming from the Department of Defence Production, which became a funding partner in 1964. This in turn became part of the Space Research Institute of McGill, including a firing range in Barbados, and an impact laboratory at Highwater, Quebec. The first director was Gerry Bull whose history has already been described.

The beginnings of space science at the University of Western Ontario (UWO) and the University of Toronto were described in Chapter 3. In 1961, Forsyth left the University of Saskatchewan to become head of the Physics Department at UWO, building on its historical roots in experimental radio to create an Institute for Radio Science that included ground-based studies but also rocket and satellite measurements. Don Moorcroft and Gordon Lyon moved with him from Saskatoon. Later, UWO expanded into atmospheric science as well, adding Bob Lowe, Bob Sica and Jean-Pierre St. Maurice, a modeler with interests in both the neutral and ionized atmospheres. St. Maurice has since moved to the University of Saskatchewan.

The sixties was a time when new universities were being created, or expanded. York University was established in 1965 and the new Dean of Science, Harold Schiff, a renowned atmospheric chemist who had come from McGill University appointed Ralph Nicholls of the University of Western Ontario as Chair of the new Department of Physics. They agreed to make "space" a priority research area, by creating what is now the Centre for Research in Earth and Space Science (CRESS), with Nicholls as its founding director. CRESS has over the years made substantial contributions to the space program through ground-based, rocket and satellite experiments as well as modeling. Space researchers who joined during the early years include R.A. (Bob) Young, Allan Carswell, Jim Laframboise, Diethard Bohme, Gordon Shepherd, John C. (Jack) McConnell, John Caldwell and Ian McDade. The results are described in subsequent chapters as are the activities of more recent appointees. After a long career of supporting space science in Canada, Ralph Nicholls died peacefully on January 25, 2008, at the age of 81.

University level teaching in Calgary began as a branch of the University of Alberta in Edmonton and passed through many names before becoming the University of Calgary in 1966. Before that happened, Don Rose had taken an interest in the region by selecting Sulphur Mountain in Banff as a site for one of his cosmic ray observatories to be operated during the IGY, with a young Irish Assistant Research Officer, Brian Wilson, in charge. Wilson also took a part-time sessional instructor position in Calgary, commuting back and forth, with the observatory providing data for the first M.Sc. graduate of the university. Another cosmic ray faculty member was added in 1960, John Prescott. During the years 1960-71 the Sulphur Mountain observatory was doubled in size to accommodate a "superneutron monitor", developed at Atomic Energy of Canada, with a comparable facility in Calgary to compare observations at different altitudes. The group was greatly enlarged by Chang Kim, Doraswamy Venkatesan, Titus Matthews and John Bland. Sulphur Mountain was closed down in 1978, but the Calgary unit continued to operate. This cosmic ray influence was enlarged to include upper atmospheric physics with the addition of Clifford Anger, Neville Parsons and Tony Harrison (from the University of Saskatchewan). When the rocket opportunity developed, Cliff Anger and Brian Wilson became heavily involved. Anger developed three-dimensional auroral scanning photometers while Wilson developed novel experiments in X-ray astronomy. Alan Clark joined as a faculty member in 1968 and was the motivating force behind establishing an optical astronomical telescope on donated land south of Calgary, now called the Rothney Astrophysical Observatory. He also developed payloads for high altitude balloon flights, observing atmospheric species concentrations through the absorption of infrared solar emission, using Michelson (Fourier transform) interferometers. While other physics disciplines grew as well, Space Research increasingly became a dominant group, with the addition of Leroy Cogger and later Sandy Murphree, leading to an Institute for Space Research. This activity revolved around the success of the ISIS II Auroral Scanning Photometer as described in the next section, as well as the later Viking and Freja missions.

The Alouette/ISIS Program

Perhaps the most dramatic project of the sixties was the Alouette-ISIS satellite program. Following the acceptance by NASA of a DRB suggestion that Canada should build a topside sounder to be launched by the US, work began at DRTE in January, 1959. The first satellite, Alouette I was launched September 29, 1962 into a 1,000 km circular orbit at an inclination of 80°. This was a remarkable achievement, accomplished in less than four years, starting with no satellite experience and no test facilities. The development of this program is described in detail by Hartz and Paghis.

A topside sounder is an ionosonde (like that used by Henderson and by Rose) that probes the ionosphere from above, rather than below. A ground-based ionosonde contains a transmitter that launches a radio wave upwards. On entering the ionosphere, the electric field associated with the electromagnetic (radio) wave interacts with the electrons and ions, exerting forces on them. The same kind of interaction occurs when a ray of light enters water, or glass (a lens, say) and is bent according to Snell's law. The amount of bending is determined by the "refractive index" of the material. Similarly, when a radio wave traveling at an angle to the vertical enters the ionosphere, its direction is altered by the refractive index of the ionosphere, which decreases with increasing altitude as the electron density (the number of free electrons per cubic meter, denoted by N) increases with increasing altitude. Eventually the direction of the ray becomes horizontal and it then propagates downward, finally reaching the ground; the radio wave is reflected by the ionosphere. This happens even if the wave is transmitted vertically. The interaction of the radio wave with the ionosphere is such that a given frequency, denoted by f in hertz (cycles per second), is reflected from the height where a specific corresponding electron density occurs. The relationship is given by $f^2 = 80.5\ N$. If the frequency is increased, the reflection moves to a higher altitude where a higher electron density is found. This continues up to the altitude where the ionosphere reaches its maximum electron density; if the frequency is increased further, the wave penetrates through the ionosphere and into space. If the radio wave is sent as a pulse, the time of travel between transmission, reflection and reception can be measured, giving the altitude of reflection for that frequency, and thus the electron density at that altitude, using the equation relating reflected frequency to electron density. By measuring the delay time for each transmitted frequency pulse, a profile of the electron density as a function of altitude is measured for the ionosphere, but only up to the maximum electron density. The ionosphere above the maximum was largely unknown in 1959, but the Alouette I topside sounder changed that, by transmitting the radio wave downward from above, allowing the electron density of the ionosphere above the maximum to be determined.

This remarkably ambitious concept was enthusiastically, if somewhat skeptically, received by NASA in an agreement confirmed in April, 1959. Other US groups had considered it possible to transmit waves of discrete (fixed) frequency from a satellite and receive the returned signal, but Canada was proposing a "swept-frequency" topside sounder, in which the radio wave would be scanned through a wide range of frequencies, a truly challenging goal. To be on the safe side, NASA authorized another mission to carry a US fixed-frequency instrument, to be built by the Central Radio Propagation Laboratory in Boulder, Colorado. The frequency range planned for Alouette I was 0.5 to 12 MHz, corresponding to wavelengths of 600 to 25 meters, respectively. To radiate effectively, the antenna must have a length that is a significant fraction of that; for Alouette I, it was determined that two dipole antennas would be required, with tip-to-tip lengths of 23 and 45 meters. The concept for their deployment in space was based on the simple tape measure. But the flat tape has no strength in the direction perpendicular to its surface, so for the satellite the 10-cm wide tape was made of heat annealed spring steel that would curl up to form a tube as it was extended from the satellite after being placed in orbit; it would be rigid in all directions. This was a modified version of an invention of George Klein, an NRC engineer, developed as a collaboration between DRTE and a division of De Havilland aircraft in Toronto that later would become known as Special Products and Applied Research (SPAR), later still to become a major Canadian space company, SPAR Aerospace. The antenna system was named STEM (Storable Tubular Extendible Member) and became one of Canada's first commercial space products. The first unit was tested in flight on a NASA Javelin rocket launched in June, 1961.

A second critical item was the high power transmitter needed to drive the antennas. The signal had to be strong enough that the sensitive receiver could detect the reflected radio wave above the weak radio noise coming from space itself, i.e. the rest of the universe. This noise is called the cosmic radio noise background, but at the time nobody knew what the signal level was. It was therefore measured with a special receiver flown on the US Transit 2A satellite launched in June, 1960. Ground-based ionosondes used vacuum tubes, but these were very inefficient in their use of power, and so only solid state electronic devices could be used for Alouette. These were new at the time, but a transmitter was developed having the required power level. The engineers at DRTE educated themselves on the use of transistors by building a computer from scratch. The last challenge was the batteries. Solar cells would be used to power the spacecraft when it was in sunlight, but as Alouette passed into the dark side of the Earth opposite the sun on most orbits, it would have to rely on batteries that would be charged again during the sunlit portion of the orbit. With the help of DRB's Defence Chemical Biological and Radiation Laboratories, the reliability of commercially available nickel-cadmium batteries was significantly improved.

It is difficult to identify individuals who made more important contributions than others, as all contributed to the ultimate success. However, one can justify the selection of Colin Franklin, manager of electrical systems for the Alouette satellites as an outstanding contributor. Franklin is a leading pioneer in Canada's space program. He played a key role in the design, construction and application of Canada's first satellite, Alouette I, in both an engineering and a program management capacity. In particular, as chief engineer for Alouette I electrical systems, the achievement of the swept-frequency sounder is arguably Franklin's most enduring contribution to space science. Later, his extensive work on Canadian research and industrial development activities in space-related research and manufacturing, have been a significant influence in establishing Canada as a world leader in these fields.

The engineering model for Canada's first satellite, Alouette I, launched on September 29, 1962; with Colin Franklin on the left, Keith Brown, manager of the spacecraft team and John Barry. Photo by courtesy of the Communications Research Centre, Ottawa.

A final crucial element was transmitting the received data to the ground. Tape recorders for on-board storage were notoriously unreliable, with their complex mechanical mechanisms, and computer memory was decades in the future. The DRTE team decided to keep it simple and simply transmit the data to ground in real time, as it was acquired. This could be done only within sight of a telemetry receiving ground station. As part of the USA-Canada agreement, Canada was to provide three ground stations in Canada and NASA would make its worldwide network of ground stations available. As described earlier, the UK had a long-standing interest in the ionosphere, and agreed to provide stations in the UK, the South Atlantic (the Falklands) and at Singapore. In return for this assistance they would be allowed to use these data for their own scientific studies. This international interaction continued to expand as it was joined by France, India, Japan, Australia, New Zealand and Norway, and later Hong Kong, Brazil and Finland, which led to the program becoming the International Satellites for Ionospheric Studies (ISIS). John Chapman was designated the Alouette program coordinator and Eldon Warren the Principal Investigator for the topside sounder. Frank Davies and Jim Scott, who alternated as Chief Superintendents of DRTE, had the difficult task of obtaining approval for the program from the government of Canada, defending it from criticism and acquiring the required funds. Although many people were involved, Chapman was considered to be the one providing the vision and driving force behind the project.

John Herbert Chapman was born in London, Ontario on August 8, 1921, son of Lt. Col. Lloyd Chapman and Kathleen Chapman. He attended the University of Western Ontario and obtained his M.Sc. and Ph.D. degrees from McGill University, the latter in 1951. He served with the RCAF as a Flight Lieutenant from 1940-45 with service in the UK and West Africa. According to Frank Davies, "John Chapman's early experience as a young officer in the RCAF was perhaps the most unexpected. He was posted to West Africa and when the CO found that he was a scientist, he was assigned to direct the building of quarters for a large number of Swahili-speaking Africans." It has already been mentioned that he worked as a summer student with Jack Meek in the Radio Propagation Laboratory, which as the Radio Physics Laboratory later became part of DRTE; Chapman became an employee in 1949. The following material taken from the Friends of CRC website (http://friendsofcrc.ca/) attests to his capability and leadership. As described later, when the new Department of Communications was created, DRTE was administratively moved there and renamed the Communications Research Centre (CRC).

"'John Chapman's vision and determination were central to the success of the space program," says Bert Blevis, former DG Space Technology, for the Canadian Department of Communications, as he describes the man often referred to as the father of the Canadian space program.

"John Chapman's career with the Canadian Government spanned from 1948, when he started with the Radio Propagation Lab as a young Ph.D. graduate, to 1979 when he met an untimely death at the age of 58 while serving as Assistant Deputy Minister of Space Technology for the Department of Communications.

"A talented scientist and an excellent communicator, Chapman provided the link between the scientific capabilities at Shirleys Bay and the government decision-makers who needed his technical understanding and judgment to provide direction to the emerging Canadian space program.

"His major accomplishments were to facilitate such programs as the Alouette/ISIS scientific satellite program, the Anik communication satellite program including the creation of the Communications Research Centre Canada and Telesat Canada, and the Hermes CTS demonstration satellite.

"'Chapman was a natural leader," says Doris Jelly a DRTE physicist who worked for Chapman in the early 1950s. "Even though Chapman was always in a hurry, he had a twinkle in his eye and a spark of fun ... I remember the time we found him demonstrating his version of the twist dance craze in the DRTE machine shop.

" 'Chapman, was a very quick and intelligent person, with little time for small talk," says former colleague Dr. LeRoy Nelms. *"Chapman would have liked to have become the first head of a Canadian space agency, but unfortunately he didn't live to see that happen.*

"Milestones in Chapman's career included:

- joining the Radio Propagation Laboratory as a Ph.D. summer student in 1948;

- his successful submission, with Dr. Eldon Warren, of the Alouette I proposal to NASA in 1958;

- co-authoring in 1967 of what became known as the Chapman Report, or the "Upper Atmosphere and Space Programs in Canada", *here he recommended the cancellation of ISIS C in favor of the Hermes satellite and pointed the direction of focus for the eventual space program;*

- leading the task force in 1968, which produced a White Paper on "A Domestic Satellite Communications System for Canada", *this was the basis for the establishment of Telesat and the formation of a Canadian Department of Communications in 1969;*

- being the primary force behind Canada's co-operative program with NASA and the European Space Agency to design, build and demonstrate the Hermes Communications Technology Satellite, which would provide Canadians in remote areas with direct-to-home television by satellite.

"Chapman did not work alone; he was part of an enormously talented technical team. He was, however, personally able to gain the support and confidence of the Canadian Government and their partners to make the investments that led to Canada's position of leadership in the aerospace and communications technology industries."

As the work progressed it became apparent that more instruments could be added. The radio receiver could be used to record the cosmic radio noise background as well as the ionospheric reflections – this became the Cosmic Radio Noise Experiment. Another receiver was added to detect Very Low Frequency (VLF) radio emissions and Don Rose's Cosmic Ray section in NRC used their rocket experience to provide detectors of energetic electrons and ions such as those that had been detected in the Van Allen radiation belts, and above auroras in the polar regions.

All these challenges were met and the spacecraft found itself at the Vandenberg Air Force Base in southern California (near Lompoc). This was an unusual site in that the Southern Pacific Railroad passed through it and the countdowns had to take into account the schedules of the fruit trains passing through. In those days, satellite launches were only 50% successful, but Alouette was successfully launched on September 29, 1962. After the Thor rocket that launched it burned out, the second stage Agena took Alouette to a higher orbit. The shroud protecting the satellite was next ejected. After the Agena's booster cut off, booster and satellite coasted half way around the world, until it reached the east coast of Africa. There, over Madagascar, at the highest point of its initial launch and a full 48 minutes after the Agena's cutoff, the second stage of the Agena fired and gave Alouette what space buffs like to call a "kick in the apogee", the apogee being the point on the orbit furthest from the Earth. A final push from a spring separated the satellite and the second stage of the booster, setting Alouette up in an orbit 994 km above the earth, to circle the earth every 105 minutes for centuries to come.

Shortly before completing its first orbit, Alouette, now spinning at 120 rpm was sighted by the telemetry station at College, Alaska, where the DRTE engineer, especially sent there for this task, commanded Alouette to extend the two sets of antennas of 23 and 45 meters, each driven out by tiny motors that had been perfected to work in the harsh space environment. As the antennas unwound, the spacecraft slowed down to its final rotation rate of one and a half rpm, keeping a constant percentage of solar cells exposed to the sun. The shape of the body, the oblate spheroid, had been designed to do this.

Within weeks of the launch, the data from the 13 ground stations showed that the topside sounder was a spectacular success, providing scientific knowledge that had never been obtained before. The engineers had designed Alouette I to last for one year, but it was also deemed that the mission would be a success if it obtained three months of data. In fact it continued to provide data for ten years, a remarkable testimony to the work of the DRTE scientists and engineers.

Based on this success, NASA and DRTE, in 1963, laid out a plan for future missions. This was to involve four more satellites, gradually expanding the exploration of the ionosphere and the magnetosphere to greater distances from the Earth. It would also include a focus on the aurora, and instruments would be added that made measurements in the vicinity of the spacecraft, so-called "in-situ" measurements. This International Satellites for Ionospheric Studies) ISIS program would consist of spacecraft built in Canada, but containing US as well as Canadian instruments, launched at intervals during the rising sunspot activity from 1964 to 1970. Tape recorders were added, but there was still a critical reliance on ground stations. The Canadian government also required that the work be contracted to industry to the greatest extent possible. The first of these four was Alouette II, a modification of the backup spacecraft to Alouette I, which happily had not been needed. Alouette II would have an elliptical orbit with its highest altitude at 3,000 km (its apogee) and its lowest altitude (its perigee) near 500 km; the frequency range of the topside sounder was adjusted to take advantage of this. A Langmuir probe instrument from Larry Brace at the Goddard Space Flight Centre was added to determine the electron density and temperature in the vicinity of Alouette II and the whole mission would be coordinated with Explorer 31, the pair of missions being known as ISIS X. One purpose in this was to find out if the topside sounder would perturb the electron densities around the spacecraft in such a way as to make the Langmuir probe data invalid – Explorer 31 had the same probe, but no sounder. Thus a comparison of the data from the two spacecraft would provide the answer. The results, following the launch on November 29, 1965, showed that there was no problem. Like Alouette I, its successor also operated for ten years.

During the building of Alouette II, 18 employees from RCA Montreal were sent to Ottawa in 1962 to work with DRTE on the project, to learn from them the experiences gained from Alouette and transfer it to Canadian industry. A similar contingent was selected from DeHavilland in Toronto, for the mechanical aspects. One of the RCA crew was Joe L. McNally who had graduated from the University of Ottawa with additional mathematics and physics courses, and a year of post-graduate study, but he was not an engineer. After a year with limited responsibility he was suddenly appointed "Payload Engineer" for Alouette II. As noted earlier it can be somewhat arbitrary to identify individuals in a team effort but McNally's career is one that stands out.

After a successful completion of this project the teams returned to their plants and prepared to build ISIS I in industry. Work on ISIS I began in March, 1964. The company RCA Limited of Montreal (in Saint-Henri) was selected as prime contractor with John Stewart as Program Manager and Joe McNally as Payload Engineer, and it was supported by SPAR of De Havilland on the mechanical systems, including the spacecraft structure. Additional US instruments were added, the Soft Particle Spectrometer (SPS) by Walter Heikkila, now at the University of Texas at Dallas; a Retarding Potential Analyser (RPA) by Rita Sagalyn of Air Force Cambridge Research Laboratory in Cambridge, MA; and an Ion Mass Spectrometer (IMS) by R.S. Narcisi, also of the AFCRL. The telemetry transmitter and antenna would serve as scientific VHF beacon. The spacecraft can be described in terms of instruments (pieces of hardware) or experiments (operations resulting in specific datasets). Usually the two are the same, but for the ISIS satellites additional experiments were derived from the basic instrument set, such as monitoring the automatic gain control of the topside sounder, which allowed separation of the swept- and fixed- frequency modes.

The launch of ISIS I, on January 30, 1969 was delayed and almost prevented by the California weather when a series of storms flooded the launch complex, immersing some of the ground support electronic equipment with water and covering the floor with a layer of mud, but all was brought back to operation in a matter of three days. ISIS I had a remarkable lifetime of more than 20 years, and ISIS II almost as long. These spacecraft outlasted their builders as the CRC felt it could no longer justify operations after 1983 (the scientist's funding for data analysis had long since run out as well). The CRC turned the two satellites over to the Radio Research Laboratories – Communications Research Laboratories (RRL-CRL) of Japan in 1983 and they continued operating the sounders to the end of their life.

One might wonder how an international program can be coordinated, with scientists from so many different countries subject to the support of their own funding agencies. This was done through a Working Group, which determined the policy decisions, into which each country had an input. John Jackson of NASA was its first chair, to be succeeded by Ted Hartz of CRC in 1974. If a country accepted a responsibility, it had to be met. The scientists had a separate committee that determined scientific priorities and made their wishes known to the Working Group that had the final say on policy and operations activities, especially with regard to cost. The scientist body was called the Experimenters Committee and it was originally chaired by Walter Heikkila of the University of Texas at Dallas. Although then a US investigator for the Soft Particle Spectrometer (SPS) carried on ISIS I and II, he was born a Canadian in South Porcupine, Ontario, near Timmins. Before moving to Texas, as was described earlier, he got his experience leading the early rocket experiments at DRTE. From 1976 onwards, the Committee was chaired by Gordon Shepherd of York University.

The ISIS II satellite, with Joe McNally on the right and Norm Harrison on the left. The small circular object at the top of the spacecraft (in the plane of the page) with a dark circle at the centre is the cover for the optical input to the Auroral Scanning Photometer, opened in flight. The dark hole is a window, used to inject calibration light without opening the instrument. The similar cover at the bottom is for the Red Line Photometer.

ISIS II was even more ambitious, but for this mission Joe McNally was upgraded to Project Manager. Then DRTE (later CRC) determined that the mission should be opened up to Canadian university scientists and made this quietly known to the community. Clifford Anger and Gordon Shepherd discussed this during a rocket launching campaign at the Churchill Research Range and agreed to each submit a proposal to image the aurora from above. Anger would concentrate on the well-known "green line", the 558 nm emission from atomic oxygen giving the aurora its well-known normal green colour, and on a violet emission of 391 nm wavelength from ionized molecular nitrogen (N_2^+). Both emissions were known to arise from normal aurora at around 100 km altitude. Shepherd would concentrate on the "red line" emission from atomic oxygen (with a wavelength of 630 nm) that causes the red colour at the top of auroral forms, or throughout the whole form, during major magnetic storms – it was associated with higher altitudes, around 250 km. These proposals were accepted by the Working Group but only then did Don Rose, who had not been fully informed during the process, point out that these experiments would have to be paid for by the NRC, as university research was their responsibility. In this sense the procedure was all wrong; NRC should have given permission before the proposals were submitted, but there was a good chance that they would have turned them down. Certainly they would have had no money available in their existing budget so there is no way now of knowing what might have happened. After expressing some frustration, Don Rose went to bat for these university optical imagers and got their acceptance approved through the NRC system and then the Treasury Board. Although Anger (University of Calgary) and Shepherd (York University) both had rocket experience they did not have adequate facilities for the design and fabrication of these experiments and so contracts were let to RCA Limited by NRC. Anger's instrument became known as the Auroral Scanning Photometer (ASP) and Shepherd's as the Red Line Photometer (RLP). As it turned out, the "one-time" money that Treasury Board placed in the NRC budget became permanent so everyone benefited from the flawed process.

Anger's concept of spin-scan imaging took advantage of the spinning motion of the ISIS II spacecraft. If a spacecraft is simply injected into orbit it will tumble randomly in its orientation, but if it is given a spinning motion the direction of its axis will stabilize as with a gyroscope. This approach was adopted for the Alouette/ISIS satellites, which with their long antennas had a good deal of angular momentum for this stabilization. With the ISIS II photometers looking perpendicular to the spin axis they measured the intensity of the auroral light at a spot on the Earth, but the spin of the spacecraft carried the spot across the Earth below, giving a "strip" image. One rotation later the spacecraft had moved forward in its orbit so that the next strip was adjacent to the first and by building up a sequence of strips an image was obtained, as with a TV scan. It would take about 20 minutes to build up one image. ISIS II contained more instruments that related to the aurora: Heikkila's SPS, that had been on ISIS I, Larry Brace's CEP (Cylindrical Electrostatic Probe), John Hoffman's IMS (Ion Mass Spectrometer) – Hoffman was also at the University of Texas at Dallas, and Gene Maier's RPA (Retarding Potential Analyzer) – who like Brace was from NASA's Goddard Space Flight Center in Maryland. All this was in addition to NRC's energetic particle detector led by Ron Burrows. Together with the topside sounder and related experiments this was a powerful auroral observatory. However, all the experiments just listed made in-situ measurements and in studying particles in space one wants to scan over a complete revolution in order to study the "pitch angle distribution", the angle at which the particles travel with respect to the magnetic field that constrains their motion. At the first Experimenters meeting therefore, Anger and Shepherd were astonished when the US investigators argued for having the spin axis perpendicular to the plane of the orbit, the so-called cartwheel configuration, moving like a wheel along its track. In this configuration the spin-scan concept wouldn't work, as the scan strips would lie on top of one another, underneath the orbit. For the spin-scan to work, the spin axis had to be in the orbit plane so that the strips were scanned perpendicular to the orbit. A compromise was worked out in which the spacecraft would operate for three months in cartwheel configuration and three months in orbit-aligned configuration. There were some advantages even for the ASP and RLP in the cartwheel configuration. One study of interest was to relate the in-situ measurements of auroral particles, at 1,400 km, the altitude of the ISIS II circular orbit, to the intensity of the light they produced lower in the atmosphere at the foot of the magnetic field line that they followed. With the repeated strips of the cartwheel motion one could always find correlated measurements made at the same time, and on the same magnetic field line. Also, in cartwheel mode the measurements were made all the way around the orbit, while in orbit-aligned configuration images were taken only in the region of the orbit where the spin axis was roughly horizontal, which happened twice per orbit, but one of these was normally in sunlight where no aurora could be viewed.

It is clear from all this that the orientation of the spin axis is crucial and had to be controlled. This was done by using "magnetic" coils of current that created a torque on the spacecraft, leveraging against the Earth's magnetic field. This was a slow process, but worked very well. Dave Boulding, head of spacecraft operations at CRC very patiently met all the scientist's requests for the control of the spacecraft and the data acquisition.

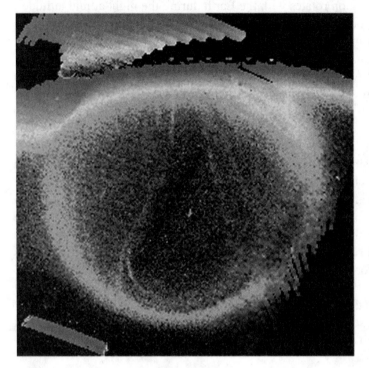

A global view of the aurora obtained with the Auroral Scanning Photometer seen as an almost perfect circle (the auroral "oval") indicating that aurora was present during all 24 hours of local time. The bands of light at the top are sunlight. Inside the auroral oval is "polar cap aurora", discovered with ISIS II. The peculiar shape for this unusual image gave rise to the name "horse-collar" aurora. This image was kindly provided by Leroy Cogger on behalf of the University of Calgary group.

It was found that the green aurora was dominant on the night side of the Earth while the red aurora dominated on the dayside. Of course the red aurora in the daytime could be seen only at high latitudes during the long polar nights when it was dark near noon, around the winter solstices—December 21 in the northern hemisphere and June 21 in the southern hemisphere. These high latitude regions are sparsely populated so this red aurora was generally unknown, but it was very evident in the ISIS-II satellite measurements that viewed all of the polar region. The red aurora related well to the low energy electrons observed by SPS, of about 100 electron volts, that is, the energy of electrons accelerated through a voltage difference of 100 volts. The nightside aurora corresponded to electrons of a few kilo (thousand) electron volts. These nightside and dayside regions connected to different regions of the Earth's magnetosphere. Magnetic field lines on the dayside turn equatorward to cross the equator while those on the nightside extend out into the tail of the magnetosphere. Charged particles can't cross a magnetic field in the perpendicular direction as they are deflected sideways; the force on them is perpendicular to their velocity and to the direction of the magnetic field. Hence it is very difficult for solar wind particles to enter the magnetosphere. However, there is a point on the Earth's dayside that separates the equatorward-directed field lines and those which turn upward and over the magnetic pole, extending into the tail. This divide allows the solar wind electrons to follow the field lines all the way down to the ionosphere, causing the red aurora (which is produced efficiently at low energies). This region was called the magnetospheric cusp (a point), but Heikkila called it the "cleft" (a slot), and the ISIS experimenters supported this. However, decades later only the cusp name has survived.

Alouette I was not just a major technological accomplishment—the science that flowed from it revolutionized the current knowledge of the ionosphere. By studying the global distribution of electron density at the F-region peak, Don Muldrew showed in 1965 that there was a depletion of ionization just equatorward of the auroral region which he called the main trough. This low density region rotates with the Earth during the night at mid latitudes. The depletion ends at the plasmapause, the boundary of the plasmasphere which is a large bulge of plasma stretching out into the magnetosphere as described earlier. Nishida (earlier Jack Jacob's student) showed that plasma outside the plasmasphere is swept into the night time tail of the magnetosphere. All of this is consistent with the magnetospheric paper by Ian Axford and Colin Hines. Muldrew's paper was named a "Citation Classic", with 148 citations by other scientists between 1965 and 1977. Alouette I produced 2 million ionograms and 300 scientific papers.

While the data processing and scientific analysis and interpretation have not been described in any detail, it is evident that dealing with such huge volumes of data is very demanding. Archiving the data afterwards in perpetuity is also demanding. The Alouette/ISIS program generated warehouses of magnetic tapes, with limited lifetimes. Without recopying these tapes at regular intervals they would not last forever, and CRC with its new mandate was unable to take the responsibility for this. Fortunately Bob Benson of NASA/GSFC was successful in gaining support for the digitization and securing for posterity Alouette II, ISIS I and ISIS II ionograms. Since 1995 over 400,000 digital ionograms have been deposited in the National Space Science Data Center at the Goddard Space Flight Center in Maryland. These data can be freely accessed on the internet, and continue to provide data for current radio-science research. The web site is http://nssdc.gsfc.nasa.gov/space/isis/isis-status.html.

The swept frequency sounder found many plasma resonances within its frequency range. A plasma resonance is a received signal that resonates or exists in the plasma near the sounder for an unusually long time after a transmitter pulse. In a space plasma the electrons that are free of their atoms are still loosely tied to them. Glen Lockwood discovered electron cyclotron resonances in 1963, related to motion around the magnetic field. The observed frequency can thus be used to deduce the strength of the magnetic field and Len Hagg used the measurements to map the magnetic field at 1,000 km (Hagg was born in Sweden and served as a radio operator with the Swedish Volunteer Corps of the Finnish Army, fighting the Russians. He later moved to Canada and joined DRTE in 1949; he passed away in Ottawa on September 11, 2007.) Ron Barrington and Jack Belrose discovered Lower Hybrid Resonance noise with the VLF receiver in the frequency range, 5 – 20 kHz. It was found at mid-latitudes 20% of the time, peaking at night, and in summer. Don McEwen and Ron Barrington studied this in more detail; the lower frequency cutoff is related to the ion mass, and so was used to distinguish the ion population; atomic oxygen ions were dominant at high latitude and hydrogen ions at low latitude (these dominate in the plasmasphere).

Ian McDiarmid and Ron Burrows used the NRC energetic particle detectors to map the boundary of trapping – the Van Allen belts – for 40 keV electrons; this was found to be near 75° magnetic latitude near noon; and 70° near midnight, lying just outside the auroral oval. This boundary was interpreted as the high latitude limit of closed magnetic field lines. Above the trapping boundary and during auroral substorms they found extremely intense spikes of particle precipitation that must have originated in the tail of the magnetosphere. Alouette I was launched three months after the US 'Starfish' nuclear test, in which a 1.4 megaton bomb was detonated 400 km. over Johnston Island. Energetic electrons from the explosion were trapped in the geomagnetic field and electron intensities were measured by the particle detectors on Alouette. A paper by McDiarmid, Burrows, Budzinski and Rose described the measurements, which showed a thin intense shell of electrons with energies greater than 3.9 Mev on a magnetic shell having an L value between 1.25 and 1.3 earth radii, roughly connected to the location of the explosion. For a dipole magnetic field the shell parameter L identifies the shell obtained by rotating a particular field line about the axis, it is the distance, in the equatorial plane, from the centre of the dipole to the field line. A similar parameter (L) can be calculated for the Earth's magnetic field and is expressed in earth radii— the parameter is useful for organizing data taken at different locations in the geomagnetic field. The particle data also indicated it would take ten years for the electron intensities to decay to the pre-Starfish levels measured by earlier satellites. The Alouette particle detectors also measured trapped radiation from three Russian high altitude nuclear explosions in October and November 1962. A paper by Burrows and McDiarmid reported the early spatial distribution of electrons with energies greater than 3.9 Mev, and tracked the temporal variations of the radiation during the three months following the explosions. Effects of geomagnetic conditions on the electron precipitation rate were measured and an estimate was made of 3 times 10 to power 21 for the total number of electrons precipitated into the atmosphere at an L value of about 2.95 earth radii.

One wonders how new things were found to be done with the fourth satellite, ISIS II, but they were. All the satellites had a magnetometer, to sense the angle between the spacecraft axis and the magnetic field; this was the direction-finder that Dave Boulding used when he moved the spin axis around. It was a relatively crude device, not intended for scientific measurements until Ron Burrows got the idea that he could use it to observe the magnetic field associated with the currents flowing up magnetic field lines. A series of papers by McDiarmid, Burrows and Wilson used particle measurements and measurements of perturbations in the geomagnetic field, due to field aligned currents, to infer plasma convection patterns in the auroral region and in the polar cap. It was found that the east- west component of the interplanetary magnetic field (supplied by the US National Space Data Centre) had a strong east-west effect on convection patterns on the day-side of the earth, particularly in the region of the cusp. In addition, changes in the north-south component of the interplanetary field could be related to some convection pattern reversals at very high latitudes. The results suggested a direct connection between the Earth's magnetic field and the interplanetary field which was influencing plasma motions in the magnetosphere. Much earlier there was a major scientific dispute about the possibility of field aligned currents, with Chapman saying there were no such currents, and Alfven saying there were. The only way to settle such a disagreement is by measurement, and by sophisticated data analysis from this simple magnetometer the NRC group was able to detect these currents, now a major field of study and accepted by everyone.

In ending this story it should be remarked that Canadian scientists got world-level results and recognition for the Alouette/ISIS program, built on the engineering success of those who designed, built, launched and operated the spacecraft. This was done in the collegial atmosphere of the ISIS Experimenters' and Working Groups, which was a rewarding experience for all those involved and was highly productive. John Jackson tried to establish the definitive bibliography of Alouette-ISIS papers, but was being frustrated because the papers kept coming in, exceeding 1,200 in total.

In looking back at this program what is most remarkable is that these four spacecraft were launched in the space of nine years. Today any scientist embarking on a single space mission had better be prepared to allocate ten years of his or her career, most likely more. Much of this has to do with going from exploratory science to science with much more specific goals. The Alouette I investigators didn't spend much time discussing the scientific requirements – anything they observed would be new and valuable. Nowadays, science experiments have very specific goals to be met with defined accuracies, and it takes time to ensure the project will deliver on these. Risk assessment is a major part of this. In a mature space program, failure is not tolerated and with the experience available need never happen. The ISIS program was dominated by the necessity to get the spacecraft built and launched; but one can't say those involved were unaware of risk when they built spacecraft that operated for more than twenty years.

A home-grown Canadian rocket program

While in some ways the ISIS program dominated the sixties, it by no means involved the entire Canadian space community as it was primarily a DRTE activity. The other major activity was the rocket program, which was an activity more easily accommodated within university laboratories. Don Rose, who guided so much of what happened in Canadian space science, convinced the government to support Bristol Aerospace in developing a series of research rockets and the results were spectacular. The Black Brant III (the smallest rocket), the Black Brant V, a larger and higher altitude vehicle, and the Black Brant IV, a two-stage vehicle that could reach beyond 500 km were extremely successful. In the US they tended to use left-over military rockets rather than develop a version strictly for scientific purposes. And UK and French rockets developed early did not survive. Thus Bristol Aerospace eventually found good customers with NASA and from Germany. Moreover it was very convenient to ship the rockets from Winnipeg to Fort Churchill.

The rocket results obtained from Churchill during the IGY were so unique in vertical probing of the polar regions that scientists from both the US and Canada wanted this capability to continue. The US Army re-opened the range in August, 1959, with access available to many other organizations. Unfortunately a fire in February 1961 caused serious damage, and it closed again. In this renovation the capability was significantly enhanced and the range opened again in November, 1962 under the responsibility of the US Air Force Office of Aerospace Research, with the operations carried out by a contractor, Pan American World Airways.

In the meantime DRTE became interested in sounding rockets as a means to greater understanding of the physics of the ionosphere in relation to improved radio communications. The first two payloads built at DRTE, managed by Walter Heikkila, were launched in September, 1959, aboard Aerobee rockets provided by the Americans. These rockets carried an experiment designed to determine the electron density of the ionosphere for comparison with ground-based measurements. In addition, "cosmic-ray" detectors were provided by Donald Rose and Ian McDiarmid of the NRC to determine the fluxes of energetic charged particles entering the atmosphere from above. The successful co-operation between the US and Canada during this early rocket program established protocols that were used in the Alouette/ISIS program. These early CARDE and DRTE flights meant that Canada had rocket data to present at the first COSPAR meeting in Nice, in January, 1960.

The following year, DRTE designed payloads for three Black Brant rockets built at CARDE. Three experiments were carried out with each rocket: (a) radio wave signal strength variation with height from the ground, using transmissions from DRTE; (b) energetic particle measurements (NRC) and (c) positive ion measurements (University College, London). Two of the firings penetrated auroral absorption layers.

For scientists at DRTE, the Black Brant sounding rockets proved to be a valuable tool to measure parameters other than those gained from ground experiments. The rockets were particularly helpful in obtaining information on the absorbing region in the lower ionosphere that interrupted high frequency radio communication, but was difficult to study from the ground. Over the next 25 years, more than 50 rockets carried DRTE/CRC experiments. A variety of experiments was devised and the payloads became more sophisticated. Some payloads were equipped with parachutes to provide a soft landing, so that instruments and/or data might be recovered. Rapid advances in knowledge were made using rockets in conjunction with satellites and ground-based measurements. In 1967 and 1968 launches were made from Resolute Bay as well as Churchill.

Two streams of activity began to converge, with Don Rose at the apex of both of them. The first was the very strong interest that had developed in Canada about the access to rocket launches from Fort Churchill and the other was the capability to build and launch Canadian rockets, based on the technologies developed at CARDE and Bristol.

Under the impetus of the second stream, a motor and rocket to test propellants was implemented in a system called the PTV, or Propulsion Test Vehicle, with no evident specific goal other than the development of capability. Solid propellant was identified from the start, in order to respond rapidly to an attack by a foreign power, and also to provide reliability in cold Arctic conditions. This program got under way about 1956, with Lionel Dickinson leading the effort. He had worked on classified British missile projects after the war before coming to Canada so was very knowledgeable. It was decided to use propellants based on a solid crystalline oxidizer and an organic polymerizable bonder, the former limited to ammonium perchlorate or potassium perchlorate. Their analysis put them at the forefront of the field, as the US was concerned with strictly military vehicles. The first PTV was 20 cm in diameter and was built just to study the propellant, and how to fill a motor. This involved mixing the fuel, a potentially dangerous procedure best done remotely, and curing it at 140° for a few days. But they were already looking ahead to a 43 cm diameter motor with a thrust of 9,000 kg. The test vehicle to be attached to this motor would be designed and constructed by Bristol Aircraft in its Canadian and British facilities; it would be capable of carrying a 45 kg payload to an altitude of 145 km. Another contract was given to Canadair so that in the end Bristol designed the vehicle casing while Canadair did the fin stabilizer system and nose fairing. This became the Black Brant II (named after a small dark goose that nests along the Arctic coast). In the meantime, the first PTV flew in September, 1959, from the Churchill range. Altogether seven Black Brant I vehicles were flown. The Black Brant IIA was flight tested at the NASA Test Range on Wallops Island, Virginia and cleared for use for high altitude experiments in 1962. After motor improvements this became the BB IIB, which was flown in a number of experiments, but later replaced by the higher performance BB V. Within a month of the first Black Brant launch in October, 1959, the Department of Defence Production received an application from Bristol to make commercially available sounding rockets exploiting the success of the development by CARDE of improved solid propellants and rocket engines, proposing to develop a Black Brant III, Black Brant IV and Black Brant V. Their optimism was based on a market survey of international scientists, done by Don Rose and Al Fia, who was to become the major figure for the Black Brant at Bristol.

Al Fia was another prairie boy whose tenacity moved forward the Canadian rocket and space program. As did many of his peers, Al Fia took his training in propellants and missile work in England. A former teacher and career military officer, Fia had joined the Army in 1940 and saw action in northwest Europe during the war. After the war there was a period of uncertainty that included even a stretch as artillery instructor in Shilo, Manitoba. But then he qualified for the Royal Military College of Science in the UK, spent two years there, then two more at the propellant research establishment at CARDE where he worked as an engineer on Velvet Glove. It was at the Royal Military College of Science that Fia's interest in rocketry was awakened when he had the opportunity to work on a captured V-2 missile. Fia was enormously interested in this, and after two years posting at CARDE, Fia returned to the United Kingdom for post-graduate work at the Royal Military College, where he now studied rocket and missile design. After ten months of study, he was loaned to the British government to work for two years as a program manager on a guided missile project, after which he returned to duties at defence headquarters in Ottawa. By this time Fia learned to guide a large engineering team as well as to control a budget: in short he learned to become a good manager. These skills were found to be of little use in the slow-paced job he found himself in at defence headquarters. After some soul-searching, he gave up the security of the military, and joined Bristol Aerospace Ltd. in Winnipeg in 1958 to build an aerospace company. Thus Velvet Glove helped pave the way for the Black Brant sounding rocket family, for which Fia is given much credit. His death on June 5, 2004 was noted in the Legislative Assembly of Manitoba.

The second stream involved a place from which to launch the rockets, most obviously the Churchill rocket range. The administration of what was called the Rocket Research Facility went through a number of changes. The Canadian Army continued to maintain overall support and control of all matters to do with the range. Following the reopening of the range in 1959, the responsibility for maintenance, operation and management of the range passed from the US Chief of Ordnance to the White Sands Missile Range, and the Military District of Washington provided the logistic and administrative support. The actual operation of the army then was supervised by a new unit called the US Army Ordnance Rocket Research Facility (USAORRF) whose task it was to maintain the range for the use of the various agencies involved in upper atmospheric research using rockets. The unit accomplished this with 13 officers and 120 enlisted men. The Canadian Army provided utilities, communications facilities, living quarters and service facilities.

The fire of February, 1961 was discovered about 3 am in the building that housed the diesel generator plant. The plant attendant tried to telephone for help, but the fire had already damaged the wires. The attendant rushed out on a snowblower to the nearest telephone two miles away, but by the time the firefighters arrived much of the rocket launch facility that could burn was in ashes. The generator building, the helium storage building, the helium compression building, the Aerobee assembly and preparation building, mess hall, launch control building, and part of the tunnel between the launch control building and the Aerobee tower had gone up in smoke. Only the boiler plant, Aerobee launch tower, Nike Cajun facilities (the assembly building, launch area, and tunnels) remained intact. The damage was estimated to be

$1 million for the facilities and three-quarter of a million for the instrumentation. The irony of the situation was that the attendant did not know how to operate the snow-blower. Had he known, he could possibly have put out the fire all by himself by blowing snow on the fire.

In March 1961, the White Sands Missile Range, its new supervisors, developed plan to reestablish the range. This plan did not provide for a full restoration of the range but only a temporary rehabilitation and some modification of the Rocket Research Facility, and the restoration of damaged equipment which could provide a limited launch capacity for a number of specific vehicles. A figure of $650,000 was named as necessary to replace the $1.75 million damage to the place. The Army no longer needed the type of experiments that Fort Churchill made possible, and wanted to put its money into Fort Wainright in Alaska instead of a facility it considered no longer useful for its purposes, and in another country at that. A joint US-Canadian group was to determine who was to use the range and in what order.

The US Army plan was sent to Canada's Minister of Defence and approved, as long as it cost nothing to Canada to reconstruct the range. Other stipulations were that the limit Canada had placed on the number of US personnel, 500, not be exceeded; that reconstruction be completed by November, 1962, and that the layout and construction be done according to the agreement drawn up between the US and the Canadian governments. Canada provided some money for administration and instrumentation.

As the US Army was eager to unburden itself of the reconstruction of the range, and as Canada preferred that it be maintained by military personnel, the US Department of Defence assigned the range to the US Air Force, to be handed over by July 1, 1962. The US Army was to keep on with rehabilitating the damaged facility, and to construct additional launch facilities for rockets such as the Nike Cajun, Javelin, Aerobee and Black Brant. The Army then was to transfer all properties at Churchill to the US Air Force. The Army had operated the range with 175 military personnel, including 36 in communications, 43 manning the motor pool, and helicopter operations, and another 27 directly in support of these. The receipt, storage, preparation and firing of missiles, what was termed missile operation was contracted out to Aerojet General Corporation of the US.

The Air Force wanted to have fewer of their own people on the range and depended more on a large contractor to do the tasks the Army had done. The Air Forces total personnel commitment to the range was limited to 10 officers and 10 airmen; of these eight would stay on the Range itself, two liaison officers were stationed in Winnipeg and two at their Washington headquarters. A transition period was allowed while some of the Army's men remained either on site or in Winnipeg, and the Air Force officially took charge of operations on July 1, 1962. For three months after that date there was a period of behind-the-scenes work while new staff was assembled, trained, the existing or damaged equipment was repaired, new equipment was installed and calibrated, and new support systems were devised. On November, 1962, the range was officially re-opened and sported another name, one which finally stayed: the Churchill Research Range (CRR).

As the Air Force was going to rely on contractors more than the US Army did, there was a serious search for qualified companies, and as many as 50 companies expressed interest in the job; of the twelve that applied, Pan American World Airways Company (New York

City) was selected in May, 1962. One of the conditions of getting the contract was that at least 72 per cent of the workforce employed at Churchill should be Canadian citizens and this quota was exceeded by Pan Am by three per cent.

By the time the range officially re-opened, there was a limited firing capability already in place to handle rockets like Arcas, Nike Cajuns, Nike Apaches, but the first rockets to be launched under the USAF's reign were two Arcas Robins, launched on the night of October 30, 1962. The range was back in operation. In two weeks, the first scientific payload was off the rails aboard a Nike Cajun. It carried a sodium vapor grenade experiment from the Goddard Space Flight Center for the purpose of measuring upper atmospheric winds. The last of the added new facilities, a Universal Launcher, which was capable of launching any type of rocket, was completed in March, 1963. The first rocket launched from the Universal was a Canadian Black Brant I which carried a nitric oxide seeding experiment from the CARDE. By this time CRR could handle a large variety of rockets, from little rockets such as the Arcas for meteorological purposes that carried 34 kg to an altitude of 50 km, as well as larger ones, such as the eight meter long (the length varied somewhat with the size of the payload) Nike Cajuns, Nike Apaches, Astrobee 200s, the liquid-fuel Aerobee 100s, 150s, and 300s as well as the twelve-plus meter long Javelins, in addition to any one of the four Black Brant Rockets by then in existence.

The Nike launcher under construction at the Churchill Research Range. The Aerobee launch tower can be seen on the left.

The interior of the Auroral launcher with a BBVIIIB rocket, also called a Nike VB, in the horizontal position.

This enormous capability shows just how ambitious Canada was, since it shared half the cost during the NASA-NRC joint operation from 1965-1970, but it couldn't match the US in its use of the range, as there were many more space scientists in the US than in Canada. It was a bit like feeding a monster as Canadian scientists came up with new ideas, new designs and new instruments for rocket payloads. This occurred on an annual cycle, as auroral observations could be taken only during the winter; there was too little sunlight in summer, and only very few flights were non-auroral. The procedure was that Canadian scientists submitted proposals annually to Don Rose who organized a panel that selected compatible instruments for a given rocket payload, so that perhaps a dozen experiments could be flown on say three rockets. One scientist was named Project Scientist for each rocket and this individual had the responsibility of choosing the right conditions, such as the occurrence of a certain type of aurora overhead, for launch. Each instrument would be supported by a professor or government employee, normally with a graduate student, and usually with technical support. After the flight they would be busy looking at the data before writing a proposal for next year's flights.

In 1960 the Radio and Electrical Engineering Division (REED) of the NRC became involved in the development of telemetry components for Black Brant rocket payloads.

Rose took advantage of this by convincing this group, under Dick Rettie, to take the responsibility for the payload integration of Black Brant rockets, to be launched, under contract, by Bristol Aerospace. University and government scientists could build their own instruments that would be integrated into the rocket payloads by REED. Integration involved building the rocket structure, installing electrical power and telemetry, and then installing the instruments and connecting them to these devices. In 1965, the Space Research Facilities Branch (SRFB) was formed to coordinate what was becoming a large operation. Under this arrangement the integration was no longer done by NRC personnel but was contracted out to Bristol Aerospace and to the new Space Engineering Division (SED) at the University of Saskatchewan as described earlier. Scientists would build their individual experiments within their own laboratories and deliver them to one of these two organizations for integration and test.

A review of the Summary of Range Activities for May-October, 1966, by Superintendent Jim Brandy is of interest. There were 19 NRC personnel on site, and 194 Pan American employees, with a capacity of 66 apartments for married quarters, but Brandy stated that 126 were needed if employees were to be retained. A total of 143 rockets were launched between January 5 and September 30. This consisted of 97 Arcasonde IA vehicles, 3 Aerobee 150, 2 Astrobee 200, 3 Black Brant IIA, 2 Black Brant III, 2 Black Brant VA, 1 Javelin, 15 Nike Apache, 10 Nike Cajun, 6 Nike Tomahawk and 2 Sidewinder Arcas. Brandy discussed the limitations of the existing facilities, and made a case for the capability of simultaneous launches of two rockets into the same auroral event. The ability to launch a number of rockets in rapid sequence was also a goal. With the large variety of rockets being launched from a limited number of launchers, it was often necessary to change the rail on the launcher to accommodate a different rocket. This could take four to eight hours and not compatible with rapid launch sequences. Brandy suggested an increase in the capability of rocket launch by a factor of two or three over the next five-year period to accommodate the coming solar maximum. He predicted the number of 337 rockets per year by 1969.

Enhanced rocket instrumentation

As the rocket program matured, the individual investigators were attracted to the development of more challenging experiments, and both SED and Bristol Aerospace enhanced their support capability for the control of the rocket during flight. Instruments were often mounted under the nosecone, which had to be removed once the rocket was above the dense atmosphere. This was normally done by splitting it into what were called clamshells. These opened on hinges and then fell away from the rocket. Instruments also were installed in the "parallel section", looking perpendicular to the rocket axis. Here, small doors in front of each experiment would be ejected after launch. Booms would be extended if the experiment package needed to be some distance from the rocket, outside the shock wave, for example. If rocket recovery was required, a parachute system would be added to the payload.

One novel experiment was developed by Alex Kavadas and a newly hired faculty member at the University of Saskatchewan, Jim Koehler, an exceptional experimentalist. They had worked together for a long time as Koehler was employed as a third-year summer student for Kavadas, who was not simply an auroral theorist, he developed radar systems as well, adding polarization capability. Koehler recalled walking into the radar room for the first time and seeing the floor covered with wires. "Don't step on the red ones", Kavadas said, "they have 250 volts on them." This impressed Koehler but they nevertheless continued to work together for a long time. Now they wanted to measure the electric fields associated with the aurora. The existence of electric fields parallel to the magnetic field was a very controversial topic, a battle between giants in the field. Sidney Chapman insisted that parallel electric fields could not exist in the ionosphere and magnetosphere as they would be shorted out by currents flowing through the plasma. Hannes Alfven said there were ways in which charge regions that he called "double-layers" could be formed and that the electric field between them in fact caused the acceleration of auroral particles. At the UWO Conference on Auroral Physics in 1951 this battle continued and Gerson arranged a special debate on the topic. As Chapman was the senior scientist, and well established in the UK and the US, his view dominated over Alfven's revolutionary new approach, based in Stockholm, Sweden. However, in the end, Alfven was proven right, and received the Nobel prize.

Kavadas and Koehler wanted to resolve this experimentally, rather than theoretically, and their concept was to release probes, 4 cm or so in diameter and about 40 cm long, with electrical contacts at either end. If an electric field existed, a voltage would be generated between these contacts. The probes were actually mounted on the outside of the rocket, with covers that protected them from the heat of the ascent through the atmosphere. Then they were spun up to high velocity and released, with gyroscopic stability. Several probes were used in order to obtain the different components of the electric field but it was necessary to know their orientation in space. This was accomplished by tiny optical sensors that could sense the difference between the Earth and space, and the horizon. The experiments were successful, but limited in their accuracy because of the short distance between the contacts. Later Sweden extended probes on cables to generate a larger sensor separation, and later in Canada a distance of 1 km was achieved using a tethered rocket body, on Oedipus C. That story is told later.

Passive auroral plasma observations at the NRC

The Cosmic Ray Section of the Division of Pure Physics, by this time headed by Ian McDiarmid, made a major advance when it appointed Dr. Brian Whalen. Here is Whalen's account of the activities that followed. Some editorial comments are added as [ ].

"Brian Whalen joined the Comic Rays and High Energy Physics section of the Physics Department at NRC in late 1965 after graduation from UBC with a degree in Nuclear Physics. He, along with Ian McDiarmid and E. (Bud) Budzinski, immediately embarked on a series of studies of auroral energetic particle precipitation associated electron and proton aurorae. This programme was stimulated by the establishment of the sounding rocket range at Churchill Manitoba (Churchill Rocket Range CRR) which was strategically situated near the centre of the auroral zone. Instruments designed to investigate the sources and energization mechanisms for auroral particles were developed which measured electron and ion energy (1 - 100 kev) and pitch-angle distributions and their relationships to auroral arcs and substorms from which the location and source mechanisms for particle precipitation could be inferred. Electron and proton energy analysers and ion mass spectrometers sensitive to particles in the auroral energy range were developed and flight tested on a series of approximately 15 sounding rockets flights launched into various auroral conditions.

"The sounding rocket programme provided this group with the opportunity to develop and flight test instruments and techniques which would later serve as the basis for a number of satellite programmes. Below we outline some of the highlights of their investigations.

Auroral Electron Precipitation and Ionospheric Electric Fields

"*Early observations of electron fluxes in and near aurora displayed magnetic field aligned features which were most easily interpreted as resulting from acceleration through a quasi-static electric fields parallel to the local magnetic field lines. Later observations suggested that these distributions may result from plasma wave interactions. The energization mechanism for these particles remains a matter of dispute to this day.*

"*It also became clear that ionospheric and magnetospheric convection electric fields (fields perpendicular to magnetic field lines) played an important role in the formation of auroral arcs and the acceleration of auroral particles, which was consistent with Gerry Atkinson's ionosphere-magnetosphere feedback theory published in the mid-sixties. To test this theory Whalen and his group designed instruments to measure, for the first time, the thermal (~ eV) ionospheric ion distribution function from which the ion drift velocity (local perpendicular electric field) could be inferred. Dennis Green working with these rocket-based ion measurements, developed the analytical techniques needed to determine the ionospheric ion density, drift velocity and temperature. Ground-based observations and in-flight electron and thermal ion distribution function measurements showed explicitly the (anti-) correlation of electric fields and electron flux in a (co-rotating) earth frame of reference.* [This means a coordinate system that rotates with the Earth].

"*One of the campaigns in which the group was involved was the Cape Parry launch into the Dayside aurora (1974). Dr. Hirao from Japan was also invited to take part in that expedition and in association with that activity Dr. Hirao visited the Space Plasma group at NRC and engaged in spirited discussions with Whalen on techniques to measure the thermal plasma environment. This was the first scientific contact the group had with Japanese space scientists and was a precursor to significant collaboration in the future.*

The "User Room" at the Churchill Research Range. Here the scientists and technicians conducted the final tests and made any necessary adjustments or repairs to the payload prior to being mated with the rocket. Two rocket payloads are evident in this photo.

Solar Wind Source for Auroral Ions

"*Early in the sounding rocket programme, observations were taken in a proton aurora where ionospheric emissions (observed from the ground by Fokke Creutzberg of NRC) seemed to be produced exclusively by precipitating magnetospheric protons. Analysis by John Miller (most recently Chair of the Department of Earth and Space Science and Engineering at York University) confirmed this speculation and led to further investigations of the sources of these energetic magnetospheric ion fluxes.*

The first energetic (1 - 100 kev) ion composition measurements in and near auroras were made by the group and indicated a solar wind origin for these particles. The instrument developed for these observations was a simple electrostatic analyser followed by magnetic deflection to determine the ion energy and mass per unit charge. Flights into and near aurorae showed that ion precipitation was dominated by H^+ and He^{++} with roughly the same relative composition as the Solar Wind, strongly suggesting a solar wind origin for these particles. Later satellite observation by the group at Lockheed Palo Alto showed that ionospheric sources may dominate at lower energies (< 1 kev) .

A complex rocket payload, (left) VB-18, integrated by SED Systems in 1968. Photo by courtesy of SED Systems.

Transverse Ion Acceleration (TIA)

"*Energetic ion observations made from a sounding rocket launched into an intense auroral sub-storm revealed that the payload had passed vertically through a very thin region located near 400 - 500 km where the local ionospheric ions were being accelerated to Kev energies perpendicular to the magnetic field. These data were analysed by Patrick Daly (now at MPE [the Max Planck Institute for Extraterrestrial Physics]) who showed that this mechanism could be responsible for a large loss of ionospheric ions to the magnetosphere and geomagnetic tail. Several rocket flights followed which confirmed that this was a common occurrence at auroral latitudes.*

"*It is notable that two dominant particle energization mechanisms had been directly observed in space plasmas: one which accelerates electrons parallel to the local magnetic field and the other which accelerates ions perpendicular to the magnetic field. Much of the effort of the group has been exerted in clarifying the mechanisms and free energy sources responsible for these two processes.*

Polar Cap Aurora

"*In spite of the extremely difficult late November and early December weather conditions, the Bristol Aerospace engineering team led by Ralph Bullock successfully launched a sounding rocket from Resolute Bay NWT into Polar Cap auroral arcs. This campaign was supported by ground-based photometric observations by Fokke Creutzberg and by Clifford Anger, who provided an early version of his auroral all-sky imager. Results from this campaign led the group to propose a unique interpretation of the source mechanism for the 'soft' (100s of eV) electron precipitation observed in these arcs, namely direct reconnection of magnetic field lines originating in the polar cap during periods of strong northward directed interplanetary magnetic fields."* [Direct reconnection means that the magnetic field lines in the polar cap actually connect to the magnetic field lines of the sun.]

This ended the first phase of Whalen's activity. The account continues in the next chapter. It should be noted that during 1966 a requirement to launch rockets near the magnetic pole led to establishing a BBIII rocket launch facility at Resolute Bay. According to government reports, after this time rockets were launched yearly. The US launched Arcas rockets there in 1967 and 1968.

Canada Centre for Remote Sensing

Yet another space development was taking place during the sixties, quite independently of the upper atmosphere, ionosphere and space, that was the focus of the rocket and satellite programs. This major space activity being created in Canada was then called "Remote Sensing."

One of the major discipline areas represented by Government Departments in the Interdepartmental Committee for Space was that of Earth Observations. Its early history has been recorded by Dr. Larry Morley. Its development went on throughout the sixties and it came to realization only in 1971, so is included in this Chapter. Besides being an important historical document, Dr. Morley provides a great deal of insight into the workings of governments (Canada and the US), international relations and the military-civilian relationship. Here is his story.

"Canada has a notable history in the early development and application of aerial photography, photogrammetry and airborne geophysics to the mapping, resource development and environmental monitoring to its very large and remote territory. Topographers, foresters and geologists used aerial photography extensively, especially in the years immediately following WW II. Their combined experience in this art was heavily relied upon to develop the Canadian Remote Sensing Program. The term 'Remote Sensing, was first used by the U.S. Military to describe a type of aerial surveillance that went beyond the use of photography into the use of parts of the electromagnetic spectrum other than the visible such as the infrared and the microwave parts.

"It was my privilege to have an appropriate scientific and technical background in order to have been in a central position as far as the launching of the Canadian Remote Sensing Program was concerned. Prior to 1963, remote sensing was militarily classified by the U.S. Department of Defence. U-2 aircraft reconnaissance flights, spy satellites and airborne infrared line-scanners were being used to great strategic advantage in Vietnam and in the Cold War in general. As a civilian with no access to classified technology, serving as a geophysicist in the Geological Survey of Canada (GSC) and active in all aspects of airborne geophysical methods, I was naturally interested in these things.

"By chance, in 1962, I happened to serve on a United Nations Mission on 'photogeology and airborne geophysics' at the Geological Survey of Japan with William Fischer, chief photogeologist with the U.S. Geological Survey. He told me that the U.S. Department of Defence would shortly be de-classifying a lot of remote sensing technology and that he was hoping to get the USGS involved. I was, from the beginning, on the lookout for information. In 1963, the Environmental Research Institute of Ann Arbor held the first unclassified international symposium on remote sensing. Steve Waskurak from the GSC attended and brought back much valuable information.

"As soon as Bill Fischer returned from this mission, he set to work on the idea of using satellite imagery for photogeology and later became a force for leading the U.S. into its National Remote Sensing Program.

"The GSC, allowed a lot of freedom to its scientists in choosing what kind of R & D they undertook. Thus the Geophysics Division was able to pursue R & D in remote sensing, even though aerial methods were considered quite 'flakey' at the time for geological applications. We set up a Remote Sensing Section with Alan Gregory as its head. He had been conducting the original experiments in the development of the airborne gamma ray spectrometer which we considered to be a type of remote sensing. He went on a fact-finding mission to several laboratories in the U.S. which we knew were doing development work on remote sensing and returned with much valuable information and contacts which were used to point the direction in which we should be moving in Canada.

"During this period, there were two persons in Canada who had been working with remote sensing in Canada within the classified area. One was Trevor Harwood, a geophysicist with DRB and the other a geologist with Mt. Allison University, the late 'Harky' Cameron, Cameron was an ex RCAF navigator who had maintained his connections with the RAF and was able to get access to PPI radar images taken over Nova Scotia by the RAF Vulcan aircraft based at Goose Bay, Labrador. They were taken at altitudes of 30,000 feet and each one covered nearly half of Nova Scotia in one image. These images showed dramatic 'linears' which Cameron interpreted as major geological faults. The geological 'establishment' did not think much of his interpretation and nicknamed him 'Faulty Cameron' which discouraged him from publishing. It is interesting to note, however, that his interpretation now stands as 'self evident', confirming Schopenhauer's words.

"Trevor Harwood and Moira Dunbar, both of DRB, studied floating ice in the Canadian Arctic using aerial photography and remote sensing. This was, and still is, a subject of great concern to Canada, both from the point of view of unauthorized passage of foreign vessels through the N.W. Passage as well as providing information for safe surface navigation. Harwood organized the first infrared line-scanning survey of a test area in the Arctic which demonstrated the possibility of ice reconnaissance during the Arctic night. The Ice Branch had been conducting ice reconnaissance flights for several years during the summer, but had no information on winter coverage as they were using visual observation methods.

[Six paragraphs have been deleted from Dr Morley's text here]

"Civilian remote sensing in Canada would have developed incrementally had it not been for the proposal brought forward to the U.S. Government by Bill Fischer of the U.S. Geological Survey. By 1966, he had managed to convince the Department of the Interior and its then secretary, Bill Pecora, to sponsor an Earth Resources Orbiting Satellite (EROS). The satellite bus would be the same one that RCA had made for NASA's experimental meteorological

NIMBUS program and the payload would consist of three bore-sighted, RCA, very high-resolution return-beam vidicons. Each vidicon would have its own optical filter, one for the near infrared band, one for the red and the other for yellow, corresponding to the three emulsions layers on camouflage film used during the war and found to be especially useful in airborne experiments to map vegetation. Bill Pecora managed to get approval for funding the satellite as an Interior initiative.

"I invited Bill Fischer to come to Ottawa and meet privately with senior EMR officials and John Chapman ADM of the Satellite Communication Branch of Eric Kieren's newly-formed Department of Communications. After the success of the Alouette and ISIS programs Chapman had also become interested in satellite remote sensing. The meeting was held at the Royal Ottawa Golf Club. Jim Harrison ADM/EMR and John Chapman, Yves Fortier, director of the GSC and Sam Gamble, director of the Surveys and Mapping Branch were present.

"Fischer, after briefing us on the EROS program showed some of the pictures of the Earth taken by the astronauts with hand-held cameras. At the time they seemed very dramatic indeed and impressed all of us. Chapman then offered the Prince Albert Radar Laboratory as a readout station for EROS. It contained a 25 meter diameter paraboloid tracking dish which was no longer needed as the ionosphere experiments, for which it had been erected, were complete. Fischer was pleased with the idea, as this station would be able to read out data for the whole of North America. At that time, no plans had been made for the readout of EROS.

"Alas, the U.S. Bureau of the Budget had EROS cancelled. NASA had complained that EROS would be an experimental space project, and as such was within the mandate of NASA. Mandates are very important in the government service. Nevertheless, NASA was honour-bound to come up with an alternative program, which they finally did in 1969. It was to be called The Earth Resources Technology Satellite (ERTS). It was to retain the three RCA return-beam vidicons, but was also to have a new sensor, the multispectral scanner designed and made by Hughes, Santa Barbara. A vibrating mirror focused the earth scene onto an array of solid state detectors arranged in five clusters, each cluster being sensitive to a different wave band, making five channels ranging from blue to the far infrared. The sensor was to have a ground resolution of 80 metres.

"Our problem now was how to get the same deal with NASA on ERTS as we had with Interior on EROS, particularly as NASA originally had no plans to share their program internationally. John Chapman, who had successful cooperation with NASA on the Alouette and ISIS programs and I visited their Assistant Administrator to propose that Canada read out ERTS at Prince Albert. We were politely told 'it would be foolish to invest in a readout and ground data handling facility as it would be too large a risk for Canada. It was costing NASA $40 million and besides, they could cover most of the Canadian landmass from their three readout stations in Alaska, Goldstone, Arizona and Wallops Island, New Jersey (sic). Furthermore ERTS would carry an on-board tape recorder which could record any data beyond the range of the three readout stations.' It was clear they did not want to have a foreign country reading out this satellite. Chapman appealed to President Nixon's Scientific Advisor but to no avail.

"Canada had a strong tradition, as did most states, of keeping the control of aerial photography and mapping within the country. We did not like the idea of having to purchase imagery of Canadian territory from a foreign country nor of having a foreign country acquire

imagery of Canadian territory without advance permission. Such a concept was in violation of the Chicago International Convention which required any state wishing to acquire air photography of another country to first get permission. NASA took the position that in this case, the U.N. Treaty on the 'Peaceful Uses of Outer Space' applied—in which any state was free to conduct any activity in space provided it did no harm to other states (particularly in the case of falling debris).

"The Deputy Minister of Communications, Alan Gottlieb and his legal advisor, Charles Dalfen, at first wished to make a diplomatic objection on the grounds that the U.S. would be able to obtain exclusive information on the location of potential mineral and petroleum deposits in Canada by means of this satellite and might give advance information to U.S. exploration companies. By orbiting a resource satellite and taking imagery of Canadian territory, the U.S. would be invading our sovereign rights. In EMR, we took the view that it would be preferable for Canada, and indeed the international community, to gain knowledge of the technology which would allow us to better control the use of the data. To that end, Harrison, Gregory and I presented a paper at the International Astronautics Federation held in Brussels in 1971, in which we recommended that an international legal regime for the operation of remote sensing satellites, including the transfer of technology, be established. We suggested the establishment of an international network of readout and ground data handling centres and even supplied a map showing the possible locations of such a network. This aroused a lot of hostility from one of NASA's assistant administrators. Its publication was delayed for four years because the editor misplaced the manuscript and failed to notify me of this.

"A break in the question of Canada's reading out the ERTS satellite occurred in March, 1969. NASA's Chief Administrator was on a tour of the countries active in space to seek technical contributions towards their 'Post-Apollo Program' which was the development of the Space Shuttle. When he came to Canada, he addressed a meeting of about 50 senior scientific administrators and politicians held in the Centre Block of the Parliament Buildings. After delivering a briefing on the re-useable shuttle concept he was asking Canada, as he had asked several other nations to consider how they might contribute. (Canada eventually responded to this request suggesting we contribute the remote manipulating system, later to be known as the CANADARM.) During the question period I had the temerity to ask if NASA would allow Canada to read out their proposed ERTS satellite at Prince Albert. To my surprise he immediately replied 'yes.' While it did take another two years to reach a written agreement, we were definitely on our way. Whatever happened within NASA, I do not know, but within a few months, it became their policy to encourage other states to read out ERTS and to pay a fee of $200K per year per station for the privilege of doing so. After a year of negotiations with NASA and the U.S. State Department, 'an exchange of notes' on the Canada/U.S. Earth Resources Agreement was finally signed.

"It was still an open question as to which department was to be the lead agency. Jim Harrison ADM/EMR instructed me to prepare a memorandum to Cabinet from EMR. When it was completed he told me to show it to John Chapman, ADM Communications, which I did. Chapman reached into his desk drawer and pulled out a draft document to Cabinet proposing that Communications be the lead agency. My reply was that he had better have lunch with Jim Harrison. After the lunch, Jim Harrison told me that they decided they would recommend that the two departments would manage the proposed centre jointly. When the document reached Treasury Board for vetting before going to Cabinet, Sid Wagner, the Board's scientific advisor,

told me in 1992 that he had questioned whether two departments could be jointly responsible for the same program. The answer was 'no.' Treasury Board decided that EMR should be the lead agency. If, at the time, there had been a Canadian Space Agency as had been recommended in the 'Chapman Report' in 1967, it would undoubtedly have been given the mandate.

"Finally the big day came on Feb. 11, 1971 for approval by the Cabinet Committee on Science, of which Bud Drury was the chairman and Bob Uffen the secretary. For any large proposal there is usually a series of approval levels to go through. The higher the decision-making level, the greater the jeopardy for project approval. Cabinet was the highest level.

"On the first presentation, it was obvious the ministers had no idea what we were talking about. 'Remote sensing' at that time was a subject that was not in every day use and was a difficult concept for the lay person to understand. I remember particularly Eugene Whelan, the Minister of Agriculture, who kept asking 'Yes, but what does it do?' Afraid that ten years of work was about to go 'down the tube', I frantically suggested that I show some of the astronauts pictures of the earth taken during NASA's MERCURY project which I had brought with me, as well as a miniature Japanese projector and screen. Bob Uffen wagged his head, so I proceeded to show pictures for the next half hour. The ministers were so intrigued they were unaware of the passage of time until a bell rang which signified they were required in the House for a vote. They all disappeared leaving the bureaucrats sitting there. I asked Bob Uffen whether or not the proposal was approved. He replied 'no problem.' The Canada Centre for Remote Sensing came into being. After eleven years of working on the project, it was one of the high points of my professional life."

Postscript

It is truly remarkable that with no evident policy to guide it, Canada assessed the appropriate level for its space activity, and virtually reached it during one decade, the sixties. This achievement was based primarily on the insights and responsibilities accepted by individuals, and the support they were able to obtain from their funding sources. The DRTE (within DRB) established a flourishing scientific satellite program, launching a satellite about every three years. The NRC created a marvelous rocket program, using rocket motors manufactured in Winnipeg, not far from the Churchill Research Range, and scientists from across the country built instruments that were integrated into these payloads and launched at a rate of about twenty per year. The personnel needed for the factor-of-ten growth required to operate this ambitious program had to come from the universities, which proved themselves able to attract good graduate students and teach them how to do space science. These came initially from the University of Saskatchewan, but during the decade other universities joined in as well; including the University of Calgary, the University of Western Ontario, the University of Toronto and a new University created during the decade, York University.

References:

Axford, W.I. and C.O. Hines, A unifying theory of high-latitude geophysical phenomena and geomagnetic storms, Canadian Journal of Physics, 39, 1433-1464, 1961.

Friends of CRC website: http://friendsofcrc.ca/

Gainor, Chris, Canada in Space, Folklore Publishing, 2006.

Hartz, Theodore R. and Irvine Paghis, Spacebound, Canadian Government Publishing Centre, 1982.

Hines, C. O., Internal atmospheric gravity waves at ionospheric heights, Canadian Journal of Physics, 38, 1441-1481, 1960.

Kirton, John, Canada, the United States, and Space, Canadian Studies Program, Columbia University, 1986.

Chapter 6

Decades of Transition

A Department of Communications

During the 1970s the Canadian space program, which exploded into existence in the 1960s, began to lose momentum at an alarming rate. In 1967, the so-called "Chapman report" was published, entitled, *Upper Atmosphere and Space Programs in Canada*, commissioned by the Science Secretariat of the Privy Council Office and authored by John Chapman (DRTE), Peter Forsyth (University of Western Ontario), Phil Lapp (Special Products and Applied Research division of de Havilland Aircraft) and Gordon Patterson (University of Toronto Institute of Aerospace Studies). This then became the basis of the new Science Council of Canada's Report No. 1, entitled *A Space Program for Canada*. This was sent by the chairman, O.M. Solandt to Prime Minister Lester Pearson on July 4, 1967. It began by referring to an earlier report by the Royal Commission on Government Organization, from January, 1963, which drew attention to the fragmentation of non-military space research and suggested "that it be consolidated into a single agency." The Chapman report, based on 112 briefs and 11 hearings across Canada highlighted three issues:

> *The need for a central organization for space activities in Canada;*
> *The need for Canadian satellites for domestic communications by 1970 or 1971;*
> *The growing need for a Canadian satellite launching capability.*

It recommended "the establishment of a central co-ordinating and contracting agency for space research and development." In order not to duplicate existing facilities it proposed that authority would be contracted back to certain agencies and departments. It also recommended "the initiation of a design and cost study for a small-satellite launch vehicle and related facilities for Canadian use." Finally, it recommended "that steps be taken to ensure continuing Canadian control of domestic communications," that Canada should obtain rights to appropriate geostationary longitudes and that reassignment of communications frequencies be negotiated to provide frequencies for direct broadcast satellites.

The Science Council report made one major recommendation:

> *"The Science Council of Canada recommends the establishment of a broadly conceived central agency responsible to the Government of Canada for the advancement of Canadian capability in the science and technology of the upper atmosphere and space; for furthering the development of Canadian industry in relation to the use of the upper atmosphere and space; and for the planning and implementation of an overall space program for Canada."*

It may have weakened its case by the sentence immediately following the recommendation: *"There is some difficulty in defining initially the full extent of the role that may be played by a central agency."* The seven functions that it recommends for this body begin with *"advising"*, *"recommending"*, *"planning and coordinating"*, *"coordinating"*, *"developing"*.

Only the last two contain any muscle: "*Control of all facilities in Canada capable of launching rockets or satellites into the upper atmosphere or space and of all launches by or on behalf of Canadian institutions*", and "*initiating and conducting by contract or otherwise such projects in the upper atmosphere or space as the Minister may approve.*"

It was noted that the current expenditures on space were 0.032 percent of the gross national product (about $17 million) and that the Chapman report proposed an increase to 0.1 percent ($60 million), not including the cost of the communication satellites. There was no government response with respect to the central co-ordinating and contracting agency (read "space agency"). Perhaps the government was afraid of creating a large organization that might assign priorities to the various space activities that did not align with those of government departments. Similarly the small-satellite launch capability never came about.

On the communications side the response was to create a new Department of Communications (DOC) in 1969, with Eric Kierans as its first minister. As part of this, DRTE was moved from DRB to DOC, although it physically remained in the same building, and was given a new name, the Communications Research Centre (CRC). To the space community this change was fairly transparent, as CRC continued the Alouette/ISIS program in conducting the launch and operations of ISIS II, launched April 1, 1971. The same faces were present around the table. But the mandate had changed, and as the seventies progressed it became clear that there would be no scientific satellite to follow ISIS II (although an ISIS C was in the original Canada-US agreement). It meant that the highly skilled expertise developed by DRTE would not be available for the launches of scientific satellites and without this how could they come about? The DOC was abolished in 1996 and replaced by the Department of Canadian Heritage, but CRC survives to this day, under the Department of Industry.

The DRTE plan called for an ISIS C mission (the missions are labelled by letters before launch, and changed to Roman numerals after) that would take the spacecraft out of the magnetosphere and use the topside sounder to probe its surface, the magnetopause. But that was not to be; the next satellite launched by DRTE (now CRC) personnel was the Communications Technology Satellite (CTS). The technology of communications satellites had reached the point of becoming economically feasible, and Canada needed satellite communications as a way of reaching its far-flung and thinly-spread Arctic communities, as well as carrying heavy commercial traffic between commercial centres such as Toronto and Vancouver. This involved sending information from the ground to the satellite at a frequency of about 6 GHz (1 Gigahertz = 1,000,000,000 oscillations per second), and from the satellite back to ground in the 4 GHz band. A corporation, Telesat Canada, was created to operate this, with shared ownership between the Canadian government and the telephone companies constituting the Trans Canada Telephone System (TCTS). John Chapman was named to the Board of Directors. It was expected that RCA in Montreal would build the satellite using existing technology, and based on its ISIS experience. The three Anik A (Anik meaning "brother" in the Inuit language, normally referred to as Inuktitut in Canada) spacecraft would be placed in geostationary orbit, 35,800 km above the Earth. Satellites in Low Earth Orbit (LEO, a few hundred km above the Earth) complete an orbit in about 100 minutes, but with increasing altitude the orbit time increases and eventually reaches 24 hr. Since this is the rotation time of the Earth a satellite at this distance hovers over a fixed point on the ground, but only if the orbit is in the equatorial plane, as otherwise it drifts back and forth across the equator. Although the equator is some distance from Canada, from this "fixed" vantage point, the geostationary satellite can "see" all of Canada, and so have communications with any ground stations located in it, although not reaching quite as far as the North Pole.

The Anik satellites were based on very conservative, and thus highly reliable, technology, but the so-called 4/6 GHz band had its limitations. One was that this was the same band as used by telephone companies for carrying telephone communications across the country via micro-wave towers (familiar to all those who have driven along the trans-Canada highway). This meant that the Aniks had to operate at low power, in order not to cause interference; this was not a problem for the satellites as the technology worked only at low power anyway, but it did mean that large dish-like antennas were needed on the ground, perhaps 30 meters in diameter, depending on the type of usage. The result was that the communications would go to a few selected locations, an Arctic community for example, or a heavily populated area in southern Canada from which the signals would be distributed over ground-based networks. Although RCA was the only expected bidder, Hughes Aircraft Company of California submitted an unsolicited bid, at considerably lower cost, and a much shorter delivery time than for the RCA bid. An intense competition ensued, with each side improving its offer in turn, but Hughes won the day by offering to subcontract to Canada not only the Anik work, but work on other communications satellites as well. Although this ran counter to the strategy of using the Alouette/ISIS development as a way of creating a Canadian space industry capable of building entire satellites it has to be remembered that RCA, even with a laboratory in Canada, was not Canadian either. By using Hughes, with Eric Kierans' approval, Telsat Canada got into space quickly, in November, 1972 with the world's first domestic communication satellite system. It occupied a "slot" in geostationary orbit, established a place for Canada in the space communications business and made itself a national entity with services in English and French across the country. Quebec had been considering an alliance with a European communications satellite company, Symphonie, but Telsat Canada made this irrelevant. This saga of satellite communications continued on into the future. Hughes provided three Anik A satellites, and three Anik Cs, but in 1979, Telsat Canada awarded the contract for two Anik D satellites to Spar Aerospace, and the prime contractor was back in Canada again, and was able to follow up with two Anik E satellites. But the worldwide competition was too keen, and although Spar built two Brazilsat satellites they never made another foreign sale, and the Anik F satellites went back to Hughes.

CRC began its research life in satellite communications in the completely unexplored territory of the 12/14 GHz band with the Communications Technology Satellite (CTS). This frequency band was still open to users; and available solely to such satellites. This meant that higher satellite power could be used and thus smaller antennas on the ground. The new CTS satellite was planned to operate at power levels greater by a factor of 20 than existing commercial communication satellites, allowing satellite dishes as small as 0.6 meters diameter, something that could be envisaged for a single household. Without having to rely on ground-based distribution systems, this could be a "direct broadcast" satellite. The challenges in going to these higher frequencies and higher powers were formidable, and were achieved through a three-way partnership between CRC, NASA and the European Space Research Organization (ESRO, later to become ESA, the European Space Agency). To manage this challenging program Joe McNally (remember him from the Alouette-ISIS days) was brought to CRC in 1973 and made Deputy Manager. One of the many challenges he remembers is learning how to work with the NASA Lewis Research Center.

This program is described in detail by Ted Hartz and Irvine Paghis, and so only a brief summary is needed here. To achieve high power density levels on the ground, the satellite antennas had a relatively narrow field of view, covering an area roughly that of the province of Ontario. So the whole country couldn't be covered at once, but the two antennas (one for upgoing signals, the other for downgoing) could be pointed towards any desired locations for particular communications experiments. For the Alouette/ISIS satellites the power was derived from solar cells mounted on the surface of the satellite, so-called "body-mounted" cells. For CTS much larger areas were needed to achieve the desired power levels, requiring two solar sails, each 1.3 x 7 meters in size. These had to be unfurled in orbit, providing 1365 watts of power and allowing 200 watts of radiated signal. The spacecraft had accurate "attitude control", with a fixed orientation towards the Earth. This was needed for the accurate positioning of the communications antennas, but also allowed, through rotation, the solar panels to be kept perpendicular to the sun. The 676 kg satellite was launched January 17, 1976, on a Thor Delta 2914 satellite, and after using fuel to reach geostationary orbit had a final mass of 346 kg. After launch, the name of the satellite was changed to Hermes, after the son of Zeus, a messenger of the Olympian gods. It was inaugurated by the Honourable Jeanne Sauvé on May 21, 1976.

Hermes succeeded admirably in achieving its technological goals. After launch, it conducted communications experiments on tele-health, community interactions and communications, tele-education, administrative and community services and technology and science. These continued until 1979. The testing of Hermes required a full-scale environmental test facility, and this was opened in 1972 on the CRC site at Shirley Bay (also often called Shirley's Bay or Shirleys Bay); the facility is now called the David Florida Laboratory. CRC turned this facility over to the CSA after the latter was created in 1989.

Impact of NASA's Space Shuttle on Canada

NRC had both concern and interest about the continuity of space science in Canada, but did not have the personnel, and no fundamental mandate. Nevertheless, NRC continued to strongly support space science within its existing structure. Their rocket program continued to perform well but with diminishing resources. But other things were happening outside the country. Human spaceflight, which was territory available only to the USA and the USSR, suddenly became available to other countries with the initiation of the NASA Space Shuttle program in 1972. NASA was looking for international partners and the European Space Agency (ESA) agreed to provide a laboratory module called "Spacelab", to ride in the cargo bay. But NASA was also looking for a robot arm for moving payloads and launching satellites from the shuttle. In Canada this idea was picked up by Frank Thurston, head of the National Aeronautical Establishment of NRC. He won the support of the Interdepartmental Committee on Space, chaired by John Chapman, and the then Science Minister, Jeanne Sauvé, and government approval came in 1974. In 1976 NRC initiated a technology program, to provide a Remote Manipulator System, later known to Canadians as the Canadarm, to NASA's Space Shuttle system. Following its first flight on shuttle Columbia in November, 1981 (the second shuttle flight), as Canadians have all witnessed on numerous television newscasts, this became a remarkable success, leading to the later provision of the Mobile Servicing System to the International Space Station. This work was initially performed by SparAerospace, allowing its maturity to continue under the management of NRC, and later by the Canadian Space Agency. However, in 1999 MacDonald, Dettwiler and Associates Ltd. acquired the

Space and Advanced Robotics Division of Spar Aerospace and this division became known as MDA Robotics, and later MDA Space Missions, located in Brampton, Ontario. But there was a second impact, perhaps not entirely forseen when, during a ceremony celebrating the 20th anniversary of Alouette I, NASA invited Canada to join its astronauts in space. In actual fact, the invitation was to fly a "payload" specialist, someone able to perform specific experiments in space—only pilots and mission specialists are considered to be real astronauts. Canada responded aggressively by establishing its astronaut program within the NRC, a year later, in 1983, as a long-term program, not just a single flight. Thus Canada was changing its space research direction, piece by piece.

It was intended that other government departments, with needs to be met through satellite missions, would coordinate their needs through an Interdepartmental Committee for Space, (ICS), chaired by DOC, which had the capability to support project development. The concept was that if a government department really needed a satellite mission, it could make its case before other relevant departments, with possible collaboration and synergism. The departments involved in 1977 were NRC, DOC, the Department of Energy, Mines and Resources, the Department of Fisheries and the Environment, the Department of Transport, the Department of National Defence, the Department of Industry, Trade and Commerce, the Department of External Affairs and the Ministry of State for Science and Technology.

A Grass-Roots Space Science Initiative

In 1975, following a meeting of the Canadian Association of Physicists (CAP) at York University, a group of space scientists sat around on the grass (pun intended) in a circle on the lawn, feeling rather depressed, but determined to do something. Their division within the CAP was the Division of Aeronomy and Space Physics (DASP), and on behalf of this division, under the leadership of Alister Vallance Jones (NRC), with Peter Forsyth and Robert Lowe of the University of Western Ontario, Don McEwen of the University of Saskatchewan, Gordon Rostoker of the University of Alberta and Gordon Shepherd of York University, a report was developed entitled *The Future of Aeronomy and Space Physics in Canada*. Following an assessment of Canadian capability and needs in this area, the document outlined a detailed space program for Canada, and recommended "That a university-government consortium be established to plan, organize and finance the high latitude program, the small satellite program and other programs and facilities as appropriate." They submitted this report to the Associate Committee on Space Research (ACSR), which was already acutely aware of the problem and responded immediately. Under the Chairmanship of Ron Barrington of the CRC, the ACSR commissioned Peter Forsyth to prepare a report and make recommendations. Together with Ted Hartz (CRC), Don McEwen and Gordon Shepherd, Forsyth formulated a highly penetrating and insightful document called *Canadian Research Opportunities in Space*, in which he beautifully categorized what space research is, according to whether the research was done: 1) in space 2) on space or 3) from space. He then developed the concept of what is different about space science in terms of 1) disciplines 2) sectors and 3) facilities. One specific project will involve a particular mix of all of these. He then stated his overall position as:

1. The quality of Canadian space research has been high.
2. The effectiveness of Canadian space research has not been as great as it could have been because of fragmentation and lack of balance. Consequently Canada's place in

the international community of space science is rapidly deteriorating.
3. The lack of effectiveness is no one's fault.

The last item is a pointer to the ultimate conclusion. It is no one's fault because there was no one in charge. Forsyth then goes on to give the reasons why Canada should do space research:

> * *because some of the problems are uniquely associated with Canadian territory;*
> * *because these problems are ripe for solution;*
> * *because Canada has the capability to do front-rank research in solving these problems;*
> * *because Canada needs better quality weather forecasts and can expect to achieve them through space activity;*
> * *because Canada needs new, more efficient means of managing its natural resources;*
> * *because Canada needs remote-sensing to monitor its environmental characteristics efficiently and for a variety of purposes;*
> * *because Canadian industry will benefit from a program of space activity provided that activity is incorporated into a logical comprehensive program;*
> * *because Canada needs access to international space science and can gain that access by contributing knowledge obtained from a Canadian-based selective research program of high quality;*
> * *because Canada is committed to a continuing program for the utilization of space in the interests of the nation, and that program needs research to guide it.*

Following this, all the areas of space science are discussed in detail, with recommendations for each: aeronomy and space physics, the space shuttle, meteorology, remote-sensing, space astronomy, the life sciences, cosmic rays, space processing, data platforms, special orbits and gun-launched vehicles, finally leading to his major conclusion. Forsyth considered that the recommendation of a Canadian Space Agency was beyond the scope of his responsibility, but there is no doubt that his goal was to recommend something that could be implemented simply and easily. Thus he recommended:

> *"That an agency of government, such as the National Research Council, be designated the lead agency for space research and that it set up an Office of Space Research."*

Although the report was submitted in December, 1975, it was aimed at recommending where Canada should be in 1985. By that date few of his recommendations had been accomplished; in fact it took about 30 years, but most of them were ultimately achieved. Any present-day manager would benefit by reading this perceptive analysis again.

His report was submitted by Barrington to Dr. W.G. Schneider, President of the NRC. This last recommendation neatly sidestepped the question of a space agency, allowing immediate steps to be taken, should the NRC agree. This essentially invited NRC to expand its mandate to satellite missions and thus take the responsibility for all the space research in Canada.

The NRC accepted the challenge, and began immediately by opening a Space Science Coordination Office (SSCO) under Forsyth's interim leadership. Forsyth developed a strategy based on "gap-filling," the concept being that the goal should be a flat level of budget and resources for space activity in the entire Canadian community. Since individual projects are like mountains, in which activity (and expenditure) rise rapidly to a peak and then decay, Forsyth's approach amounted to filling the valleys between the successive peaks of applied satellite missions with scientific ones. His perception in terms of government programs progressing through the ICS was that such a gap was on the horizon and this could be filled by a scientific satellite. A group was formed to propose such a mission, coordinated by Gordon Shepherd. The pent-up energy of the academic community was released in this long-overdue opportunity, and they immediately focussed on the Canadian polar region. Two satellite missions were designed: 1) Energetics of the Atmosphere, Ionosphere, Magnetosphere System and 2) Energization Processes Above the Polar Cap, the first with an extensive suite of instruments – later called POLAIRE, and the second a smaller mission with a more specific focus of understanding processes of producing the aurora. The enthusiasm engendered by POLAIRE was extremely stimulating, but also became very ambitious in its technical requirements. The first of these arose in the difference between the way space plasma and atmospheric measurements were made, intended to be contemporaneous in space and time on this comprehensive mission. Space plasma scientists like to make measurements of auroral, or other energetic electrons coming from all directions; this is most easily done from a spinning spacecraft, like ISIS, which samples over 360° of direction. Atmospheric scientists normally prefer to look at a fixed point on the Earth, or in the atmosphere just above the Earth's edge (called its limb), so the atmosphere is viewed tangentially. If this is done from a spinning satellite, the fraction of the time spent in looking at one particular location is very small, making the measurement inefficient. For atmospheric observations what is needed is a non-spinning satellite, with a fixed orientation with respect to the Earth. After prolonged discussion, a solution was found—to have the satellite in two parts, one spinning, and one not spinning, connected through a bearing. The space engineers were attracted by the challenge, saying that it would be no problem. But it did mean higher costs. As the POLAIRE concept progressed, the supposed funding gap disappeared, as another applications mission appeared on the horizon. Eventually these collided, with the result that both proposals were taken off the table. On October 3, 1978, Peter Forsyth wrote a letter to the scientific community, saying that *"I am writing to tell you that on the 29th of September the National Research Council formally notified the Interdepartmental Committee on Space that it was withdrawing the proposal for the POLAIRE program and did not intend to proceed with it. This decision was similar to one taken by the Department of Communications with respect to the MUSAT proposal. These decisions were made in the light of a recent assessment of industrial impact which concluded that the spacecraft fabrication segment of Canadian industry will be adequately loaded for the next several years without these two spacecraft."* Thus neither mission took place, and scientific hopes were dashed again.

The National Research Council takes charge

The NRC recognized that within its large structure it had a number of groups with strong synergism, located within both "pure" physics and also its Radio and Electrical Engineering Division (REED). Building on the award of Gerhard Herzberg's award of the Nobel Prize in 1971, they created the Herzberg Institute of Astrophysics (HIA) on April 1, 1975. This combined the cosmic ray group, high energy physics and spectroscopy from Physics; upper atmosphere research and radio astronomy from REED; the Canada-France-Hawaii astronomical telescope on Mauna Kea, along with other astronomical activity in the form of the Dominion Astronomical Observatory (DAO) near Victoria and the Dominion Radio Astrophysical Observatory (DRAO) in Penticton, as well as the space work in space plasmas and optical measurements, including scientists supporting the Churchill Research Range.

In parallel the NRC persevered in its support of space science and by 1980 had obtained a significant increase in funding. A new division of the NRC was established called the Canada Centre for Space Science (CCSS) under Ian McDiarmid's leadership, and later under Roy Vankoughnett. McDiarmid had been successor to Rose as leader of the Cosmic Ray section of the Physics Division of the NRC. He had inherited the vision of Rose, but given the recent history of space science in Canada, he was very much a realist. He knew that from the government perspective the development of a healthy space industry was a priority, and if this could be done in the context of excellent space science, then it would be supported. In implementing a space science program the CCSS was responsible for planning and evaluating future science projects proposed by scientists in university and government laboratories. The Centre was also responsible for providing space engineering expertise for the scientists, funding major expense items, and acting as an interface between scientists and Canadian industry which also played a major role in the program. To carry out these activities, the CCSS included a scientific planning and evaluation group led by Gerry Atkinson, who had moved to the CCSS from CRC; a satellite instrumentation group which provided project management for the development and integration of satellite instrumentation led by Bob Gruno, who had also moved from CRC; and a facilities group, headed initially by John Aitken and later by Bob Hendry, which had previously been the Space Research Facilities Branch (SRFB) created in 1965 within REED. This group was responsible for the payload integration and launch of rockets and balloons at various locations in Canada, as well as the operation of facilities at Fort Churchill and Gimli, Manitoba, the latter for balloon launches.

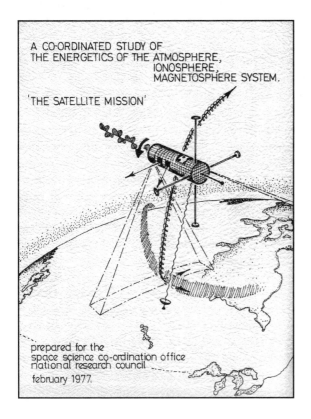

The cover of the POLAIRE study document of February, 1977, an attempt to resurrect scientific satellites in Canada.

McDiarmid came from nearby Queen's University in Kingston, where he obtained B.A. and M.A. degrees in Physics and Math. He worked with their 70 Mev particle accelerator (called a synchrotron) to study electron pair production. From Einstein's famous $E = mc^2$ equation it is known that particles of mass "m" can be converted to an amount of energy "E", where "c" is the velocity of light, and the reverse, that energy can be converted into particles. When an electron is created in this way the result is not one electron but two, one normal negatively charged electron and one positively charged electron called a positron; hence the term "electron pair production." McDiarmid also studied some photo nuclear reactions, where energetic photons change the structure of the nucleus. He then went to England, to P.M.S. Blackett's laboratory at the University of Manchester where he obtained his Ph.D. working on the scattering of mu-mesons in lead and iron plates in a cloud chamber. McDiarmid's career path was remarkably similar to that of Don Rose.

They both went to Queens, worked there for a year or so (Rose after his Ph.D., McDiarmid after his M.A.) and both went to the UK on 1851 scholarships for their Ph.D.s. Both had military experience, Rose during the war and McDiarmid after (during summers spent during his undergraduate years in the RCAF as a pilot). Both joined the NRC, both headed the Cosmic Ray section (at different times) and both were Associate Directors of Divisions (Rose of Physics, McDiarmid of the Herzberg Institute of Astrophysics), and they chaired some of the same committees, including the Associate Committee on Space Research. This long continuity of expertise served NRC and the space science community extremely well.

Working in Blackett's laboratory must have been a stimulating experience. After serving in World War I, Blackett entered the world renowned Cavendish Laboratory, then under Ernest Rutherford, the discoverer of the atomic nucleus, who had done much of his work at McGill University between 1898 and 1907. Rutherford gave Blackett the task of perfecting the cloud chamber, which he did, leading to a number of fundamental discoveries including the first photograph of a nuclear transformation. A cloud-chamber is a chamber containing supersaturated vapour, in which the passage of a charged particle causes condensation and thus visible tracks. Blackett later moved from this field to the study of the magnetic fields of the sun and the Earth, for which he had to invent a very sensitive "magnetometer." This magnetometer was used by his colleagues to establish the field of paleomagnetism, determining the direction of the Earth's magnetic field in ancient rock samples. This in turn contributed to the concept of moving continents, called continental drift. Blackett was both brilliant and controversial and was, for a period, blacklisted from entering the US. In a single issue of his local newspaper, the *Manchester Guardian*, in 1948, one page was devoted to his recent award of the Nobel Prize while the facing page contained a critical review of his book, *The Military and Political Consequences of Atomic Energy*. His ideas, considered then to be highly radical, are entirely conventional in the present day.

After Manchester, McDiarmid accepted a Post-doctoral Fellowship at the NRC and after a year he was appointed to the staff. For several years he used scintillation detectors (in which particles produce light), in a rather large multi-plate cloud chamber, to study various interactions of cosmic ray mu-mesons. While doing this, he slowly got into space physics. This illustrates the interdisciplinary nature of space science, in that while McDiarmid began his career as a nuclear physicist, exploring the interiors of atoms, there was a clear connection to space through cosmic rays that influenced his career much later. Moving from the highly energetic cosmic rays to the much lower energy auroral particles was his final step in becoming a space scientist. McDiarmid became successor in 1966 to Rose as leader of what was then the Cosmic Ray Section and was later named the Space Physics Section.

A Maturing Rocket Program

In spite of all these perturbations, including reduced budgets, the rocket program continued to be strong and strongly supported by the CCSS. After 1970, with the NRC covering the entire cost of the Churchill Research Range (CRR), funds were tight, and they had to learn to be very clever, and they did. At one point it was said that the SFRB could launch a rocket with the same number of people required to staff the firehall at CRR during the Pan Am days. There had always been rocket experiments that could not be conducted at Churchill. For example, rocket launches during solar eclipses have to be flown from under the eclipse tracks. These experiments were organized into campaigns; one eclipse campaign was conducted at East

Quoddy, Nova Scotia on March 7, 1970 and another from Red Lake, Ontario on February 26, 1979. Don McEwen from the University of Saskatchewan led a campaign to study pulsating aurora that required latitudes lower than Churchill – this was conducted from Rabbit Lake in Northern Saskatchewan (north of Lac La Ronge). The ISIS II Red Line Photometer and Soft Particle Spectrometer showed that the green nighttime aurora that was overhead at Churchill moved north to the Arctic coastline and turned red during the daytime where low-energy electrons from the magnetospheric cusp entered the ionosphere directly. The Space Research Facilities Branch (SFRB) was asked if it could locate a site where rockets could be launched into the daytime aurora and after some study identified Cape Parry as a suitable location. This peninsula is to the far west along the Canadian Arctic coastline and juts into the ocean extending towards Banks Island. What made it feasible for a launch site is that it was a DEW (Distant Early Warning) line station with a radar and operators constantly on the watch for Soviet bombers coming over the pole towards Canada. Cape Parry was the geographical location, but this DEW line station was known as PIN Main. What it had, besides the radar, was a runway (gravel), a large hangar, living quarters and an ample kitchen. In fact it had duplicate living quarters, separated by some space. If one building should catch fire, it was essential that the crew be able to move to alternate quarters to wait for a rescue, which could thus be made available to guests intending to launch rockets into the aurora.

It's of interest to look back at "Operations Requirement No. 7450, prepared by the SRFB of NRC, dated October, 1974, for which the abstract is as follows:

"This O.R. is for the launching of two Black Brant vehicles from a temporary site at Cape Parry, N.W.T., during the period 1 to 15 December 1974. The vehicles are a Black Brant IVB Mod. 1 designated AAF-IVB-32 and a Black Brant VB designated AMF-VB-41.

"AAF-IVB-32 is to be launched at 1430 hrs., (magnetic noon) local time, 1430 – 1600 hrs. approximately, to take auroral measurements with a solar depression angle of five to ten degrees. AMF-VB-41 is also to be launched at magnetic noon and is to take auroral measurements under quiet magnetic conditions.

"The Project Scientist for AAF-IVB-32 is Dr. B.A. Whalen of the National Research Council Division of Physics. Experiments are carried for Dr. Whalen, Dr. A.G. McNamara, and Dr. G.G. Shepherd.

"The Project Scientist for AMF-VB-41 is Dr. G.G. Shepherd of York University. Experiments are carried for Dr. G.G. Shepherd, Dr. W. Evans, Dr. Hirao/Dr. Oyama, Dr. D.J. McEwen, Dr. A.G. McNamara, Dr. J.C. Gerard and Dr. B.A. Whalen.

"The payloads have been instrumented and manufactured by Bristol Aerospace Ltd under contract to the Space Research Facilities Branch of the National Research Council of Canada."

The introduction gives a little of the scientific rationale for the flights.

"The prime object of these tests is to probe the dayside ionosphere precisely at the foot of the field lines presumed to extend to the magnetospheric dayside cusp. Recent observations from several spacecraft have stimulated interest in the dayside aurora. Direct measurements of particle precipitation in the dayside aurora suggests that the magnetosheath plas-

ma, which is effectively excluded elsewhere, may gain access to the magnetosphere through the dayside neutral points, or lines, producing among other things dayside auroral precipitation. Some authors have speculated that this is the dominant mode of particle entry into the magnetosphere. The rocket borne measurements will be complimented by ground based observations."

Going on into the "Operations Command" section of the document, it notes that a High Altitude Diagnostic (HAD) launcher was installed at Cape Parry by SRFB personnel during the summer of 1974, at 70.1666° North, 124.6° West. A map shows the projected impact areas for the two vehicles, with exclusion regions marked for "oil rig operations, airfields, Eskimo hunting parties and trapper and hunting areas." The open areas included the McClure Strait and Viscount Melville Sound. Jack Tarzwell of SRFB was Facilities Coordinator, responsible for coordinating accommodation and "messing arrangements" for the launch team with the Station Supervisor. Normal telephone communications were available, but NRC arranged for use of the DND "Hot Line" to Air Traffic Control in Edmonton during countdowns. Hal Roberts of SRFB was the Mission Controller (MC).

The charter flight out of Winnipeg was scheduled for November 21, 1974. The personnel list includes the two persons from SRFB named above, two from York University, seven from Bristol, two from the NRC Division of Physics, three from NRC Radio and Electrical Engineering, one each from the University of Saskatchewan and the Atmospheric Environment Service and six from the Churchill Research Range. It is remarkable that this crew of 24 could conduct such an operation from a remote site, but of course a site with excellent basic facilities offered by the DEW line had been chosen.

The Cape Parry launcher inside its tent, which was pulled back for launching. The tall person at the head of the group on the right is John ? from Bristol Aerospace and behind him is Hal Roberts of SRFB. At the end (farthest right) in the dark toque is Jack Tarswell of SRFB. Photo by Gordon Shepherd.

The Black Brant IVB was a two-stage vehicle. The first stage motor MT-147 with nozzle FSN-48 and the second stage motor SSLW-157 with nozzle IVBN-12 were moved to Cape Parry during the summer of 1974. It was planned to have a maximum flight altitude (apogee) of 650 km to take it out of the major ionospheric influence in order to detect the incoming magnetopheric particles before their interaction with the ionosphere. The nosecone was split into two clamshells to expose the particle and plasma detectors, while an optical photometer to detect the red daytime aurora was contained in the parallel section below, with a door opened in flight to allow viewing.

The larger Black Brant VB was planned for an apogee of 300 km, enough to penetrate the red dayside aurora. York University provided a four-channel auroral photometer to measure four prominent auroral wavelengths. Wayne Evans of the Atmospheric Environment Service of Environment Canada provided two spectrophotometers to measure water vapour through the absorption of solar radiation, viewed as the rocket emerged from the Earth's shadow, while Don McEwen of the University of Saskatchewan provided two experiments, one to measure the incoming low-energy cusp electrons and the other to observe Lyman-alpha radiation from incoming protons. Brian Whalen of NRC measured both electrons and ions over a wide energy range, 1 electron volt to 100 kilo electron volts. But it was an international payload, with Kunio Hirao and Koh-Ichiro Oyama of the University of Tokyo providing an instrument to measure the temperature of the ambient electrons and Jean-Claude Gerard of Belgium providing an instrument to measure nitrogen emission in the aurora.

The airstrip, hanger and the rocket launchers were located at the bottom of the steep hill on which the radar and living quarters were located. The Black Brant IVB countdown began 8 hours before launch, with a review of weather data, establishment of a "NOTAM" to alert aircraft, firing line checks and a decision to proceed. At 7 hours the payload was moved to the vehicle assembly building and road blocks were set up in the launcher area. The payload assembly was mated to the second stage motor, the first stage installed on the launcher and the second stage then moved to the launch pad for mating, after which the protective tent was moved over the launcher. The telemetry antennas were installed and the umbilical cable connected. At T-200 minutes (T stands for "test"; for some reason all rocket launches are called "tests") the horizontal checks were performed (payload testing with the rocket in the horizontal position). At T-160 the launch area was cleared for the arming of the vehicle by the designated personnel, with RF (radio) silence in effect. At T-130 minutes the tent covering the launcher was pulled back and the rocket elevated to the preliminary launch direction. At T-110 minutes the final arming was completed and the crew returned to base. At T-90 minutes the vertical checks were begun and the backup generator was started. At T-60 minutes, with the vertical checks complete, the payload was declared "GO." The final launcher settings were then set, based on the latest wind data from radiosondes. At T-30 minutes the Range Safety Officer (RSO) was required to confirm that the countdown was clear to proceed. At T-20 minutes the ground station Doppler (for tracking the rocket) was synchronized, payload temperature checked and the Test Conductor (TC) declared the telemetry station a "GO." At T-15 minutes the RSO again confirmed a "GO" and the Vehicle Manager (VM) declared that the second-stage ignition batteries were fully charged. At T-5 minutes all motor vehicles near the launch site were turned off, the Doppler azimuth started recording, and payload temperatures and batteries were again checked, as well as voltage calibration. T-3 minutes was the long-term hold position, allowing a wait for "geophysical conditions."

The tent has been pulled back and the rocket elevated, awaiting the geophysical conditions for launch. Photo by Gordon Shepherd.

This is where the Project Scientist (PS) came in. To achieve the scientific objectives the rockets had to be launched into the red daytime aurora. Because Cape Parry was far to the west of the magnetic pole, magnetic noon occurred a few hours after local noon, by which time there were twilight conditions, just dark enough that the red aurora could be seen—not by eye which is not very sensitive in the red, but by photometric instruments. Fokke Creutzberg of NRC had brought a magnificent ground station, which scanned the sky to locate this feature. The instruments were outdoors, but the recorders inside, in the furnace room of the hangar, so it was comfortably warm. Here the scientists gathered to watch for the red aurora to appear as the sky brightness faded. This was all new to them, as very little was known about the red daytime aurora except for ISIS II, so they couldn't be sure what they would see at Cape Parry. During the preparation of the launch site they had a number of days to practice, and soon were able to identify the feature they were looking for, on Creutzberg's oscilloscope screen.

But during the afternoon of the first countdown for the BBVB the red aurora simply didn't appear, and the Project Scientist held at T-3 until the end of the window. The crew had to pack everything up without launching. This wasn't unusual at the Churchill Research Range, because the green nightside aurora was much more unpredictable, but the dayside aurora was thought to be more stable. Hal Roberts, the MC, was visibly disturbed. "I thought you said this would happen every night, he upbraided the PS." "Sorry, we thought so too, but apparently there are exceptions; don't worry, we will launch next time." The concern was certainly valid, given the enormous investment of budget and effort, and Christmas approaching. The consequences of failing to launch these two rockets were unthinkable.

The next night they were holding at T-3 when the red aurora appeared on the northern horizon, close enough for a launch and the PS told the MC to pick up the count. Then things got very busy. They checked the commutators, Doppler, tape playbacks and then asked the PS if he wanted a hold at T-10 seconds, an option in case a further wait was needed. This time there was none and they counted straight through. At T-2 minutes they switched the payload to internal power, did calibrations, voltage checks, confirmed the tape recording. At T-20 seconds they asked the PS again about the T-10 second hold, but this was waived so at T-10 seconds they turned on the tape recorders, confirmed payload status, switched payload to flight condition, confirmed the tape recording was OK, set the strip chart recorders to 5 mm/sec, and pointed the antenna trackers to the launch site. At this point the RSO could still call a hold if it was not clear to fire, and at T-0 the TC pressed the fire button for three seconds. For the BB IVB, the booster burned out at 18 seconds, with vehicle separation at 20 seconds, second stage ignition at 22 seconds and second stage burnout at 36 seconds. The nose fairing deployed at 42 seconds and the payload was de-spun at 45 seconds. The plasma probes were released at 50 seconds, the door released at 55 seconds; experiment high voltages were then turned on with apogee reached at 113 seconds. The antenna trackers relayed the final position on impact, the recorders were turned off, RF silence was lifted, the crew headed to the launcher to inspect for damage, power supplies were turned off and the launcher was lowered to the horizontal position. The test termination was announced and the last item on the list was to open the road block. It was over.

There was little scope for entertainment at Cape Parry, although it was a beautiful location, with the radar located on a promontory; around noon it was light enough to view the vast expanse of the Arctic Ocean. The runway, hangar and launch pad were down the hill, from where one could walk out to an abandoned aircraft that had clearly missed the runway, but longer walks were discouraged because of polar bears. The meals were a form of entertainment because they were huge, with many choices of main course, and especially desserts. The one thing that management does offer a crew in such an isolated location was good and abundant food. In the evening there was a bar, where the DEW line crew could clink glasses with the guests. There was also a beaten-up red piano where the Project Scientist would provide accompaniment for sing songs. These usually began with bar-type ditties, but with the approach of Christmas would eventually move to songs and carols of the season. Those workers, hardened to long periods away from home, lost their dry eyes at such times, as everyone thought of family far to the south.

With the flights over, and everyone home again, the data were processed. Yes, the BB VB had flown right through the red aurora and the cusp precipitation, and the BB IVB had flown over it, observing the particles coming in, and the red aurora below. The Belgian experiment observed atomic nitrogen emission, from a so-called metastable state called $N(^2D)$ with a very long lifetime, and what was seen was a plume of emission, coming out of the excitation region where the red aurora was observed and flowing northward. This was later modelled by graduate student Richard Link as a strong wind, flowing from the sunward direction and carrying auroral plasma into the polar cap, a completely unexpected result.

Active Experiments at the NRC

In an attempt to apply new techniques to investigate the problem of auroral particle acceleration, Brian Whalen's group at the NRC undertook a series of active experiments involving ionospheric perturbations and remote probing. Passive experiments are ones in which the rocket instruments simply respond to the characteristics of the surrounding environment. Active experiments are ones in which the rocket releases energy or material to stimulate the environment; other instruments measure the response to that stimulus. What follows is the continuation of Whalen's story, begun in the last chapter.

"Waterhole" Auroral Perturbation Experiments

"During a Los Alamos Laboratory Barium ion release experiment from Cape Parry (Morris Pongratz, Los Alamos Laboratories, PS), Whalen's thermal ion sensors detected a sudden drop in E and F region ionospheric ion densities, simultaneous with the release, followed by a slow recovery to pre-release levels. It was concluded the release had created a large ionospheric hole. The presence of this artificial hole, which was attributed to the presence of molecules, mostly water from the explosive, led Whalen's team to propose a series of perturbation experiments first undertaken in 1980.

"The objective of these experiments was to test current theories of the formation of auroral arcs by modifying the ionospheric ion density and thereby interrupting or perturbing the ionospheric currents associated with the arc formation. With the collaboration of Los Alamos Laboratories (Morris Pongratz) and the U of Western Ontario (Peter Forsyth), a series of explosive releases in and near auroral arcs were undertaken. These experiments were referred to as "Waterhole."

"Observations from instruments on the rockets showed that large electron density depletions were produced in the E and F regions of the ionosphere and ground-based photometric observations (Fokke Creutzberg) indicated that the arcs may have been perturbed by the release, suggesting that the holes had indeed effected the mechanism(s) responsible for arc formation."

ECHO Auroral Probes

"One of the most popular theories of auroral electron acceleration involves the acceleration of magnetospheric electrons through electric fields parallel to magnetic field lines. To test this theory William (Bill) Bernstein (Rice University) proposed to launch electrons of various energies up field lines and to look for 'Echoes' of these particles reflected by the parallel electric field. He invited Klause Wilhelm of the Max Planck Institute for Extraterrestrial Physics (MPE), Germany, and Whalen's space plasmas group as co-investigators. MPE provided free flying sub-payloads to look for the electrons reflected by the electric field and the NRC space plasmas group provided the electron gun and particle instrumentation for the main payload.

"The electron gun was developed at NRC and tested in the large vacuum chamber in NASA Houston facility. During these tests it became evident that the experiment might turn out to be 'Active' as well as simply probing. In the chamber the electron beam demonstrated an instability referred to as a 'Beam Plasma Discharge' which under certain conditions filled the large chamber with a glowing hot plasma. It is not certain at this time whether this same instability occurred during flight.

"The ECHO rocket was launched successfully into an auroral display and time delayed 'echoes' were observed by the MPE sub-satellites. However, analysis of the particle 'echoes' indicated that simple reflection by parallel electric fields could not account for the broad energy range of the reflected particles."

Firewheel

"Gerhard Haerendel of MPE invited the space physics group (with Brian Whalen as Project Scientist) to provide one of the four sub-satellites for a combined active and passive observation programme called Firewheel which involved deep space chemical releases. The invitation was accepted and SED Systems was contracted to develop the NRC sub-satellite. To support the observations required for the release, a variation on the Bennett ion mass spectrometer was developed at NRC and flown on Firewheel. The NRC sub payload also included a battery of standard charged particle detectors. Firewheel was launched from the ESA range at Kourou in French Guayana in 1980. A failure of the launch vehicle soon after lift-off led to a sudden unfortunate end to this mission. However, not all of the instrument design and development effort was wasted since the mass spectrometer concept initiated in this program served as the model for the EXOS-D SMS instrument which followed soon after."

Termination of the Churchill Research Range

The rocket community was stunned to learn in 1984 that when the Canadian rocket program was cancelled. Following the change of government at that time, Federal departments were required to offer up programmes to be terminated and NRC offered the Churchill Research Range. The CRR was an Arctic research centre and an important element in the economy of Manitoba, but to the surprise of many, the government accepted its offer, and the rocket program ended. Canadians can, and still do, conduct rocket experiments, but not from CRR; they launch from other locations such as White Sands, New Mexico; Poker Flat, Alaska; or Andoya in Norway. This served to reinforce the vision that the Space Shuttle was the vehicle of the future for Canadian space science. The building of the Canadarm also provided some potential synergism between the Canadian technical and scientific elements of the space shuttle, although these were never realized.

The Apollo-Soyuz Docking – Another Kind of Active Experiment

Robert Young came from Stanford Research International (SRI) as a faculty member to York University in 1968, a university just three years old, situated on a what had been a farmsite in the north-west corner of Toronto. He brought with him an idea, embodied in the following United States Patent, Number 4024431.

"A controllable evaporation source of parent metal species, produced either from a volatile metal or a metal halide, and under some circumstances a chemical getter sink in a sealed RF excited discharge. This discharge occurs in a second, extremely pure gas which is present in great excess over the gas produced by evaporation which may be followed by chemical decomposition. Excitation of species whose emission is desired occurs by electron impact or energy transfer from the major species which are in turn, excited by the electron impact."

What this meant was that Robert Young brought to Canada a concept for making a lamp that would emit spectral radiation from specific atomic species. He also brought the designs of an RF lamp exciter and VUV (Vacuum Ultra Violet) xenon line source from SRI, which soon became a working light source in Canada, in about 1970. But Young also had the idea of measuring atomic oxygen in the atmosphere using an oxygen lamp flown in a rocket, although an atomic oxygen lamp would be much more difficult to build. Young's approach was to mount the lamp in a rocket, illuminating the atmosphere just outside the rocket body. The atomic oxygen radiation would first be absorbed by the atomic oxygen and then emitted in all directions—a process simply called fluorescence. The light coming back towards the rocket could be measured with a detector, and since this light was proportional to the amount of atomic oxygen present, this constituted a measurement of the concentration of atomic oxygen in the high atmosphere – atomic oxygen is a significant species above 100 km altitude.

Oxygen is extremely reactive, so making a miniature lamp containing a source of atomic oxygen and keeping it from destroying itself on the walls of the tube was a challenge. Fortunately there was a mix of researchers at York University at that time who all contributed ideas in one way or another. It was Bill Morrow, who had been hired by Young, who finally realized that one could create a dynamic equilibrium between an atomic oxygen source and a sink (absorber of atomic oxygen) that would maintain a steady concentration. Working with Molly Morrow, a suitable source was finally identified as potassium permanganate, and the sink as powdered uranium. Heinz Sammer built the miniature version of the SRI exciter and by 1973 a version suitable for flight in a rocket had been achieved. A graduate student, Mike Walton appeared about this time, and was given the research task of making the overall design for the rocket experiment, with assistance from Bill Morrow, flying it on a Canadian rocket and analyzing the results, which was successfully done. This is the only method of measuring atomic oxygen concentration in the atmosphere that was truly successful, and the technique was copied by others who made successful measurements as well. One of these was Paul Dickinson of the Appleton Laboratory at Slough, England, who got the basic information from Young and then proceeded to build his own lamps. This led to a remarkable campaign called ETON (Energy Transfer in Oxygen Nightglow) with the launch of seven Petrel rockets from South Uist in Scotland. The campaign was supported by Ted Llewellyn and Ian McDade, in which the latter led the data analysis and determined definitively the reaction rates of the processes leading to the oxygen airglow. This campaign just celebrated its 25th Anniversary at South Uist.

The team of Fred Kaufman, Tom Donahue and James Anderson at the University of Pittsburgh had a different version of this idea, to measure atomic oxygen and also atomic nitrogen in the thermosphere, around 200 km where the concentration is very low. To do this it was proposed that absorption be used, rather than fluorescence. To do the absorption experiment one shines the lamp towards a detector with the atomic oxygen in between and the loss of light owing to absorption by that atomic oxygen is measured (this was in fact the method used by Walton, who extended a mirror at the end of an arm out from the rocket to reflect the absorbed light back to the detector). The advantage of this method is that its sensitivity could be increased by lengthening the path over which the light is absorbed. The basis of Kaufman's idea was a unique space mission that had just been approved. The US and the USSR had agreed that there would be an international link-up in space, in which a Soyuz vehicle would dock with an Apollo module. By placing the lamp on one spacecraft and a detector on the other, the absorption could be measured over a range of distances as the two approached one another. However it turned out to be simpler to have the lamp and the detector on Apollo and a retro-reflecting mirror on Soyuz to reflect the absorbed light back to the detector. But the opportunity arose so quickly and the time line was so tight that the Kaufmann team did not have time to learn the technique themselves, so they looked towards Canada where the fabrication of lamps had already been successful.

In fact the experiment is not so simple because the spacecraft are moving at about 7 km per second, which causes a large Doppler shift of the lamp wavelength, so much so that the lamp line no longer overlaps the atmospheric line and is therefore not absorbed. Thus the Apollo module had to move out of its joint orbit plane with Soyuz so the light beam would be perpendicular to the spacecraft velocity and thus not Doppler-shifted. This involved some very fancy flying by the Apollo crew. Apparently they failed on the first attempt at 150 meters range because they mistakenly locked onto a light on the Russian spacecraft rather than the retroreflector. However, on the second pass at 500 meters they were successful, and fluorescence data were acquired as well.

When Young was approached by NASA in 1973 to provide the lamps for the Apollo Soyuz mission, he formed a company, Intra Space International (ISI) with Bill and Molly Morrow as its first employees, and three others added later, then a contract was put in place. After the work was under way NASA discovered that the American Morrow was in Canada because of the Viet Nam war. Rather than jeopardize the project, they decided to keep this a secret within the Johnson Space Center (JSC) where the work was being coordinated. Lockheed Electronic Corporation built the space instrument in which the Canadian lamps were mounted while ISI provided the lamp bulbs and integrated them in a Canadian designed exciter system fabricated by Lockheed. The lamp systems were integrated and VUV tested in thermal vacuum in Canada at ISI and shipped to JSC for integration into the flight instrument. The detector was a lunar VUV spectrometer flight spare that is thought to have come from Johns Hopkins University. Part way through the project Young set up a company in Boulder, Colorado and tried to move his team there, but they wouldn't budge. Nevertheless the technology then became NASA property and was to some extent lost to Canada. In fact NASA ordered Young to keep out of his Canadian company. With Dick Leveson as leader the Canadian group continued until Leveson left to form his own company, Photovac; Photovac eventually became a $60 million company, manufacturing lamps to detect hazardous gases. Other companies entered the field, which is now a one billion dollar market, with lamps not so different from those first made at York University. Young closed the Canadian company in 1975 after the end of the contract, but just before it closed Bill Morrow received notice that the State of Virginia had granted him amnesty, so NASA brought him with Molly to witness the thermal vacuum test of the experiment at the Applied Physics Laboratory in Maryland.

Morrow moved to Barringer Research in Toronto, an innovative company about which more stories could be told, where he learned about correlation spectroscopy. Later he moved to an environmental monitoring company called Moniteq where he revived the oxygen lamp technology, but in 1980 formed his own company, Resonance Ltd. that provided lamps for two Canadian rocket payloads, OASIS (1981) and ARIES (1984). Later Morrow got a contract from N. Iwagami and T. Ogawa from Tokyo University to build a rocket experiment (WINDO) for the measurement of atmospheric atomic oxygen. The flights began in 1992 and continued until 2004, with six flights altogether, four of them very successful. The work was done in a York University laboratory as an incubator company then became Resonance.

Resonance Limited is now a successful company located in Barrie, Ontario. Lamps made by Resonance Ltd. have flown on a number of NASA spacecraft including the WIND and POLAR missions, the WF/PC II Hubble project (part of the Hubble Space Telescope rescue) and the Gravity Probe B mission described later. However the company product line has expanded considerably beyond these unique lamps. Currently Resonance is a leading supplier of thousands of photo ionization detector lamps used in hazardous vapour detection by firemen and environmental workers. In their small niche they are the most successful company in the world, thanks to the inspiration of Bob Young and the help of so many in the York University community.

Space Shuttle Science Initiatives

Brian Whalen was appointed to the space shuttle study group, whose mandate was to promote the shuttle as a platform for space science activities. Informal meetings between the Chair of the working group on Active Experiments William Bernstein and Whalen led to a productive long term scientific collaboration on transverse ion acceleration (TIA) studies and the electron ECHO project.

When an Announcement of Opportunity for future missions was released in 1978 the NRC Space Plasmas group responded by submitting a proposal with the Lockheed Palo Alto group headed by Ed Shelly for an Energetic Ion Mass Spectrometer (EIMS). After selection for shuttle flight, design and development of this instrument was undertaken. Approximately one year into the project NASA indicated that they no longer could support the Lockheed effort. Simultaneously negotiations were under way for participation of the NRC group in the Japanese EXOS-D Ion Mass Spectrometer (SMS) program. Whalen suggested to the CCSS that since NASA was not able to support their science team, that the program should be cancelled and that the efforts of the group should be put into the SMS project, described later. This proposal was accepted and the EIMS project was terminated.

There were still scientists at CRC interested in the fundamental physics of space plasmas, and Gordon James led a second shuttle experiment called WISP (Waves in Space Plasmas), which was based on the Alouette/ISIS topside sounder concept, but much more sophisticated in concept and execution. Finally, Gordon Shepherd of York University led a team for a third experiment called WAMDII (Wide Angle Michelson Doppler Imaging Interferometer) for the measurement of atmospheric winds from space. These three experiments were to be the next wave of Canadian space science activity, and with the full support of the CCSS, a rapid program of development was initiated. WISP was designed and built by CAL Corporation (originating in the earlier name of Canadian Astronautics Limited, a home-grown Canadian company) while EIMS was designed and built mainly within HIA, in collaboration with Lockheed, but as already noted, this experiment was short-lived. The contract for WAMDII development was won by SED Systems in Saskatoon, the company that had grown out of the prairie University institute, under Alex Kavadas' guidance, as we have seen.

But again, the gap between optimism and realism began to grow. Between NASA's first operational shuttle mission in November, 1982, and the launch of Canada's first astronaut, Marc Garneau, on October 5, 1984, the US program had grown farther and farther from the vision of von Braun. Shuttle missions would not take place twice per month, but more likely every two months. NASA was clearly not going to be able to honour all the commitments it had made to fly experiments, and accurate schedules became difficult to implement. WISP was ready to fly, but on the shelf, where it remains today, never finding a place in the mission sequence. So of the three Canadian shuttle experiments it was two down, and one to go.

The WAMDII and WINDII Missions

WAMDII fared better than the other Canadian space shuttle experiments, receiving a booking on an actual mission, but in the end was the victim of its own success. In 1978, following the "two eggs" approach again, Shepherd had submitted two proposals to NASA, the WAMDII proposal, and another with the same name, to a competition for the Upper Atmosphere Research Satellite (UARS). This mission, conceived by a blue-ribbon team of scientists, was to study the energetics, composition and dynamics of the middle atmosphere, from the stratosphere to the thermosphere, in a far more comprehensive way than had been conceived or implemented before. There were to be no compromises, each instrument would have the resources needed, and there would be redundant measurements, different instruments measuring the same thing to ensure that the results were valid. Like POLAIRE, but on a much grander scale, this added up to a 7,000 kg satellite, and a price tag approaching one billion $US. It took some time for this to be accepted by the US House and Senate, but the project finally came to life in 1983. However, WAMDII was not to be part of it; an instrument of similar concept by the French, submitted by Gerard Thuillier, called WINTERS, had been accepted instead. While UARS was awaiting its approval, WINTERS began to run into problems. The original plan was to have the instrument built "in house" at the Service d'Aeronomie (of the Centre National de la Recherche Scientifique, CNRS), near Paris, but as time passed, other experiments gained priority over it. Thuillier decided that collaboration with Canada was the only solution, and came to Ottawa to propose it. He met with Shepherd, Ian McDiarmid and Gerry Atkinson and at the end of the discussion McDiarmid said simply, "Let's do it." No proposal was ever written. A Centre National d'Etudes Spatiales (CNES) delegation came over from France, and there was a return visit to France, but in the end the only realistic scenario was for Canada to take the lead role, with Shepherd as Principal Investigator, and France contributing a smaller but important role for the experiment. It was also agreed that the design would be based on WAMDII, which had already been proven in the laboratory and with field measurements. Once the CNES and the CCSS had approved this approach, NASA had to be convinced. Happily, they agreed that since this was replacing a proposal already accepted, namely WINTERS, that the change would not require a re-competition which would have involved an impossible delay. But they were uncertain about the Canadian capability; nobody in the UARS team seemed to know anything about Canadian capability in space. But on one of Shepherd's visits to NASA he met Douglas Broome, who had worked on the Alouette/ISIS program. Broome had no doubts about the matter: "Of course Canada can do it", he said. And that settled it. WINters combined with wamDII became WINDII, which for simplicity was said to stand for the WIND Imaging Interferometer.

WINDII's problem was that it was now 1984, and the launch was scheduled for 1989, making for a very tight schedule. Bob Gruno, a skilful strategist with Hermes experience, and now with the CCSS, was called upon to develop an approach. The result was a WASP (WINDII Advanced Studies Program) team, drawn from the major space industries in Canada that worked intensively during the summer of 1984 to reach a conceptual design. The CCSS then solicited bids from competing consortia to design and build the instrument to this design. Two groups responded. CAL Corporation teamed up with AIT (Advanced

Information Technologies), headed by Allan Churgin, who had worked in the US space program before coming to Canada. AIT would be prime contractor, and also be responsible for the "ground-segment", the computers and software on the ground, including instrument testing, while CAL would be responsible for the "flight-segment", namely the WINDII instrument itself. SED Systems had teamed up with COM DEV. The AIT/CAL consortium won the selection and the work began with Tom Darlington, another Hermes graduate now with the CCSS, as the Project Manager. Once into the detailed optical design, shortcomings were found by Donnalee Desauliers in the WASP team design. These had to be corrected, and that took time. This wasn't allowed for in the "success oriented" plan, the only one that would meet schedule. The team began to fall behind. NASA managers came regularly to Ottawa to check on progress. At once such meeting, the CCSS and AIT representatives took the UARS project manager, Peter Burr, and his team to a congenial dinner, where the interactions were very friendly. Next day at the meeting, Burr stunned the Canadians by telling them that they were behind, would never catch up, and that they might as well save their money and quit. This kind of psychological warfare is part of space activity. Burr had ten experiments to worry about, and if one was allowed to fall behind, then others would too—the whole schedule would get delayed, and the program costs would get out of control. So the strategy is that no group should ever admit that it is behind any other, and be prepared to demonstrate it. Tom Darlington did not withdraw, but a catastrophic change of schedule occurred soon after.

Everything changed on January 28, 1986, when the Challenger vehicle exploded just after lift-off. Now there would almost certainly be an enforced delay in the launch date, but the work on UARS didn't let up. In the end, they launched in 1991, two years late, and although the Challenger accident was routinely blamed for that, the work on the UARS spacecraft and its ten instrument teams never slackened; all that time was really needed. WINDII would not have achieved a 1989 launch, but neither would the other investigator teams, or the spacecraft itself. For example the CLAES instrument had to be redesigned because it was designed with liquid hydrogen for cooling which was no longer considered safe. Meanwhile, the work on WAMDII was continuing, but the value of continuing with both experiments was being questioned by both NASA and by the CCSS. The WAMDII flight would be for ten days, out of which perhaps 20 hours of data would be obtained. WINDII would fly for 18 months for sure, and UARS had an optional extension to 30 months – and this was for 24 hours of WINDII data every day. The choice was clear, so the WAMDII project was cancelled. But it had served a valuable purpose: without WAMDII there never would have been a WINDII.

COSPAR in Canada—1982

In the middle of this transition to a new way of doing space science in Canada, COSPAR came to Ottawa, from May 16 to June 2, 1982, for what was then called its 24th Plenary Meeting. This was a combined meeting with SCOSTEP (Scientific Committee for Solar Terrestrial Physics), each meeting occupying one week. As well, one session was devoted to celebrating "A Quarter Century in Space." Doris Jelly (1988) has described those 25 years for Canada. The keynote speakers were N. Hinners of the National Air and Space Museum in Washington, D.C., R.Z. Sagdeev of the Space Research Institute in Moscow and P. Morel of the Centre National d'Etudes Spatiales of France. In the larger panel discussion that followed, Canada was represented by Alex Curran, listed as the "Minister for Space", Department of Communications. It's purely coincidental that the "half-century" Anniversary of COSPAR will be celebrated in Montreal, in 2008.

Of the nearly 1,800 attendees in 1982, 38 were Canadian, about 2%, a rather small contingent for a host country at a meeting dominated by the USA, the USSR, the major European countries, Japan, and to a lesser extent, India and China. As already explained, 1982 was a low point for Canadian space science, but the hosting of the event was likely seen by the CCSS as a stimulus for what was expected to emerge on the Canadian scene. There were only two ISIS papers, indicating that this project had truly come to an end. A relatively new field in Canada, Remote Sensing as described in Chapter 5, was quite active, with 10 papers; the Earth's space environment including the aurora accounted for 11, solar planetary topics numbered 8 and there were 7 atmospheric papers. There was also a session on "Instruments and Techniques" organized by Ian McDiarmid, looking forward to the future in which papers on the NRC mass spectrometers, new auroral radars, the Ultra Violet Imager and WAMDII were given. One paper counted as Canadian, but really not, was an invited paper by Donald Hunten on "The Atmosphere of Titan" (Titan is the largest of Saturn's moons); Hunten was no longer in Canada but at the Planetary Sciences Institute in Tucson, Arizona. His contributions to Canada were nevertheless recognized in 2003 by election as a Foreign Fellow of the Royal Society of Canada. The final paper was given by Ron Barrington, on "Communication Satellites: The Experience Gained in Under-Populated Regions of Canada", given in the Symposium on "The Role and Impact of Space Research in Developing Countries." Barrington was also chair of the Local Organizing Committee.

Nuclear winter and the Defence Research Establishment Valcartier (DREV)

In the 1970s, it was recognized that nuclear explosions could inject large amounts of nitrogen oxides into the stratosphere; here they would act as a catalyst to reduce ozone levels with the effect on skin cancer of which Canadians are now well aware. Some of the material in this section is taken from an article by Brian Martin.

In the 1980s another possible consequence, severe climatic effects, became the subject of intensive scientific investigation and extensive political discussion. A major nuclear war would lead to vast amounts of soot and dust being injected into the atmosphere, mainly from the burning of cities. This material would absorb incoming solar radiation but continue to allow infrared heat from the Earth's surface to escape to outer space, causing a significant drop in surface temperatures. This temperature drop could cause the destruction of ecosystems on a vast scale. The popular term for this was 'nuclear winter', and it was predicted to last for several years.

Paul Crutzen was asked in 1982 to provide his analysis of nuclear winter for a special issue of Ambio, the journal of the Swedish Academy of Sciences. His expertise goes back to his Ph.D. thesis on the role of nitrogen oxides in regulating the amount of ozone in the stratosphere, an issue at the time, because commercial supersonic jet transport (SSTs) were being promoted. Crutzen drew attention to the dangers inherent in this because of the nitrogen oxides in the exhaust emissions from these aircraft. In fact, Crutzen and Birks' model-runs for this special issue showed no significant effect of a nuclear war on the ozone layer. However, they did suggest that the soot injected into the atmosphere could cause a climatic effect.

Those supporting SST's used the argument that they would be no more dangerous than nuclear weapons, thinking that the level was minimal. This just served to broaden the debate. Jonathan Schell's book, *The Fate of the Earth* (1982) brought this into the public sphere.

The analysis of Crutzen and Birks was taken up by Richard Turco, Owen Toon, Thomas Ackerman, James Pollach and Carl Sagan, the so-called TTAPS group, which predicted severe consequences on climate. The atmospheric science was rather well known; what was not known was just how many megatonnes of energy would be released in a nuclear war, and how much smoke and dust this would produce. Those arguing for or against the dangers of nuclear winter could adjust this number to suit their argument. There is no doubt however, that the overall argument was in favour of a reduced stock of nuclear weapons. It also seems true that the military had not taken account of nuclear winter in their own studies. A point of primary concern is that up to this point all of the knowledge was in the hands of the military, who apparently were not aware of all the consequences of a particular strategy to be followed in the initiation of or in response to a nuclear attack.

It is interesting that a Canadian, John Hampson, then a scientist at the Defence Research Establishment Valcartier (DREV) was in the front line of this debate. CARDE had changed its name to DREV in 1969. The DREV group, involving a large number of very capable scientists had been studying the stratosphere for some time, using high altitude balloons. This group was way ahead of its time in understanding the possibilities of the destruction of ozone in the stratosphere and what this would mean for humankind. Hampson in particular was extremely concerned about nuclear winter. Eventually he left DREV and returned to a quieter life in England. With the end of the cold war and some reduction in nuclear weapons this seems no longer to be an issue, replaced now by concern about global warming.

STRATOPROBE High Altitude Balloon Flights

STRATOPROBE was a high altitude balloon project of Environment Canada (EC) to investigate the chemistry of the ozone layer and its depletion by chlorofluorocarbons (CFCs). Under the leadership of W.F.J. (Wayne) Evans of the Experimental Studies Division of the Atmospheric Environment Service scientists from EC and several universities were involved (the University of Saskatchewan, York University, University of Calgary). It began in 1973 with the design and construction of a large balloon payload to loft six instruments to over 35 km in the stratosphere. Large balloons of up to 700 million liters volume were used and the payload had a mass of over 900 kg. The payload engineering was conducted by SED Systems and later by SIL of Saskatoon. The first three flights were conducted at Churchill, Manitoba in July, 1974 with the balloon launches performed by the US National Scientific Balloon Facility (NSBF) from Texas. In the summers of 1975, 1976 and 1977 launches were conducted at Yorkton, Saskatchewan. Extensive flights of ozonesondes and ground measurements of the ozone layer were conducted to support the balloon flights. The first NO_2, HNO_3 and odd nitrogen measurements were reported in several papers by Evans and colleagues in 1976. In the fall of 1977 launches began to be flown from the NSBF base in Palestine, Texas. Launches were conducted for the NASA BOIC campaign and for validation of the NIMBUS 7 LIMS (Limb Infrared Monitor of the Stratosphere) and Stratospheric Aerosol and Gas Experiment (SAGE) II/ satellite instruments. Hand launches of small balloons were described earlier. For a 900 kg payload something more substantial was required and a modified earth mover called "Tiny Tim" was used to carry and release the payload (gondola). Since balloons are carried with the high altitude winds, the recovery can be some distance from the launch point. This landing point is to some extent controlled by cutting the balloon from its payload by radio command, allowing the latter to descend by parachute. A light plane is used to spot the payload and in Saskatchewan it could normally be accessed with a small truck and winch. From Churchill a helicopter was necessary for recovery.

In 1984, a launch facility was established at Vanscoy, Saskatchewan, near the home base of SIL Ltd of Saskatoon, with the launches conducted by SIL personnel. Flights continued until the summer of 1988. In a winter campaign in 1988-89 flights were conducted at North Bay, Ontario and in February at Alert, NWT, with a much smaller payload, to investigate the suspected Arctic ozone depletion similar to that occurring in the Antarctic ozone hole. The first measurement of Arctic ozone depletion was made during this campaign (Evans, 1990). The STRATOPROBE project was terminated in 1990 but concern about ozone depletion in the Arctic led to the establishment of an ozone observing laboratory called ASTROLAB at Eureka, 80° N. STRATOPROBE was later resurrected as a project called MANTRA (Middle Atmosphere Nitrogen Trend Assessment) by Kim Strong from the University of Toronto, with EC support; the same experiments and balloon payload technology were used again, supplemented with newer instruments and with SIL as the payload contractor. MANTRA flights were made in 2001 to 2006 at Vanscoy. By comparing with STRATOPROBE measurements made a decade earlier, trends on stratospheric chemistry were investigated.

The ASTROLAB later closed because of lack of funding, but was re-opened in 2005 by a University consortium under the name PEARL (Polar Environment Atmospheric Research Laboratory), using funds from the Canadian Foundation for Innovation (CFI), NSERC and the Canadian Foundation for Climate and Atmospheric Science (CFCAS).

A high altitude balloon lifting off. The gondola, which is far below the balloon itself has just cleared the ground. Photo by courtesy of Wayne Evans.

Canada begins an Astronaut Program

It has already been mentioned that, consistent with its new focus, the NRC initiated an Astronaut Program. On July 14, 1983 it placed advertisements in Canadian newspapers, inviting "Canadian men and women to fly as astronauts on future space shuttle missions." Some 4,400 people applied and six were selected: Ken Money (from the Defence and Civil Institute of Environmental Medicine), Roberta Bondar (physician and medical researcher), Bjarni Tryggvason (NRC researcher), Bob Thirsk (physician), Steve MacLean (York University Ph.D. Physics graduate then holding a post-doctoral fellowship at Stanford University), and Marc Garneau (engineer and naval officer). This activity fared well because it was fully integrated into the NASA schedule and system, and it immediately captured the fascination of the Canadian public. The early part of this captivating story has been described in some detail by Lydia Dotto and it is brought up to date by Chris Gainor. Here the intent is to place the story in its science context.

According to Forsyth's categories, the astronaut program belongs in category 1), research done "in space", and involves either a study of the astronauts themselves, and their reactions to the space environment, or the conduct of experiments conducted by astronauts, such as doing experiments under "zero" gravity, generally called microgravity, or biological experiments such as growing plants in space.

Garneau was seconded to the Canadian Astronaut Program from the Department of National Defence in February 1984 to begin astronaut training. In March, 1984 he learned that he would be a payload specialist on a flight in October the same year, with the largest crew that had ever flown, five men and two women. The orbit would have an inclination with respect to the equator of 57° compared with the normal 28.5°, meaning that he would be able to see Canada. The altitudes would be 350 km on the first day, 275 on the second and 225 for the remaining days. Mission STS 41-G was primarily an Earth observation mission, hence the higher inclination and lower altitude. For this reason the cargo bay would be facing the Earth most of the time. The largest instrument was a synthetic-aperture radar (whose concept is described later in connection with RADARSAT) called SIR-B (Shuttle Imaging Radar). A large format camera would record images on a huge film format of 23 x 46 cm.

The mission took place October 5-13, 1984 aboard Space Shuttle Challenger, with a remarkably short training period. During 133 orbits of the earth (5.4 million km) in 197 hours and 23 minutes, the crew deployed the Earth Radiation Budget Satellite, conducted scientific observations of the earth with the OSTA-3 pallet and Large Format Camera (LFC), performed numerous in-cabin experiments, activated eight "Getaway Special" canisters, and demonstrated potential satellite refueling with an EVA and associated hydrazine transfer. The crew also filmed with the Canadian IMAX Camera.

One aspect of the astronaut program is that it often crosses disciplinary lines. On Marc Garneau's first flight he remarkably conducted ten different Canadian experiments, two of which were observations of the atmosphere. One of these was SPEAM, the Sun Photometer Earth Atmosphere Measurements, a joint experiment of the Atmospheric Environment Service of Environment Canada and York University.

The SPEAM instrument was based on earlier work at the Atmospheric Environment Service (AES) of Environment Canada, now called the Meteorological Service of Canada (MSC). It was based on a widely used technique for the measurement of the concentration of atmospheric constituents in the atmosphere, one in which Canada has gained considerable expertise. This consists of measuring the spectrum of sunlight that has passed through the atmosphere and is imprinted by the loss of light by absorption of the particular species, seen as a dip in the intensity of the solar light at the corresponding wavelength. The depth of the dip yields the concentration. However, the solar spectrum is already imprinted by absorption lines in the sun's atmosphere so that must be taken into account. The ideal way to do this is to measure the depth of the dip outside the atmosphere, and then again inside. This is not possible from the ground, but Garneau's flight offered the possibility to do this in space. Holding the battery-powered SPEAM instrument, Garneau would first point it at the sun when it was high in the sky. Then as the sun set, the rays from the sun would pass through a much longer path in the atmosphere, yielding the concentrations there. During the mission, the concentrations of ozone, nitrogen dioxide (NO_2), water vapour and aerosol (small particles in the atmosphere) were measured.

The experiment was led by Wayne Evans of the Atmospheric Environment Service (AES). That the device could be put together on such short notice was a result of an earlier design that had been developed by AES for aircraft flights. During the flight Garneau had to position this hand-held device to take advantage of the view of the sun available through various shuttle windows, both for high-sun measurements, and "limb" views just before sunset. Sunrises were not successful as Garneau was unable to determine with sufficient accuracy the position of the sun before it came up. Altogether, 21 measurements were made, 5 for high sun, and 16 sunsets. These were pioneering observations that provided a foundation for later Canadian space measurements using the same technique.

Marc Garneau inspecting the SPEAM instrument during the STS-41G mission, taken October 8, 1984. He is located on the Flight Deck, on the aft station. Photo by courtesy of NASA.

Marc Garneau conducted another experiment, this one on the high atmosphere. The Orbiter GLOW (OGLOW) experiment was concerned with photographing the light glow that formed on shuttle surfaces impacting the atmosphere at 7 km/sec, using different filters in order to determine the emissions at different wavelengths. OGLOW also photographed the same airglow and auroral emissions that were to be observed by WAMDII, to assist in the planning for that mission. It was led by David Kendall, then of the NRC, now at the CSA, along with a team of 10 Canadian and 2 US investigators. A secondary goal was to determine whether the shuttle glow would contaminate the WAMDII measurements; it was found that they would not. Because there was no Canadian equipment on which to base this experiment, camera technology from Stephen Mende, then at Lockheed (now Lockheed Martin) in Palo Alto was obtained. This consisted of a commercial Nikon film camera, with an intensifier in front (like that in the Viking ultraviolet auroral camera to be described shortly), to enhance the weak airglow emissions. As well, optical filters were used in front to isolate the different emissions to be viewed, and a mechanism was provided for tilting the filters, to move the wavelength off the primary band. The changes of filters and the filter tilting was all done by hand, while viewing airglow, aurora and shuttle surfaces as the opportunity arose – a truly demanding task for Canada's first astronaut. Nearly all of the ten scientific objectives were realized during the 19 observation sequences, and excellent data were obtained.

In addition, Garneau conducted another three experiments, the first on medical experiments, the second on the exposure of composite materials to space, and the third on Canada's space vision system; the entire package including SPEAM and OGLOW was known as CANEX. Descriptions of all these experiments as well as Garneau's report on the mission, and a description of the astronaut program by Karl Doetsch are given in Volume 31, No. 3 of the *Canadian Aeronautics and Space Journal* as referenced. Looking back, it was an enormously ambitious package of observations for a brand-new mission specialist, with limited time for training. For all these reasons it still stands as a memorable and highly successful mission.

After the momentous success for Canada of the Garneau flight, the disastrous Challenger launch on January 28, 1986 came as a great shock, and delayed the flights of the next Canadian astronauts, Roberta Bondar and Steve MacLean until 1992. But in 1984 President Reagan authorized NASA to build a space station with permanent human occupation, a concept that went back to the early days of the agency. The development of the international scenario, and the agreements with international partners as to what elements of the International Space Station (ISS) they would provide took several years. By 1986, NRC was engaged in strenuous negotiations with NASA. The Canadian effort was led by Karl Doetsch, who also headed the astronaut program. Mac Evans, Director-General of the space policy sector of the federal Ministry of State for Science and Technology, was Canadian co-chair of the Canada/NASA program co-ordination committee. The technical work was led by Jim Middleton, space station project manager for Spar Aerospace.

The negotiations were difficult because the Canadian effort had led to an excellent concept, but both NASA and US politicians were reluctant to let Canada dominate such an advanced robotics package on ISS. A successful agreement was reached, but then the Canadian cabinet needed to allocate the funding. It was a difficult time for the Mulroney government, both politically and economically, but a package of $800 million over 15 years was approved, for what was called the Mobile Servicing Centre (MSC). The entire package is known as the Mobile Servicing System (MSS). The approval was complicated by two competing, though smaller proposals, one for a RADARSAT from the Department of Energy, Mines and Resources, the other from the Department of Communications for the next communication satellite, MSAT. (Joe McNally moved to CRC to become Project Manager for MSAT in 1981, and then moved back to Montréal and the CSA to become Program Manager for RADARSAT in 1990.) Both projects were given lower priority then, but a highly successful RADARSAT flew in 1995 as a collaboration with NASA. NRC has to be given enormous credit for so successfully bringing so many elements of the new space program together, the Canadarm, the Astronaut Program, participation in the International Space Station and a strong space science program as is described in the following sections. However, they took a significant hit for their effort, as the government required that they take $74 million over five years out of their own budget, a considerable sacrifice for Canadian space research, but an extremely valuable one. MSAT was launched as one of a pair of joint Canadian US satellites in 1996, providing mobile telephone service. Because a commercial product was being supplied, the cost to the federal government was not great for this satellite that moved communications into the 40-60 GHz L-band.

A Canadian Auroral Imager on the Swedish Viking satellite

During all this upheaval, and without the capability of launching their own satellites, Canadian space scientists continued to be active and to take advantage of opportunities as they might arise. One of these came to fruition in 1986; on February 22 in that year the Swedish Viking (pronounced "Veeking" in Swedish which helped to distinguish it from the US Viking spacecraft that had gone to Mars) spacecraft was launched carrying the Canadian Ultra Violet Imager. In 1978-79 Gordon Shepherd was on sabbatical leave at the Royal Institute of Technology in Stockholm. His Swedish colleagues were thinking about their first Swedish scientific satellite that would exploit their expertise in plasma physics. The Royal Institute of Technology had taken Kavadas' electric field measurements one step further – instead of using small released free-flying probes they unleashed cables wound around the rocket with probes at the far ends. Centripetal force associated with the spinning vehicle made the cables taut, and allowed the measurement of electric potential difference from probe to probe, yielding accurate electric field values across this relatively large distance. The Kiruna Geophysical Observatory had the capability of measuring electron and proton fluxes, but Sweden wanted an auroral imager of the ISIS II type as well, a technique with which they were not familiar. To accomplish this, they would need support from another country. Denmark was invited to provide this, but they had no experience either, so Shepherd went to Copenhagen to meet with them, but the POLAIRE problem appeared again in a different form. The ISIS technique was simple, because it used the spinning motion of the satellite, combined with its forward motion to provide the TV-like scan across the Earth. But this provided only one image per orbit. Now, CCDs (Charge Coupled Devices, as used in digital cameras) had become available, and were to be used in the WAMDII instrument. A new auroral imager should also use a CCD imager, not the ISIS II photomultiplier detector. From a spinning satellite, the CCD images would be blurred unless the exposure times were very short, too short to acquire images of the relatively weak aurora. Shepherd remembered POLAIRE, and suggested something simpler; the camera could look into a rotating mirror that would "freeze" the image during each exposure, but this didn't make the Danish scientists any more comfortable because of the several new technologies they would need to master so they declined to become involved.

On returning to Canada in the summer of 1979 Shepherd discussed this idea with Clifford Anger of the University of Calgary and Alister Vallance Jones of HIA. They had been busy working on a similar proposal for a US satellite mission called ISTP (International Solar Terrestrial Physics), where the CCD would work easily, because the satellite was to be oriented to face the Earth. The sensible choice seemed to be to give up on the Viking mission, but Anger had a better idea. With CCD's, images are shifted out of the exposure area into memory area and the rate of shifting can be controlled electronically. Anger's idea was to move the captured electrical charges resulting from the exposure, row-by-row, across the CCD at the same rate as the image was moving across the CCD as a result of the spacecraft rotation. The images are thus effectively frozen to the moving charges which continue to accumulate electrons as they move across the CCD during the exposure. This method is called TDI (Time

Delay and Integrate); it replaces the rotating mirror with electronics, making a relatively simple experiment again. Although Canada could not afford to do both the Viking and ISTP missions, it followed the "two eggs in one basket" philosophy and proposed for both. When NASA declined the Canadian imager for one provided by a US scientist, one egg was left, and the CCSS approved a Canadian Ultra Violet Imager (UVI) for Viking, to be led by Clifford Anger.

But more technologies had to be learned. It was decided that the imager should operate in the ultraviolet region where sunlight is so weak that the aurora could be observed in full daylight. Glass absorbs ultraviolet light, so the camera would have to use mirrors, rather than lenses. Vallance Jones proposed a type used by astronomers, a Cassegrain design, but with the primary and secondary mirrors reversed – he called it an "inverse Cassegrain" system. But there was more. The CCD responds only to visible light, not ultraviolet, and in any case the signals are very weak, so some kind of "gain" is desired. A microchannel plate (MCP) was used, a thin layer of silica with a grid-like pattern of holes drilled through it with the walls of the holes treated to make them slightly conducting. Then a high voltage, around 1,000 V was applied across the two sides of the MCP. When an ultraviolet photon hit the photoelectric layer on the front of the MCP it disappeared and an electron was emitted. This electron was captured by the electric field of the applied voltage, and popped down the hole, gaining energy from the voltage as it went, like Hunten's photomultiplier described in Chapter 3. Each time it struck the wall it had enough energy to produce several more electrons so that for every electron entering on the front end, a cloud of electrons came out the back. The surface of the MCP is curved, to solve another problem, that the focal plane of the Cassegrain system, otherwise without aberrations, is spherical, not flat. Finally, the cloud of electrons struck a phosphor, producing a tiny burst of visible light that was captured with the CCD, connected through a fibre optic bundle. This technique was developed at CAL Corporation in Ottawa, with help from Vallance Jones, Harvey Richardson (of HIA in Victoria) and Anger and it worked extremely well. An image of the aurora is taken in just one rotation of the satellite, 20 seconds; the exposure time is 1 second. Having an image every 20 seconds allowed the dynamical motions of the aurora to be followed. Canada scored its second huge success in global auroral imaging, following on the earlier ISIS II mission, and clearly using the earlier experience. Thus while new experiments were emerging, they still had roots going back into the earlier history.

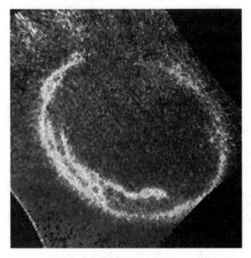

The auroral oval, about 3,000 km across, imaged by the Canadian Auroral Ultraviolet Imager on the Swedish Viking satellite. The nearly circular ring is nearly complete, except for a break at the top, which is the magnetospheric cusp, where high energy electrons excite the aurora less efficiently. Inside this ring (lower left) there is an arc extending from roughly 21 hr to midnight. Smaller "hot spots" can be seen around the oval in the early afternoon, and evening. Image by courtesy of Leroy Cogger, on behalf of the University of Calgary group.

By the time Ian McDiarmid retired from the Directorship of the CCSS in 1986, the year that Viking was launched, it had matured into an effective organization, re-established Canadian scientists in satellite measurements and laid the groundwork for the scientific accomplishments that were to follow. He also retired from the HIA at the same time. Roy Vankoughnett became Director of the CCSS and continued to guide it with a steady hand.

SMS on Akebono

Brian Whalen's story begins again, continues through this section and on through the following section.

"In the early 1980s, the Institute of Space and Astronautical Sciences (ISAS) of Japan started the development of its fourth "exospheric" scientific satellite, EXOS-D under the leadership of Professors Hiroshi Oya and Kochiro Tsuruda. The scientific objective of EXOS-D, which was renamed 'Akebono' after its launch, was to study the structure of the near-earth space environment (i.e., the magnetosphere and ionosphere) and the physical processes that take place therein, and to understanding of the flow of energy and momentum from the sun through interplanetary space to the near-earth region. At the invitation of Professors Hiroshi Oya in the early 1980s, Brian Whalen and his team at the HIA along with co-investigators from the Radio Research Laboratories of Japan led by I. Iwamoto and E. Sagawa made a proposal to the Institute of Space and Astronautical Science (ISAS) of Japan to provide an ion mass spectrometer for the Akebono mission, which was subsequently accepted by ISAS. The Canadian SMS team was led by Brian Whalen, as PI, and included Ron Burrows, Emil Budzinski, Andre Pilon and Andrew Yau from HIA; Bob Hum was CSA Project Manager. The Japanese team was led by I. Iwamoto and E. Sagawa, and several of their RRL colleagues. SED Systems was the prime industrial contractor.

"The proposed elliptical polar orbit of EXOS-D (275 km perigee and 10,000 km initial apogee) made it an ideal platform to study the ambient ionospheric and magnetospheric plasmas as well as ion acceleration processes in the high latitude auroral zone and polar caps. To meet these scientific goals the instrument requirements were set at: energy per unit charge 0 - 4 Kv, mass per unit charge range 1 - AMU/Q, and an angular resolution < 10 deg for energetic ions. [Here the Kv stands for energy in kilo electron volts (Kev), divided by the charge in electrons, so becomes just Kv. AMU is atomic mass units, the scale for which the mass of hydrogen is approximately one and Q stands for the charge in electrons.] *To measure the low-altitude major ions as well as the high-altitude energetic minor ion species, the instrument was required to have a very large geometric factor (entrance aperture) and dynamic range.*

"The instrument proposed, the Supra-thermal Mass Spectrometer (SMS), was based on the Firewheel design but had an extended energy range which included thermal ions. [Thermal ions are ones whose energy exists solely on account of their temperature; this low energy is a challenge to measure.] *SMS consisted of an entrance aperture in which planar grid electrostatic energy analysis (ion energy per unit charge) was performed. This was followed by ion pre-acceleration (or de-acceleration depending on the energy range selected) and injection at a fixed energy into a planar folded three-stage electrostatic radio frequency (rf) acceleration region. Ions of specific velocities were then preferentially (resonantly) accelerated. Varying the rf frequency (at a fixed amplitude) provided the needed mass selection. After*

exiting the rf field the ions were passed through a cylindrical electrostatic energy analyser and focused onto a planar Micro-Channel Plate (MCP) [as already described for Viking]. To achieve the large dynamic range while still avoiding MCP saturation, an electrostatically controlled variable aperture was incorporated into the design. The complex computer modelling required to finalize the instrument design was performed by E. Budzinski and most of the sensor head and rf generator design was performed by the NRC Space Plasmas group.

The Supra-thermal Mass Spectrometer (SMS) flight model for the Akebono spacecraft. Photo by courtesy of COM DEV, with the assistance of Brian Whalen.

"Akebono was successfully launched in 1989 from the Uchinoura Space Center in Japan and data collection from the spacecraft began soon thereafter. From the spinning Akebono spacecraft, SMS sampled over the full pitch-angle range, and routinely measured the 2-dimensional velocity distributions of various ion species in the iono-sphere and high altitude magnetosphere. The unmatched sensitivity, mass resolution and longevity (coverage over an 11-year solar cycle) of SMS instrument produced a unique data set which is still relevant to the science community.

The Akebono spacecraft being integrated for launch at the Institute for Space and Aeronautical Science in Tokyo. Andrew Yau (now at the University of Calgary) is on the left and Bob Hum, the Project Manager from the Canada Centre for Space Science is on the right. Brian Whalen is speaking to an NEC engineer. Photo by courtesy of SED Systems.

"The baseline analytical software required to extract the moments of the ion distribution function (Density, Temperature and Drift Velocity) from the SMS data was developed by S. Watanabe (Hokkaido University). He extended the calculations initiated by Dennis Green for the sounding rocket data but also included conditions of both positive and negative spacecraft potentials. This analysis formed the foundation for the important observational program to follow. Some of the more important results include studies of: the Polar Wind by Andrew Yau, Takume Abe (currently at ISAS Japan) and E. Drakou where ionospheric molecular ions, with unexpectedly high abundance, and other 'minor' ion species were observed flowing out of the ionosphere at high latitudes and high altitudes; TIA events in the Dayside Cusp by David Knudsen where ionospheric ions were found to be energized perpendicular to the magnetic field direction to Kev energies in very narrow magnetic field-aligned sheets; and plasmaspheric responses to geomagnetic storms by S. Watanabe where time-delayed dynamic effects were observed in the plasmasphere after the onset of major storms.

"The SMS was the first foreign instrument to fly on a Japanese scientific satellite. The program's success hinged on the close working relationships established between the Space Plasmas group and the CRL and ISAS scientists early in the design phase. This open and cooperative interaction continued throughout the analysis and publication period. These interactions also stimulated both groups to propose exciting new space plasma investigations and eventually led to further collaboration on the Nozomi and other missions.

The Swedish Freja and Russian Interball Missions

Freja CPA

"It became clear to the Space Plasmas group at NRC, from previous sounding rocket and satellite observations (SMS), that to further our understanding of the energization of magnetospheric plasmas the future lay in investigations of the small scale structure of these events. For example, studies of one of the major sources of ionospheric ion energization (TIA) indicated that the process occurred in very thin sheets. Similarly, optical images of the fine structure in auroral arcs showed that electron precipitation patterns had spatial dimensions near 100 meters. Therefore, much better temporal resolution (or spatial resolution when viewed from a moving spacecraft) was required for direct particle observations. Preliminary conceptual designs of the Cold Plasma Analyzer (CPA) were in progress to meet this goal when an invitation to participate in the Swedish satellite program Freja was presented to the group.

"The major scientific objective for the Freja mission was to investigate small scale (less than 1 km) plasma structures associated with high-latitude ion and electron heating. A science team, with Whalen as the PI and composed of members from the NRC, the University of Saskatchewan (D. McEwen, J. Koehler, G. Pocobelli), Sweden (R. Lundin, L. Eliasson) and Hokkaido University (S. Watanabe), presented a proposal for the Freja CPA which was accepted. COMDEV served as the prime contractor while PAKWA Ltd. (Dennis Johnson) was subcontracted to develop the boom deployment system and G. S. Campbell served as the Project Manager. The NRC team took responsibility for the sensor development.

"Previously, charged particle sensors had relied on relatively slow techniques to derive energy (electrostatic energy analysers voltage scans) and angle (spacecraft spin) infor-

mation on particle distributions. CPA was designed to provide near instantaneous energy/pitch-angle images. This imaging capability provided samples of thermal electron and ion distribution functions on millisecond time scales (100s of meters satellite travel distance) and was capable of 0.1 second resolution measurements of selected parameters (e.g. Density, Temperature). The instrument was composed of a spherical plate electrostatic analyser, which dispersed particles (in energy and angle of arrival) onto a high-speed Microchannel Plate (MCP) and could be operated in either ion or electron observation mode. An electrostatic gating system was inserted in the entrance aperture to increase the instrument dynamic range. To reduce spacecraft perturbations to the incident particles, the sensor was deployed on a boom.

"Freja was launched in 1992 from the Jiuquan launch site in the People's Republic of China. Successful CPA boom deployment and instrument turn-on occurred soon after launch. As planned, in-flight CPA data were made available to the science team on an almost real-time basis which facilitated a highly interactive observational program. The (anticipated) approximately one year lifetime of the MCP ended the data taking aspect of the project.

"Dave Knudsen (now at the U of Calgary) and Tai Phan (now at the U. of California) undertook the major effort of analyzing the CPA data. High time-resolution ion and electron distributions were derived and routine scientific analysis of the data set was commenced. Some initial analysis by Knudsen showed that the most intense ion heating was caused by broad-band extremely low frequency plasma waves (as opposed to VLF waves previously thought to dominate), which occurred in regions of intense, low energy (100's of eV) magnetic field-aligned electron precipitation commonly found in the dayside as well as near the poleward boundary of the auroral oval. This heating was shown to be concentrated in narrow, intense Alfven waves hundreds of meters across. The analysis carried out by Tai Phan demonstrated the ability of the CPA to make valid observations of the elusive thermal electron distribution function. Unfortunately, these analysis activities were cut short by the decision to terminate the activities of the NRC Space Plasmas Group [this is described in the next chapter].

"Although analysis of the CPA data had evolved only to a very preliminary stage, the success of instrument design had been clearly demonstrated. Indeed the awareness of foreign scientists (European and Japanese) of the potential of this approach led to invitations to provide similar (or advanced) versions for of this device for Nozomi and other programs initiated at the University of Calgary."

Freja and Interball Auroral Imagers

In parallel with the NRC CPA instrument described above, another ultraviolet auroral imager from the University of Calgary, led by John S. (Sandy) Murphree was provided for Freja. The ultraviolet auroral imager made lower altitude measurements than on Viking, as low as 600 km, allowing the imaging of small detailed auroral structures rather than concentrating on global images of the entire oval. On this scale, auroral spirals and vortices were observed; these provide a great deal of information about the plasma instability processes in the aurora. With this detailed information it was possible to relate these visible auroral structures to the plasma measurements made on board the spacecraft.

The reputation of the Canadian auroral imagers was now well established, and led to a further invitation, from Russia, for a Canadian experiment on one of their two Interball satellites; the Interball-2 satellite was launched in 1996, with the Canadian ultraviolet imager experiment led by Leroy Cogger of the University of Calgary. It was an ambitious mission, with two main satellites and two subsatellites. The role of the Canadian imager was to provide the auroral context for the plasma measurements made on both the main and the subsatellites, viewing from about 20,000 km above the Earth. A second camera viewed 67° off the spin axis, increasing the global coverage. The overall mission was very complex, and encountered a number of problems. One was that the spacecraft wobbled slightly around its geometric axis, blurring the images to some extent. While some of this blurring was removed by data processing, only about three months of data were obtained overall—very useful but less than the potential.

Since that time, the Canadian advantage in auroral imaging has been overtaken by larger and better-funded groups in the USA. However, the Canadian expertise continued to be recognized, and University of Calgary scientists were invited to participate in the USA IMAGE mission, containing a University of California (Berkeley) experiment. In fact, the imager consisted of the Freja flight spare camera with a Berkeley detector. The satellite was launched in March, 2000 and has worked at the forefront of auroral as well as other types of imaging.

Postscript

The seventies, and much of the eighties have been described as an interval of transition. For some it was a period of depression as well, but it was also a period of consolidation. The seventies began with the launch of ISIS II in 1971, the last Canadian scientific satellite to be launched until the re-establishment of Canadian science missions in August, 2003, with the SCISAT mission. However, the excellence of the ISIS II data kept scientists busy throughout the decade until CRC terminated its operations on March 13, 1984. They then transferred the operations to Japan, who operated the spacecraft until January 24, 1990. It is also true that with no new mission to move to, the effort and quality of the data analysis and scientific interpretation were enhanced. With more analysis the scientists became more experienced, were able to extract more and more information from the data. The members of the ISIS Experimenter's Group got to know each other very well, learned much from each other and collaborated closely, combining the results of different instruments. As the decade advanced, the skills were honed further and further; new results were obtained and new papers published in the journals.

As noted, rocket measurements became more mature and sophisticated as the program continued during the seventies. Some remarkable rocket payloads, such as Oedipus-C, described in the next chapter continued to be flown after the closing of the Churchill Research Range in 1984.

Thus in a time of dwindling resources the CCSS managers, the industrial engineers and the university and government space scientists had shown that it was possible to conduct new and innovative space research in unexplored locations with modest funds.

References:

Dotto, Lydia, Canada in space, Irwin Publishing, 1987.

Evans, W.F.J. et al., The SPEAM sunphometer experiment on Mission 41-G, Canadian Aeronautics and Space Journal, Volume 31, No. 3, page 240, 1985.

Evans, W.F.J., Ozone depletion in the Arctic vortex at Alert during February,1989, Geophysical Research Letters, 17, pp 167-170, 1990.

Gainor, Chris, Canada in space, Folklore Publishing, 2006.

Hampson, John, Photochemical war on the atmosphere, Nature, 250, 19 July, pages 189-191, 1974.

Hartz, T.R. and I. Paghis, Spacebound, Ministry of Supply and Services Canada, 1982.

Jelly, Doris H., Canada: 25 years in space, Polyscience/NMST, 1988.

Kendall, D.J.W., et al., OGLOW – An experiment to measure orbiter and Earth optical emissions, Canadian Aeronautics and Space Journal, Volume 31, No. 3, page 227, 1985.

Martin, Brian, Nuclear winter: science and politics, Science and Public Policy, Vol. 15, No. 5, pp. 321-334, October 1988.

Nicholls, R.W. and D.J.W. Kendall (Eds.), 2001: A Space Odyssey, Foreward: 2001: A space millennium begins, Physics in Canada, Vol. 57, No. 5, Sept/October, 2001.

Schell, Jonathan, The fate of the Earth, Alfred A. Knopf, New York, 1982.

Chapter 7

Birth of the Canadian Space Agency

The CSA moves to Saint Hubert

As part of the funding package to support the International Space Station (ISS), the government of Canada agreed to create a Space Agency, at long last. Thanks to the wisdom and persistence of the NRC, along with elements from some other government departments, all the components were in place when the announcement was made on March 1, 1989. Dr. Larkin Kerwin, then President of the NRC, became the first CSA President, serving until February, 1992. All of the employees of the NRC Canada Centre for Space Science would become employees of the new agency, but there was no immediate space available so these employees would stay where they were for the time being, at 100 Sussex Drive.

The battle over its location began earlier, in 1987, when Montreal organized an aggressive campaign. Ottawa mounted a defence, but was unsuccessful in attracting as many, or as effective, supporters. Although legislation was expected that year, nothing happened except that Art Collin, who had been given the responsibility of establishing the agency, resigned. Later it became evident that funding concerns about the ISS were giving the government cold feet about an increased commitment to space. This delay was followed by an election, but with both the funding and election crises over, the formation of the Canadian Space Agency was announced, and its location was given as Montreal. It soon became evident that it would not really be in Montreal, but in Saint Hubert, as far away from Ottawa as it could be, and still be considered to be part of Montreal. Of course it had to be on Federal land, and that was available in Saint Hubert. But it was unfortunate that it was not located closer to the academic activity of Montreal and, in addition in the grand scale of planning for the future, there was a failure to take full advantage of the NRC heritage. It is abundantly clear from what has been written so far that space research is an international activity, and so needed to be located in or near Ottawa, where all the international linkages take place, not to mention the linkages with the government itself. If the agency had been located on the side of Montreal closest to Ottawa it would not have been such a problem, but the decision may have been intentional, to make it impossible for employees to commute from Ottawa.

The agency would be located in a beautiful new building, across the river from Montreal. The rationale was that Saint Hubert would become the aerospace centre of Canada with the CSA as a focal point, but this is not where the space industry of Montreal is located. Saint Hubert may be close to Montreal, but it is not conveniently close, nor does the Saint Hubert airport have any commercial flights. Instead it is a long drive or taxi ride from the Pierre Elliott Trudeau International airport, or a long trip by bus, then metro, then bus again. This has meant considerable inconvenience for countless numbers of visiting scientists, engineers and diplomats, not to mention the employees, and the public. Certainly there is merit in distributing the research and development across the country as NASA does with its Centers, but NASA headquarters is definitely in downtown Washington, D.C.

The major impact came after the building was completed and the employees began to move there in 1993. The most experienced individuals, who had honed their talents on ISIS and Hermes were approaching retirement. The prospect of re-locating themselves and their families at this stage in their life held no joy; the attraction of working in this new organization was simply not enough to make this demand on one's family and personal life. Nearly all left the organization. Karl Doetsch moved to Europe, to the International Space University in Strasbourg, France (although he has since returned to Canada as a private consultant) and others simply retired or moved to other organizations. The David Florida Laboratory at CRC was too large to be moved, so it stayed in Ottawa. The Canada Centre for Remote Sensing, closely related to the CSA, also decided to stay in Ottawa.

The new CSA fell under Industry Canada as shown in the figure, which shows how the space activities within the NRC were gradually transferred to what became the CSA, and how CRC also became part of Industry Canada.

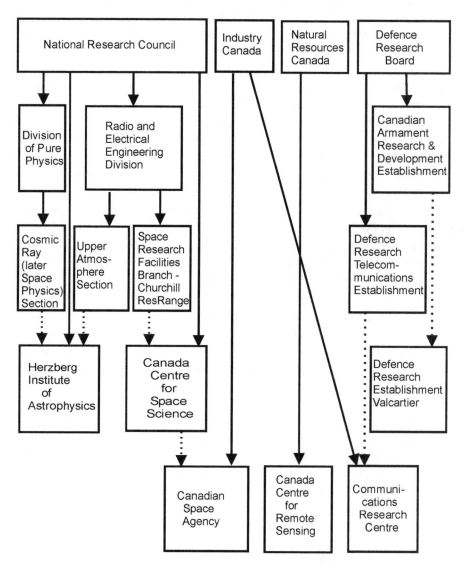

A simplified organizational chart, showing the evolution of the NRC and the DRB laboratories into what became the CSA. Solid lines indicate lines of authority and dotted lines indicate a transfer of of one organization into another.

While there are in principle some advantages in building a new organization from the ground up, in a new environment, it also is extremely challenging. The move began in 1993 and was completed in 2004, when the Space Science Programs moved from Sussex Drive in Ottawa. On October 2, 1996 this new building was named the John H. Chapman Space Centre. Dr. Kerwin stepped down in February, 1992 to be replaced by Roland Doré. In 1994, Doré also moved to the International Space University, and was replaced by Mac Evans, the first President with a deep knowledge of, and experience in, Canadian space activity. His interest in space began when as an amateur radio operator he detected signals from Sputnik 1. Evans graduated as an Electrical Engineer from Queen's University and this was followed by an M.Sc. degree as an Athlone Fellow at Birmingham University in England. On his return to Canada he joined the DRTE and over the years accepted positions of increasing responsibility in the government space sector, including the Department of National Defence, the Department of Communications, the Ministry of State for Science and Technology and the Department of Industry. Evans developed the 1985, 1994 and 1999 Space Plans for Canada and forged the broadly-based consensus among government, industry and academia that led to approval of these plans with an allocation of over $3 billion of new funding. For more than two decades he led the development of the unique partnership between the Federal Government and the Canadian Space Industry that has been responsible for the successful growth of the industry in the international marketplace.

Mac Evans saw international cooperation as a fundamental component of the Canadian Space Program. He led the negotiations with NASA for Canada's participation in the International Space Station, and obtained approval for two extensions (for twenty years) of the cooperation agreement with the European Space Agency (ESA). ESA has an interesting structure in which member countries pay a membership fee, based on the size of their economy. However, except for the cost of administration, the money that they contribute comes back to them in the form of contracts. This means that ESA has to juggle its contracts and sub-contracts in such a way that each country gets its share to within a percent or so. Canada couldn't make such a large commitment, and it's not clear that ESA would accept Canada as a full member anyway; but Canada enjoys "Associate Member" status, at a cost approaching $30M per year, of which most comes back to Canadian industry. During Evans' tenure as President of the CSA eight astronauts flew in space, RADARSAT-1 was launched and the Mobile Servicing System (MSS) was attached to the International Space Station (ISS).

In 2001 he was replaced by someone with a first-hand knowledge of being in space, Marc Garneau, Canada's first and highly respected astronaut, one who has spent 29.08 days in space according to the NASA records. Dr. Joseph Jean-Pierre Marc Garneau was born in Quebec City. He received a B.Sc. in engineering physics from the Royal Military College of Kingston in 1970, and first as an Athlone Fellow (like Mac Evans) and then an NRC bursary holder took a doctorate in electrical engineering from the Imperial College of Science and Technology, London, England, in 1973. Garneau was a combat systems engineer in HMCS Algonquin, 1974-76. While serving as an instructor in naval weapon systems at the Canadian Forces Fleet School in Halifax, 1976-77, he designed a simulator for use in training weapons officers in the use of missile systems aboard Tribal class destroyers. He helped develop an aircraft-towed target system for the scoring of naval gunnery accuracy. Promoted to Commander in 1982 while at Staff College, he was transferred to Ottawa in 1983 and became design authority for naval communications and electronic warfare equipment and systems. In

January 1986, he was promoted to Captain and retired from the Navy in 1989. In 2003 he was named Chancellor of Carleton University in Ottawa.

In November, 2005, Prime Minister Paul Martin announced that Marc Garneau would run for the Liberal Party in Vaudreuil-Soulanges, a rural riding just west of the Island of Montreal that was held by the Bloc Québécois; in order to do this he resigned from the CSA. His political bid was not successful, but he plans to run again in the next election in the more favourable Westmount-Ville-Marie riding. Very few scientists and engineers see a Parliamentary role for themselves in Canada. Garneau's decision indicates his belief that he can do more to support science and technology in Canada as a politician than as a bureaucrat inside the government system.

The Astronaut Program under the CSA

Garneau was named Deputy Director of the Canadian Astronaut Program in 1989, providing technical and program support in the preparation of experiments to fly during future Canadian missions. He reported to NASA's Johnson Space Center (JSC) in Houston, Texas in August 1992 and, after completing a one-year training and evaluation program, was qualified for flight assignment as a mission specialist, now a fully-fledged astronaut. He initially worked technical issues for the Astronaut Office Robotics Integration Team and subsequently served as capsule communicator (CAPCOM) in Mission Control during Shuttle flights.

The first astronaut to fly under the new CSA was Roberta Bondar. She obtained a B.Sc. in zoology and agriculture at the University of Guelph in 1968, an M.Sc. in experimental pathology at the University of Western Ontario in 1971, and a Ph.D. in neurobiology at the University of Toronto in 1974. An M.D. followed, at McMaster University in 1977. Bondar was selected to be prime Payload Specialist for the STS-42 mission, in orbit from January 22 to 30, 1992, that carried the prestigious first International Microgravity Laboratory Mission (IML), flown on the Discovery vehicle. This was a surprise to many as Ken Money was expected to be selected. Together with Doug Watt of McGill University he had proposed the original Canadian experiments to be performed on humans in space that laid the basis for the inception of the astronaut program. Money was selected as one of the original six (Watt was not) but now he was over 50 years of age and time was beginning to run out for him. The choice of Bondar as Canada's second astronaut in space (with Ken Money as backup) meant that he would never get to fly. During the mission, Roberta Bondar conducted 43 experiments on behalf of 13 countries. The distinguished Dr. Money is the only Canadian selected astronaut who has never flown but his contributions to the program were significant (this statement does not take into account two astronauts that were selected but later withdrew from the program).

In fact, Steve MacLean was the intended second Canadian astronaut to undertake a package of space science, space technology and life sciences called CANEX-2, for which he was selected in 1985. The primary experiment was a test of NRC's Space Vision System (SVS). This was long delayed by the Challenger catastrophe of January 1986. Steve MacLean was born in Ottawa in 1954 and attended school there. He then obtained an Honours B.Sc. from York University in physics, and a Ph.D. in physics in 1984. During his time at York University he competed on the Canadian National Gymnastics Team. His research was in lasers, so he then moved to Stanford University, working under the renowned laser physicist

Arthur Schawlow, a graduate of the University of Toronto who was a co-recipient of the Nobel Prize in physics in 1981 for his work on laser science. Although MacLean was expected to be the first Canadian aloft after the resumption of the shuttle program, scheduling considerations placed this mission behind the IML. However, he did fly not long after, from October 22 – November 1, 1992, on Columbia, as a Payload Specialist for STS-52, where he performed the seven CANEX-2 experiments, including the evaluation of the Space Vision System.

MacLean was the Chief Science Advisor for the International Space Station from 1993 until 1994, when he was appointed Director General of the Canadian Astronaut Program for two years. In August 1996 MacLean began mission specialist training at the Johnson Space Center in Houston, Texas. After successfully completing basic training in 1998, he continued with advanced training while fulfilling technical duties in the NASA Astronaut Office Robotics Branch. More recently MacLean served as CAPCOM for both the ISS Program and the Shuttle Program, at the JSC. In February 2002 Steve MacLean was assigned to his second space flight. As before, this flight would be long delayed after space shuttle Columbia disintegrated on re-entry some 60 km above Texas, February 1, 2003, as a result of problems with foam de-lamination that had somehow escaped notice for so many years. MacLean did not fly until 2006, but that story is deferred to the end of this volume.

The early years of the CSA were busy with astronaut flights, with one (or more) each year from 1995 to 2001. Colonel Chris Hadfield flew on STS-74 in 1995 and was the only Canadian to board the Russian Mir Space Station. Hadfield was selected in the second astronaut round, in 1992. He was born in Sarnia and raised in Milton, then graduated with an honours B.Sc. in mechanical engineering from the Royal Military College in Kingston. He had joined the Canadian Armed Forces in 1978, and had extensive flight experience, including flying CF-18s for NORAD. With flight testing it is said that he had flown in 70 different types of aircraft. From 1996 – 2000 he was Chief Astronaut for the CSA. From 2001 – 2003 Hadfield was at the Yuri Gagarin Cosmonaut Training Centre in Star City, Russia, coordinating work between Russia and the ISS, where he qualified as a Soyuz cosmonaut. STS-74 was NASA's second space shuttle mission to rendezvous and dock with the Russian Space Station Mir. STS-74 brought along the docking module it would need to connect with Mir and Hadfield operated the Canadarm to position it on the shuttle docking module. It was left attached to Mir for future dockings. Hadfield flew as the first Canadian mission specialist, the first Canadian to operate the Canadarm in orbit, and the only Canadian to ever board Mir, all memorable achievements.

Marc Garneau's second flight followed, on STS-77 during May 19-29, 1996. He participated in materials processing, fluid physics and biotechnology experiments in the Spacehab laboratory module carried in Endeavour's payload bay, along with other duties. This included a long list of experiments including Canada's Aquatic Research Facility (ARF) described in more detail later under "Space Life Sciences", and Canadian microgravity experiments described shortly under "Microgravity in Space."

One month later Bob Thirsk was in orbit, in June, 1996, on STS-78. Thirsk was born in British Columbia and went to school in B.C., Alberta and Manitoba before receiving a B.Sc. in Mechanical Engineering from the University of Calgary, an M.Sc. in the same field from the Massachusetts Institute of Technology and a Ph.D. in medicine from McGill University in 1982. He was at the Queen Elizabeth Hospital in Montreal when he was selected as an astronaut by the CSA in 1983. STS-78 was the Life and Microgravity Spacelab (LMS) mission, on which during 17 days aboard Columbia he and six crew members performed 43 international experiments relating to the life and materials sciences. In 2004, Thirsk trained at the Yuri Gagarin Cosmonaut Training Centre near Moscow and became certified as a Flight Engineer for the Soyuz spacecraft. He served as backup Flight Engineer to European Space Agency (ESA) astronaut Roberto Vittori for the Soyuz 10S taxi mission to the ISS in April 2005.

Next year it was Bjarni Tryggvason's turn, in August, 1997 on STS-85. Tryggvason was born in Reykjavik, Iceland but went to school in Nova Scotia and British Columbia before obtaining a B.Sc. in Engineering Physics from the University of British Columbia in 1972. He then did post-graduate work at the University of Western Ontario (UWO). He was a cloud physics specialist at the Meteorological Service of Canada, and worked in aerodynamics at UWO as well as spending time at Kyoto University in Japan and James Cook University in Australia. Like Thirsk, he was selected in 1983 and worked on the Microgravity vibration Isolation Mount (MIM) which operated on the Russian Mir space station. On STS-85 his role was to test MIM-2 and perform fluid science experiments in relation to the effect of vibrations on the experiments to be flown on the ISS.

Dave Williams flew in April, 1998 on STS-90 as Mission Specialist 3 aboard Space Shuttle Columbia. He was born in Saskatoon, and went to school in Beaconsfield, Quebec before graduating from McGill University with a B.Sc. in biology. He obtained M.Sc. degrees in physiology and in surgery, and a doctorate in medicine at the same institution. In 1988, he became an emergency physician at the Sunnybrook Health Science Centre and was later their medical director of the Advanced Cardiac Life Support Program. He was selected by the CSA in the 1992 competition and became manager of the Missions and Space Medicine Group with the astronaut program. During the 16-day STS-90 flight, called Neurolab, the seven-person crew served as both experiment subjects and operators for 26 individual life science experiments. These experiments, dedicated to the advancement of neuroscience research, focused on the effects of microgravity on the brain and the nervous system. Williams also functioned as the crew medical officer, the flight engineer during the ascent phase, and was trained to perform contingency spacewalks. From July 1998 until September 2002, Dave Williams held the position of Director of the Space and Life Sciences Directorate at the JSC. With this appointment, he became the first non-American to hold a senior management position within NASA.

Julie Payette was in orbit a year after Dave Williams, on STS-96. She was born in Montreal and went to school there before obtaining an International Baccalaureate from the United World College of the Atlantic in South Wales, UK, a B.Sc. in Electrical Engineering from McGill and an M.Sc. in Computer Engineering from the University of Toronto in 1990. Selected by the CSA in 1992, she worked as technical advisor for the MSS and served as a technical specialist on the NATO International Research Study Group on speech processing. She was in orbit from May 27 to June 6, 1999 on STS-96, which performed the first manual docking to the ISS, and delivered four tons of supplies to the station. She operated the Canadarm and was the first Canadian to participate in an ISS assembly mission and to board the ISS, during a mission of 9 days and 19 hours.

Marc Garneau's third mission in November/December 2000 on STS 97 was a significant one for the ISS. Endeavour's five astronauts including Marc Garneau were on the 101st mission in space shuttle history, launched November 30, 2000, with a total cargo of 18,740 kg to deliver the first set of US solar arrays in order to significantly increase the power generation capabilities of the ISS. The main activity was to install a 72 x 11.4 meter, 65 kW double-wing solar panel on the Unity module of the ISS.

A year later, in April 2001 Hadfield served as Mission Specialist 1 on STS-100 International Space Station (ISS) assembly Flight 6A. The crew of Space Shuttle Endeavour delivered and installed Canadarm2, the new Canadian-built robotic arm, as well as the Italian-made resupply module Raffaello. During the 11-day flight, Hadfield performed two spacewalks, which made him the first Canadian to ever leave a spacecraft and float freely in space. In total, Hadfield spent 14 hours, 54 minutes outside, travelling 10 times around the globe.

As noted before, there was a long gap following the Columbia disintegration on re-entry in 2003, so that Steve MacLean's flight was delayed until September 9, 2006, on STS-115. This challenging flight is described in Chapter 9.

A year later Dave Williams was in orbit again, arriving at the ISS on August 10, 2007. It was planned that he conduct an experiment designed by Barry Fowler, a neuroscientist in York University's Faculty of Health. This experiment is described in the "Space Life Sciences" section later in this chapter.

However, the science was overshadowed by his spacewalks, as part of installing the stations's new Starboard (S5) spacer truss for delivery, handing off the 1,814-kilogram segment of ISS to its robotic arm. The mission successfully activated a new system that enables docked shuttles to draw electrical power from the station to extend visits to the outpost. Williams took part in three of the four spacewalks, the highest number of spacewalks performed in a single mission. He spent 17 hours and 47 minutes outside, a Canadian record. A medical doctor had become an assembler of the largest space structure ever built.

An astronaut standing at the end of Canadarm2 during the STS-118 mission, with Dave Williams on board. Photo by courtesy of the Canadian Space Agency.

Microgravity in Space

Background

One's gravity force is displayed by simply being weighed on a bathroom scale. While the gravity force is always present at the Earth's surface it is easy to become weightless here. If one were to stand on a scale in an elevator and the cable broke (two unlikely circumstances) the scale would read zero during the flight to the bottom of the shaft. The person becomes weightless because in free-fall the gravity force equates to the acceleration (force = mass x acceleration, Newton's second law) so no additional balance force is required. Scientists use "drop towers" to test the results of weightlessness on samples, in which the tower is evacuated to simulate space. The University of Bremen has a drop tower 123 meters high; by firing the experiment upward from below with a catapult the weightless time is doubled to about 9 seconds. An astronaut in an orbiting spacecraft is weightless for the same reason. Gravity is still present, to keep the spacecraft in orbit – if the gravity force on the spacecraft were removed, the spacecraft would fly off in a straight line, soon leaving the Earth. The circular orbit corresponds to free-fall just as in the elevator, with the spacecraft falling continuously towards the Earth's centre. The astronaut does the same, so there is no force between the spacecraft and the astronaut and hence no weight.

Practically all physical processes are influenced by their weight. The opportunity to study such processes in orbit has some remarkable consequences. Even though gravity is still present, as explained above, the environment is called zero-gravity, or more correctly microgravity (literally one millionth of gravity), since the spacecraft is subject to perturbations that prevent the perfect balance between gravity and acceleration. Many space-based experiments are sensitive to disturbances from such things as the movement of astronauts, the firing of the orbiter thrusters, and a vibration called "g-jitter" which is common to all large orbiting spacecraft. While these vibrations are small, they can have undesirable effects on sensitive space experiments and equipment has been designed to remove this effect almost completely. A good example of a microgravity experiment is that of liquid motion and the interaction of gases and liquids. Mixtures of liquids which separate in the presence of gravity—the oil and vinegar in salad dressing, for example—remain mixed during space flight.

During Bjarni Tryggvason's 11-day space Discovery flight (STS-85) in August, 1997 he conducted five experiments, collectively known as FLEX, that used the Microgravity vibration Isolation Mount (MIM), a unique piece of Canadian technology. MIM-1 was a version that was used on the Russian Mir space station from 1996 to 1998; the STS-85 version was known as MIM-2. Scientists working through the Microgravity Sciences Program of the Canadian Space Agency discovered that the FLEX experiments helped to identify the precise effects of vibrations on certain scientific phenomena in space, as well as confirm the capabilities of MIM to isolate experiments from those vibrations. As well, the FLEX experiments provided researchers with indications of how some commercial processes are influenced by various chemical and physical factors.

Researchers are looking ahead to the International Space Station, where vibrations from equipment and astronaut activity can affect the results of scientific experiments. What's needed is an advanced version of MIM, and Canada has collaborated with the European Space Agency to provide MVIS (Microgravity Vibration Isolation System) to ESA's Fluid Science

Laboratory (FSL) that will be part of the Columbus module to be launched to the International Space Station in 2008. This program is currently on hold pending results from MVIS and potential utilization of partner facilities onboard ISS.

STS-77 Experiments

Marc Garneau flew with several Canadian microgravity experiments aboard Endeavour on flight STS-77, from May 19-29, 1996. One was the Commercial Float Zone Furnace (CFZF), a materials science crystal growth facility developed by the Canadian Space Agency (CSA) and the German Space Agency (DARA). The CFZF payload was designed around a German ELLI Mirror furnace that melted down sample materials, allowing them to resolidify in the weightless environment of space into a purer, crystallized form. This CFZF payload on mission STS-77 offered an opportunity for scientists from Canada, Germany and the United States to process four materials that had great commercial potential for the optical communications and electronics industry.

In this experiment bismuth germanate (BGeO), one of the experimental materials chosen by Canadian scientists, was examined. This has the potential to be a key element in the manufacture of the next generation of high-capacity laser disks for computer storage. Cadmium germanium arsenide ($CdGeAs_2$), the other material Canadian scientists chose, is an extremely efficient laser material with potential applications in laser technology used for medical treatment. The two materials chosen by the American and German researchers, gallium arsenide (GaAs) and gallium antimonide (GaSb), are semiconductor materials with applications in fiber optics and microwave cellular communications.

Also on STS-77 was an experiment called Atlantic Canada Thin Organic Semiconductors. ACTORS was designed to produce enhanced quality organic thin films by using the Physical Vapour Transport (PVT) method. These films could find applications as gas detectors, in computers, in lasers, and in high-performance electronic equipment. The PVT method consists of heating material so that it vaporizes, and deposits a thin film onto a plate at the top of the ampoule. ACTORS was sponsored by the CSA's Microgravity Sciences Program, the Atlantic Canada Opportunities Agency (ACOA) and the Government of New Brunswick. This payload processed a new type of semiconductor organic material prepared by a Université de Moncton research team, led by Dr. Vo-Van Truong. Under microgravity conditions, more uniform thin films are expected to be formed; this improvement in uniformity is due to the absence of convection, a phenomenon which influences terrestrial processes.

On board STS-77, ACTORS flew as a Get Away Special (GAS) along with other GAS experiments from the United States, Germany, and China. GAS experiments are a low cost means of putting experiments on board the Shuttle because they are located in the cargo bay and require very little intervention from astronauts. Essentially the experiments fit in canisters to which no services are provided. GAS experiments are self-contained, fully automated, and supply their own power source, data collection, and processing.

NANOGAS (Nanocrystal Get Away Special) was a second GAS payload flown on STS-77, designed to grow high-quality crystals of an advanced new class of materials called *nanoporous crystalline semiconductors* whose atomic structures give them unique adsorption characteristics. Potential applications for this new class of materials include high-precision lasers, computers and other high performance electronic devices. NANOGAS processed 38 samples prepared by a University of Toronto research team, led by Geoffrey Ozin. By studying microgravity-produced crystals, scientists hope to eventually grow improved crystals on Earth.

Soret Coefficient in Crude Oil (SSCO)

An unusual recent mission for Canada began when a Soyuz U rocket lifted off from Site 1 in Baikonur Cosmodrome, on Sept. 14, 2007, at 11:00 GMT. About 9 minutes later the Foton-M3 spacecraft separated from the rocket's upper stage, entering a 262-304 kilometer orbit with an inclination of 63° from the Equator. The 6,535-kilogram Foton-M3 capsule, developed by TsSKB Progress in the city of Samara, Russia, carried 27 scientific payloads with a total mass of 688 kilograms, among them 16 experiments provided by customers outside Russia. As part of the 27 scientific payloads, the Foton capsule carried 43 ESA experiments in a range of scientific disciplines, including fluid physics, biology, crystal growth, radiation exposure and exobiology. The capsule landed in an uninhabited area about 93 miles south of Kustanay in Kazakhstan, close to the Russian border after 12 days in orbit and was successfully recovered. The CSA provided coverage of the satellite from Saskatoon and from CSA headquarters in Longueuil during the flight to monitor the progress of the experiments. The experiments were retrieved and are undergoing analysis, with science results expected 6-12 months after the mission.

Two of the experiments on board were Canadian, one being the Soret Coefficient in Crude Oil (SCCO). The SCCO experiment will make it easier to determine the quantity and quality of crude oil in a given well without even breaking ground thus exploring more efficient methods that might help reduce the environmental impact of oil prospecting.

This experiment is being used to validate mathematical models to improve oil-prospecting techniques and facilitate oil extraction. With his team of 14 students, Ziad Saghir, researcher and professor at Ryerson University in Toronto, developed the digital models. The SCCO experiment examines the influence of temperature gradients on oil wells. Saghir uses thermodiffusion (heat diffusion) to measure the separation that occurs when a mixture of hydrocarbons is submitted to various temperature gradients. This phenomenon is called the Soret effect. To successfully quantify the Soret effect – and produce efficient digital simulators – it helps to eliminate the influence of gravity. In microgravity, the Soret effect is the main factor in oil dispersion and thus can be observed with precision. Usually, gravity causes the heavy components in a fluid to fall and the lighter ones to rise. But thermodiffusion has a strange effect on the distribution of petroleum components in an oil deposit—the heavier ones tend to rise, while lighter ones fall to the bottom of the well. In summary, in order to interpret data taken from real oil wells, sophisticated models are needed and the information on which these are based come in part from space measurements.

SCCO is a partnership of private and public organizations. Newfoundland's C-CORE is one of the main project partners. The petroleum industry and the scientific community will be providing fluid samples for analysis. The electronic modules and control software required for the experiment were provided by the CSA.

Other types of microgravity experiment are those involving the life sciences, and these are described in the following section, including another Foton experiment.

Space Life Sciences

Much of the material in this section is taken from the CSA website, which is gratefully acknowledged.

Acquatic Research Facility (ARF)

The Canadian-designed and built Aquatic Research Facility (ARF) is a space laboratory facility that also flew on STS-77 and which allows scientists to study and better understand early birth defects, bone calcium loss and ocean ecology. ARF is a joint CSA/NASA project – Canada built the experimental equipment and NASA provided flight opportunities. Both agencies share the results of the experiments carried out on ARF.

The ARF facility itself, built by MPB Technologies of Montreal, is used for many experiments. On board the shuttle are six separate containers, each holding two miniature aquariums, allowing up to twelve experiments to be conducted simultaneously. To ensure that all conclusions are a result of microgravity, a second set of six identical containers is mounted on a rotating plate producing an acceleration that perfectly simulates the gravity conditions on Earth. By comparing the results of the experiments conducted at "0g" (space) with those conducted at "1g" (Earth), gravity effects can be identified. By comparing the results of space experiments with experiments conducted in exactly the same manner on Earth, other aspects of spaceflight (radiation, vibration, magnetic fields) can be studied.

The three ARF experiments selected for mission STS-77, two Canadian and one US, provided an integrated international investigation of early embryonic development, ocean ecology and bone calcium loss. These areas of investigation were selected due to the importance of future applications here on Earth. ARF, apart from being an ideal laboratory for aquatic experiments, is also equipped to capture video and microscopic recordings of animals both in microgravity and under simulated gravity conditions without disturbing the sample. The equipment is designed to allow all of these functions to be controlled from the ground at the CSA's Payload Telescience Operations Centre (PTOC) or by an astronaut. For the first flight, only astronaut control was used with scientists providing ground support from the control centre.

The small aquatic species such as those studied on ARF are a major part of the food chain and an important link in the ocean ecosystem, which converts 40% more sunlight and CO_2 (Carbon Dioxide) than the rain forests into living material. Knowing how these small yet ever so important animals distribute themselves in our coastal regions will enable us to make more precise predictions as to the life cycle of coastal species. The Principal Investigators have designed experiments to investigate the effects of gravity on aquatic species such as sea urchins, starfish and scallops.

Dr. Bruce Crawford from the University of British Columbia studied the development of starfish embryo through all stages, from birth to feeding, in the aquatic environment. He studied how these animals locate the surface of the water using gravity, as well as how the lack of gravity affects bone and calcified tissue development. Dr. Ron O'Dor from Dalhousie University studied loss of calcium in the tissue as well as the feeding patterns of giant scallop larvae. Knowledge of the calcification process will enhance understanding of bone loss, a genuine problem for lengthy space travel, and closely related to osteoporosis here on Earth.

e-OSTEO

Humans and animals lose bone calcium and mass during spaceflight. These changes resemble the effects of osteoporosis, however in space the entire process occurs over several months rather than over several years, as when aging. The initial changes start occurring within the first few days. This allows researchers to develop treatment and management strategies which could not be discovered on Earth. Research into the causes of bone loss in space may lead to a better understanding of osteoporosis and other diseases on Earth, and advance the search for countermeasures or possibly even cures. The e-OSTEO experiment is an enhanced version of OSTEO, a similar study performed in the shuttle in 1998. Astronauts, including veteran NASA astronaut John Glenn, carried out the experimental tasks during that mission.

While the shuttle program was suspended following the loss of Columbia and its crew, ESA and the CSA developed an idea to use a modified version of OSTEO in a Russian recoverable satellite called Foton M3. Known as eOSTEO (for "enhanced OSTEO"), the result is a fully automated version of its predecessor. This mini science lab was successfully launched on September 14, 2007 on a Foton M3 satellite from the Russian Baikonur Cosmodrome as described earlier. eOSTEO was part of 43 life and physical sciences experiments sponsored by ESA that orbited Earth for 12 days. There were two versions of eOSTEO on board: one Canadian and one known as eRISTO from the European Space Agency.

Canadian Protein Crystallization Experiment (CAPE)

With support from the CSA's Microgravity Sciences Program, scientists from seven cities across Canada prepared biotechnology flight experiments that could help lead to advanced treatment of and possible cures for various debilitating diseases, as well as bacterial and viral infections. In the microgravity environment of space, it is possible to grow larger, more perfect protein crystals through the CAPE (Canadian Protein Crystallization Experiment) research opportunity through a launch to the Russian Space Station Mir in 1997.

Protein crystallization is a process used to determine the structure of proteins. When proteins in high concentration are put in a liquid solution, they usually form small regular structures called crystallites or nuclei and it is from these crystallites that a single crystal protein will grow. The protein crystals are subsequently analyzed by X-ray diffraction to determine their structure, but this is only possible when the crystals are large and pure. Because protein crystals are fragile, it is difficult to grow adequately large or perfect protein crystals in earth-based laboratories. However, in the microgravity environment of space, there are no gravity-induced effects such as sedimentation and convection to disrupt the growth of these fragile protein crystals and therefore, the chance of growing larger, more perfect crystals is greatly improved.

The Protein Crystal Growth Mission Support Centre has been operational since the fall of 2000 at the John H. Chapman Space Centre in Longueuil, Quebec. It is equipped for basic manipulation of protein samples. During the course of the CAPE mission, the centre served 15 research groups from the Canadian science community in protein crystallization experiments. Its facilities were used to prepare for the ill-fated STS-107 mission of the Space Shuttle *Columbia*, which broke up on re-entry on February 1, 2003. Five groups of Canadian researchers participated, and the knowledge acquired will help prepare the centre for future missions. However, this program has now been terminated.

Perceptual-Motor Deficits in Space (PMDIS)

This experiment is intended to understand whether and why astronauts have reduced hand-eye coordination in space. This is an extremely important aspect of working in space, such as in docking manoeuvres and in the assembly work on the ISS, including Dave Williams' own three EVAs, a record for Canadian astronauts. In the experiment Williams and other astronauts were asked to mark targets on a computer screen, both by tapping on the screen and by using a joystick, the level of difficulty varying with the size of the targets. To involve multi-tasking, Williams was asked to push a button in response to a musical tone while marking the targets. These tests were performed both while floating, and when attached to a support. The outcome is expected to shed light on whether the brain fails to adapt fully to the weightlessness of space, or the mechanical difficulty in pressing a button while floating in space, or by distractions from the working environment. Once this is understood, measures can be designed to counteract the problem.

It is an important issue as manoeuvring a satellite worth tens or hundreds of millions of dollars with the shuttle's Canadarm, or handling a truss weighing close to 20 tons with the Space Station's Canadarm2, are very delicate operations requiring very precise hand-eye coordination. The results of the PMDIS experiment will help determine if the loss in hand-eye coordination is due to one or a combination of the following:

*the brain not adapting fully to the weightlessness of space;
*the difficulty of performing mechanical tasks like pressing a button while floating in space; or
*distractions caused by flashing lights, floating objects, co-workers and some very spectacular scenery.

Barry Fowler of the School of Kinesiology and Health Sciences at York University is the leader of this experiment and what follows is his story of how the fundamental problem was developed and executed, beginning with the earlier Neurolab mission.

"I became involved with the Neurolab mission through Otmar Bock, who at that time was associated with the Center for Vision Research at York University. He had designed an experiment, sponsored by the CSA, to look at eye-hand coordination, since it was generally believed that this function was degraded under micro gravity. The thought was that this degradation occurred on initial exposure and then adaptation took place over days or perhaps even a week or so. This seemed reasonable since it was known that astronauts suffered from perceptual illusions due to disturbances of the vestibular system and these disturbances could very well be implicated in these putative eye-hand coordination problems. Reinforcing this belief were some earlier experiments by Doug Watt, suggesting problems with pointing at targets in space.

"When Otmar received an appointment to a Professorship in Germany and needed to move there at short notice, the CSA wanted to keep a Canadian connection to Neurolab (the costs were being shared between the CSA and the German Space Agency) and I was invited to take part. I designed an experiment that would complement Otmar's experiments by looking at the problem from a 'workload' perspective in the belief that this might be a more sensitive approach to uncovering eye-hand coordination problems (Otmar was coming at it from a purely perceptual point of view).

"In the meantime, other Germans (Dietrich Manzey and colleagues) reported further evidence for eye-hand coordination problems with a tracking task using two Russian cosmonauts on board Mir.

"Neurolab was a wonderful chance for life science experimenters because the mission was dedicated to science and data could be obtained from all six crew members – a far cry from the one or two usually available under less than ideal conditions. Imagine my surprise when our experiment showed very little evidence for any eye-hand coordination problems. Since this experiment was quite clean, I questioned the whole notion of a eye-hand coordination problem (Fowler, 2000) and resolved to try and settle the question with a definitive experiment which I proposed to the CSA in 1999 and which was funded in 2000. This experiment sought to discriminate between three hypotheses that could account for the decrements in performance that have been observed in previous experiments – this experiment is groundbreaking in the sense that previous experiments were largely data gathering exercises with post hoc explanations for the results offered after the event.

"The experiment will have taken 8 years to complete (the disintegration of Columbia held things up for 2 years). It has involved an enormous amount of time at JSC, since the subjects had to be trained for 5 months prior to the mission. I am currently waiting to go to JSC one last time to obtain post-flight control data from one of the crew of the ISS who is returning after 5 months on the current mission (STS-120). With data from 5 subjects I should be able to settle the question once and for all (neither Dietrich Manzey or Otmar Bock are persuaded that the deficits observed in previous experiments have a simple if embarrassing explanation - failure to control body movement in the micro-gravity environment – something that occurred to me due to my background in diving research, where this problem has long been acknowledged). Unfortunately, I don't know the outcome of PMDIS yet and we will not be finished analyzing the data until the end of winter. The five subjects came from crew members of STS-117, 118 and 120 and include the Canadian astronaut Dave Williams. I ran a subject on an earlier mission, STS-116, but due to teething problems with the apparatus, the data were unusable."

RADARSAT

At about the time that ESA began exploring remote sensing projects in the 1970s, Canadian officials began looking for ways to exploit Arctic resources and one of those involved a surveillance radar. When oil prices dropped in the 1980s, the project seemed to lose its relevance, but officials determined that a Synthetic Aperture Radar (SAR) could be used for more than strictly oil exploration. After study of the movement of sea ice fell out of favour there were many other needs to be satisfied including disaster management, interferometry, agriculture, cartography, hydrology, forestry, oceanography, ice studies and coastal monitoring. Overall, the monitoring of global climate change and Earth resources were the primary objectives, but many scientific and commercial needs were to be met as well.

In 1979 when it became evident that NASA was not going to launch SEASAT-2, saying that "they did not know enough about how to interpret the data and wanted to do more research on this before launching the next radar satellite", the Working Group on Satellites and Ground Station Engineering recommended that Canada now had enough knowledge about SAR radar to be able to make its own radar satellite, and that we should do so. This recommendation was taken up to the Interagency Committee on Remote Sensing which, to Larry Morley's surprise (the above is his quote), approved it and sent it on to Cabinet.

A special RADARSAT Project Office was set up in 1980. Ed Shaw took over as manager, Keith Raney as chief scientist and Bob Warren from DOC was in charge of spacecraft engineering. The project office was supported by secondees from DOC, Environment Canada and several other agencies. Spar Aerospace was selected as prime contractor supported by MDA and COM DEV, who were responsible for the SAR sensor.

The UK had promised to supply the spacecraft bus and later withdrew. After arrangements were made for another bus, the Government decided the whole project was too expensive. Phil Lapp led a group from the contractors who made a proposal for a 'stripped-down' version that finally got approval. This satellite rose to the top of the priority list for the CSA, which became responsible for the space segment and CCRS for the ground segment. As mentioned earlier, Mac Evans moved Joe McNally (remember him, from Alouette/ISIS, Hermes and MSAT) to the CSA to become RADARSAT Program Manager. It took a brave person to lead a $620 million dollar project. McNally recalled recently that "I was looking at a TV program not long ago on the Queen Mary II. I realized that for another $200 Million I could have Queen Mary III." McNally retired in May, 1998 with the title of Vice-President Engineering. He contributed much to the Canadian space program, one (of a number) of individuals who were willing to take responsibility, and had the capacity to carry a project through to a successful conclusion.

RADARSAT International, a consortium of Spar Aerospace, MDA and COM DEV won the bid to handle the marketing and sales of the RADARSAT data. Later they also took over marketing for all remote sensing satellite data in Canada and made agreements with the two other international distributors of remote sensing data, SPOTIMAGE of France and EOSAT of the U.S to market their data in Canada.

The first work began as early as 1980 and a number of studies and measurements were conducted before the parameters of the mission were determined. There were a number of similarities to the Alouette/ISIS topside sounder in that radio wave pulses were to be transmitted from the spacecraft and reflections received, not from the ionosphere, but the Earth's surface. A large antenna would be required but not for the same reason. For Alouette/ISIS the antennas were long because the transmitted and received wavelengths were long, while for RADARSAT-1 much higher frequencies were to be used, 5.3 GHz compared to around 5 MHz, that is about 1000 times greater. At these frequencies the radio wave penetrates the ionosphere with no effect so that the reflection takes place with the Earth's surface, but it is still an electrical interaction, depending on the conductivity of the surface. In fact the surface is not simple, but in the Arctic involves a layer of snow lying upon a thickness of ice. The conductivity of the snow and ice is modified by its water content, so that one can distinguish between freezing ice in the fall and melting snow in the spring. Water itself has yet a different reflection characteristic.

The RADARSAT-1 antennas had to be large because they were to form an image; just as with optical telescopes, larger telescopes (in comparison with the operative wavelength) give images of higher resolution. Similarly a large antenna compared with the radio wavelength was needed to obtain the desired resolution, in this case, 30 meters long in order to obtain a resolution of 10 meters on the ground. However a fan beam could be used since the motion of the spacecraft would carry the beam forward along the track (as with the optical imagers on ISIS II), so the other dimension for the antenna needed to be only 1.5 meters. In order to image, the beam would be transmitted downward and to one side of the spacecraft with incident angles ranging from 20° to 49°; the incident angle is the angle between the direction of the ray striking the Earth and the vertical direction at the incident point. The horizontal distance between these points is about 500 km, called the "swath width" because as the spacecraft moved along in its orbit an image was formed of this width and a length extending around the entire orbit. Because radio waves of this frequency also penetrate cloud, and sunlight has no influence whatever, the images are obtained both day and night in all kinds of weather, a tremendous advantage. In fact the antenna can be operated in different modes, with different resolutions and swath widths. The swath widths can be from 45 to 500 km, with resolutions of 8 m to 100 m.

Doppler information can also be obtained from the signal – this involves the frequency shift of the returned beam compared with that of the transmitted beam resulting from the velocity component of the spacecraft, depending on whether the reflection came from "ahead" of the spacecraft position along the track compared with the "behind" signal. The fundamental image, obtained from the basic data is a two-dimensional one, but the additional Doppler information allows the third dimension to be obtained, resulting in altitude information.

RADARSAT-1 image of the Montreal area, with the gray scale indicating the altitude. Mount Royal is clearly evident as a grey spot one-third of the way up the image, and about a quarter of the way in from the right, overlooking the bay in the St. Lawrence River that forms the Montreal Harbour. Higher (darker) elevations are seen to the west and to the north. Photo by courtesy of the Canadian Space Agency.

The concept for RADARSAT-1 was complex in the administrative/operational sense as well as from the technical standpoint. There are five distinct user groups and each of these has its own "order desk" that is coordinated through the integrated order desk. The first of these is dedicated to the Canadian Ice Service, the second to the Canadian Government Order Desk that involves all government needs other than the Ice Service, the third to the Alaska SAR facility to meet US government needs (NASA was a partner, and launched the spacecraft), the fourth was for commercial users and was handled through a corporation set up for that purpose, RADARSAT International (RSI). The fifth desk was for mission control, for internal usage. Each day, mission planners in Satellite Operations at the CSA prioritize client orders that come in from these desks and develop plans for data acquisition. From these, the planners generate command files and transmit them to the satellite's mission control facility at CSA in Saint Hubert, from where they are sent up to the spacecraft. A second antenna for receiving and transmitting data is available at Prince Albert, Saskatchewan, that dates back to the "exponential" sixties.

Although there were some perturbations associated with the establishment of the CSA at Saint Hubert, the Canadian space community in terms of industrial partners and scientists at universities and government are spread across the country and there was no loss of continuity there. The spacecraft was built by Spar Aerospace while payload subsystems were built by COM DEV. The ground receiving stations and processing facilities were built by MDA so that approximately 100 Canadian and international organizations were involved in the design and construction of the space and ground segments. The total cost of the project, excluding launch, has been estimated at $620 million, with the federal government contributing about $500 million. The four participating provinces (Quebec, Ontario, Saskatchewan and British Columbia) contributed about $57 million, while other provinces contributed by pre-purchasing data (all Canadian provinces participated in planning the RADARSAT program). The private sector contributed about $63 million.

On November 4, 1995, RADARSAT-1 was successfully launched and deployed on its orbit. It had a launch mass of 2750 kg, and derives 2.5 KW from the solar arrays. The transmitter peak power is 5 KW during a pulse, and averages 300 W allowing for the time interval between pulses. The design lifetime was 5 years, but is still operating 12 years later. It covers the Arctic daily, and most of Canada every 72 hours. Global coverage occurs every 24 days. The launch took place at the Vandenberg Air Force Base, California. NASA and NOAA provided the launch and launch vehicle, a Delta II rocket, in exchange for RADARSAT data.

RADARSAT-1 has been used for a wide variety of applications: for example, providing raw geological data to mining companies in tropical areas where weather is often a factor for remote sensing, and monitoring floods and other natural disasters in the US. SAR, which is very sensitive to water, is also used to map ice floes in the St. Lawrence waterway during the winter and in the Arctic during the summer. The Canadian government's Ice Services in Ottawa uses SAR data to determine whether these waterways are navigable and what types of ice are present. Updates are issued to vessels within hours of receiving the satellite data.

Although RADARSAT-2 was to be a repeat performance, it had new challenges, as it was designed to measure with greater spatial resolution, 3 meters rather than 10 meters; that is, the ability to see a car, instead of a house. This was not acceptable to the US which was not inclined to support any country in taking images over its territory with this resolution. As a result they withdrew their support, including the launch and its costs. Approaches were made to European agencies, and although partnerships were discussed, in the end these did not materialize and the CSA had to absorb the cost of the launch, which took place on December 14, 2007 from the Baikonur Cosmodrome on a Soyuz vehicle.

Because RADARSAT-2 was developed much later, it will benefit from the many technological advances and concepts that took place in the meantime. RADARSAT-1 was designed to look to one side, the side that favoured Arctic viewing. However, the US had a strong desire to map the Antarctic which was on the other side. In order to accommodate this request, RADARSAT-1 was reversed in its orbit for the required observing period, a tricky maneuver that turned out well. But RADARSAT-2 will be able to look on both sides, allowing coverage of the Antarctic on a regular basis. It also will be able to employ more than one polarization. Looking along the radio beam from the spacecraft, the plane containing the oscillation of the wave in the electric field can be vertical, or horizontal, or a combination of the two. The reflectance characteristics of water, ice and snow are different for different polarizations, so measurements with different polarizations allow more information about the terrain to be obtained.

The data rates for imaging data from RADARSAT-1 were very high and whenever the spacecraft was not within range of a ground station the data were recorded on an analog tape recorder, which has a limited capacity. Fortunately because of world-wide interest, about 30 ground stations were available around the globe. For RADARSAT-2, solid state memory of higher capacity and superior reliability will be used. In order to reference the acquired images against the correct Earth locations it is necessary to know the spacecraft location accurately at all times. In the early days of space research this was done with a ground-based radar tracking network while for RADARSAT-2 the new highly accurate GPS technology will be used.

The International Space Station (ISS)

President Ronald Reagan's decision to build a space station was based in part on the lead that the USSR was building up in human presence in space, but the name he gave it, "Space Station Freedom", had a political connotation as well. The European Space Agency, Japan and Canada became early partners; Canada joined in 1985. But with the loss of Challenger, and rising costs, this approach seemed destined for cancellation. During Bill Clinton's presidency, Russia became a major partner and the project became the "International Space Station" (ISS), with a rather different outlook.

Canada's ISS participation proved a challenge, particularly after the fiery and devastating end of Columbia in the high atmosphere on February 1, 2003. This led to a gap of three years in the program. The rationale of Canada's participation in ISS is that the contribution of the Mobile Servicing System (MSS) was its ticket to the utilization of this impressive space laboratory. But to use the space involved requires considerable expenditure. It is extremely difficult to design an experimental user program when the future isn't known with certainty. Is it worth the investment? Until the ISS reaches stable operation, sometime in 2010, one can't be sure. In 2005, NASA designated the ISS as a "National Laboratory", and invited other government agencies to participate in the future science and applications activities. The ISS is planned to be operated until 2016, although this could be extended. The end of the shuttle flights in 2010 means that the Russian Soyuz will serve as transporter for the operational years of ISS. Following the flight of Dave Williams in August, 2007, Canada played one further role when DEXTRE (the dexterous manipulator, with two hands) was installed on ISS during the STS-123 mission launched March 11, 2008. This was the last Canadian space robotics component to be delivered to the ISS.

On February 11, 2008 it was announced that Julie Payette will be one of the crew on the Space Shuttle Endeavour for mission STS-127, with launch planned for April 2009. During mission STS-127, the crew will deliver the last elements of the Kibo laboratory contributed by the Japanese space agency, JAXA, to the International Space Station. Julie Payette will serve as a mission specialist. At the same time it was stated that in May 2009, Dr. Robert Thirsk will be taking part in Expedition 19, becoming the first Canadian to remain on the International Space Station for a long stay, living there for 4 to 6 months. He will also be a member of the crew for the transition to a team of six astronauts living together continuously aboard the Station. Since 2000, the Station has been inhabited permanently, and the astronauts and cosmonauts occupying it change upon the arrival of crew replacements. Robert Thirsk will fly to the ISS aboard a Soyuz craft that will be launched from Baikonour, in Kazahkstan. This information was provided in a news bulletin from the CSA.

For the future, NASA is developing two new rockets, the Ares I and the Ares V, and two Orion capsules for carrying passengers. The Ares I will carry a six-person Orion module to the ISS, although it is not scheduled for operations until 2014. The larger Ares V is intended to carry a four-person Orion module to the moon, by 2020.

WINDII on UARS

The first science mission launched under CSA auspices was the Wind Imaging Interferometer (WINDII) on NASA's Upper Atmosphere Research Satellite (UARS) on September 12, 1991. Of course this too was inherited from the NRC period. The beginnings of WINDII were described in the previous chapter.

The WINDII story goes back to the early days of space science in Saskatoon, and the Wide Angle Michelson Interferometer (WAMI) concept discovered by Gordon Shepherd while visiting at the Laboratoire de Bellevue near Paris in 1961. Ové Harang was building in Pierre Connes' laboratory an instrument to measure the spectrum of the aurora back in native Norway and he explained Connes' concept. It was then implemented in a different configuration by Ronald Hilliard for his Ph.D. thesis at the University of Saskatchewan, for the measurement of upper atmospheric temperature. The fundamental instrument was the Michelson interferometer, invented by Michelson at the turn of the century in order to measure the velocity of the Earth through the "ether" that was thought to pervade the universe. The fact that Michelson found this velocity to be zero, no matter which direction the Earth was moving in its orbit around the sun meant that no such fixed frame of reference existed, a wonderful confirmation of Albert Einstein's Special Theory of Relativity, with its remarkable consequences, although Einstein became aware of this result only after his theory had been completed. Michelson's instrument turned out to have many other remarkable capabilities and various versions of it have been used by Canadian space scientists, as described here, and elsewhere.

The NASA Upper Atmosphere Research Satellite being held above the space shuttle on the Canadarm, before being released into orbit, September 15, 1991. There are ten instruments on board, but the WINDII instrument is not visible in this photo. Photo by courtesy of NASA.

Michelson's interferometer splits the incoming beam of light into two parts that travel different paths by bouncing off two different mirrors before combining again at the exit. This allows two different points on the light wave to be compared. One point acts as a reference and by moving one mirror to change the length of the other path the pattern of the combined wave form is traced out. Since the emitting atoms are carried with the atmosphere, and thus with the wind, wavelength shifts produced by the Doppler effect through the motion of the atmosphere could be interpreted as a wind. To do this the mirror positions have to be extraordinarily stable, as otherwise mirror movement caused by thermal expansion or other reasons would be misinterpreted as a wind. The stability problem was solved only when sufficient funds were provided as part of the WAMDII initial study (one of the shuttle instruments described in the previous chapter). The key was a genius in working with glass, named Jeff Wimperis, who had polished the mirrors for the first gas laser constructed in the USA (a forerunner of the laser pointers used so commonly today), and later moved to Canada where he founded his own company, Interoptics. The solution developed was to make the Michelson interferometer completely solid. The two different light beams travel in different glass blocks, silvered on the outside to form the mirrors. To trace out one wave profile, one mirror need only be moved one-half wavelength, amounting to one wavelength for a round trip of the beam. This was accomplished by mounting the mirror on piezoelectric pillars, which change their length slightly when a voltage is applied to them. Since the piezoelectric pillars are cemented to the blocks as well, the system is solid. But this solid interferometer must be manufactured to an accuracy of about a tenth of a wavelength of light, and Wimperis had to devise methods of polishing the individual blocks to this accuracy, and then maintaining this accuracy in the assembled instrument. He did this by using ultraviolet curing cements. The interferometer was assembled, with the cement in place, and then adjusted until it was aligned with interferometric accuracy; the ultraviolet light was turned on and zap!, there was a solid interferometer. This approach made the instrument stable, but also robust as required for the launch into orbit. The piezoelectric pillars allowed for slight changes in shape during the launch, as they could be adjusted once in orbit.

Underlying this novel approach to a solid interferometer was the new "wrinkle" of

WAMI, the "wide-angle" effect. In Michelson's original interferometer if the angle of the light passing through the interferometer is changed, the path difference (the distance between the two points on the wave) also changes. Previous experimenters dealt with this by restricting the light to be close to the axis, but then the field of view is very restricted. In WAMI, air was used in one path and glass in the other; in the glass path a "swimming pool" effect occurs, in which the bottom of the pool appears to be closer than it actually is. For WAMI the glass path is made to appear to have the same length as the air path, even though they are actually different and that allows a wide angle. To make WINDII solid, two different blocks of glass, with different refractive indices were used, in order to achieve the solid design already described. Also WINDII had to work at several different wavelengths, without a readjustment of the mirror position. Two blocks of glass allow an achromatic system, just as with achromatic lenses.

The wide angle effect was crucial because WINDII was to be an imager; a CCD camera was placed behind the interferometer so it had to operate over a wide range of angles. Each pixel would record its own value of the wind, so the wave had to be separately traced for each. To reduce the amount of data to be telemetered to ground, this had to be done efficiently, and so the pattern was sampled at just four positions. An image was taken and transmitted, the piezoelectrics stepped, another image taken and so on. The resulting four images were combined in a computer on the ground to create images of the wind. The WINDII pictures were taken at the Earth's limb, just above its surface, with the bottom of the image about 70 km above the Earth's surface and the top at about 300 km. Other instruments on the UARS satellite did not image in this fashion, but used pointing mirrors to step up and down in altitude. Eliminating this mechanical motion and taking the whole profile simultaneously was considered a strength of the WINDII concept.

The WINDII instrument hanging down from the UARS spacecraft, at the left side of the picture. The inverted "V" structure whose apex is practically touching the person on the left is the baffle; Charlene Lau of AIT is the person on the right.

But winds have direction as well as speed, so to determine the direction, two wind measurements had to be made, at right angles to one another. Thus WINDII had to have two fields of view. Two images were taken simultaneously through two input telescopes, but the two images were combined into one CCD image. Since the volume of atmosphere viewed by one field was seen only eight minutes later by the other field of view, the combined images were those that were taken about eight minutes apart. These then were resolved into meridional (north-south) and zonal (east-west) winds.

WINDII had to be large enough to collect enough light to measure winds from the weak night airglow that has been described earlier. There is also a day airglow (dayglow) that cannot be observed from the ground because of the bright (blue) daytime scattered sunlight. WINDII in principle could also observe this dayglow, but there is a problem in that the bottom of the field of view is at 70 km altitude, and just below that there is normally an extremely bright cloud layer at about 10 km, that is only one degree in angle outside the field of view. This bright light would be easily scattered off specks of dust on the lenses, or bounced off the walls of the lens holders and get into the field of view. The solution is a "baffle" to screen out the unwanted light. The meter-long baffle was designed and constructed by Matra in France, as part of the French contribution, and it worked extremely well. The WINDII instrument is described in detail by Shepherd (2002).

The importance of measuring winds day and night is that they could be measured over 24 hours of local time per day, which is what is needed to observe atmospheric tides. These are not like ocean tides, which are driven by the gravitational pull of the moon, but rather are solar tides driven by the heating of solar radiation. Two kinds of tides were strongly apparent in the wind measurements: diurnal tides with a period of 24 hours, which were dominant near the equator, and semi-diurnal tides with a period of 12 hours that became dominant at higher latitudes.

The project was guided by a Science Team whose members spanned the range of expertise required for this experiment. Leroy Cogger (University of Calgary), a Saskatchewan graduate, had worked on the ISIS II data with Cliff Anger and had a strong interest in the F-region of the atmosphere. Dick Gattinger (of NRC), another Saskatchewan graduate had a deep understanding of the photochemistry of OH. Bob Lowe (University of Western Ontario) was involved in Canada's first (CARDE) rocket measurement of the altitude of the OH emission and had a special interest in determining temperatures from the OH emission. Ted Llewellyn (University of Saskatchewan) had long studied molecular oxygen emissions, and so was particularly interested in the O_2 Atmospheric band. Wayne Evans (Trent University) was another Saskatchewan graduate who had a particular interest in observing the noctilucent clouds mentioned in Chapter 4. Although WINDII was not designed to observe this phenomenon it did so extremely well, observing the clouds in full daylight because of the high quality baffle. Three more Saskatchewan graduates were employee team members, working full-time on the project. Bill Gault was the Instrument Scientist, the expert on the optical system, its calibration and its performance in orbit. Brian Solheim was the data processing and archival expert who supervised the instrument operations and data processing system after the launch, while Rudy Wiens coordinated the daily science objectives. Charlie Hersom, who had supervised the WINDII characterization prior to launch became Chief Operator after the launch.Colin Hines was not an official team member, but acted as an advisor on atmospheric dynamics and attended many team meetings. Others who made major contributions to the data analysis after the launch were Charles McLandress (now at the University of Toronto) who first displayed the tides, William Ward (now of the University of New Brunswick) who studied among other things the unusual "two-day" wave, Ding Yi Wang (now of the University of New Brunswick) who studied planetary waves, Marianna Shepherd (York University) who derived temperatures from the scattering of sunlight in the daytime atmosphere (something else the instrument was not designed to do) and Shengpan Zhang (York University) who studied the influence of tides on the emission rates, the dayglow and daytime aurora. This list does not include the many students who obtained their degrees using WINDII data. The French team was led by Gérard Thuillier and included Marie-Louise Duboin, Alain Hauchecorne, Michel Hersé, Chantal Lathuillère and Francois Vial, ably supported by Patrick Charlot of the Centre National d'Etudes Spatiales (CNES), the French space agency.

Other Science Missions under the CSA

Observations of Electric-field Distributions in the Ionospheric Plasma: a Unique Strategy (OEDIPUS)

OEDIPUS was a sequence of rocket flights led originally by Brian Whalen and later by Gordon James of CRC. The story here is related by Gordon James.

"Direct measurements of magnetic-field-aligned electric fields in the magnetospheric plasma, presumed to be responsible for auroral electron energization, using standard electric field double probes had proven to be a difficult task perhaps due to the limited sensitivity of the technique. To overcome this limitation the NRC space plasmas group devised a scheme to vastly increase the sensitivity of the technique by increasing the baseline for the measurements (double probe separation) by several orders of magnitude using a space tether approach. A payload and sub-payload (each with carefully prepared exterior surfaces) configuration was used as plasma contacts (probes) and a very long tether (a thin insulated wire) connecting the two monitored the inter-probe potential difference. Brian Whalen was PI for the flight. Each payload contained thermal ion detectors to monitor the local plasma (payload potential) as well as a battery of energetic particle detectors. For a prototype of active radio-wave experiments to be undertaken on the NASA Space Shuttle, Gordon James (CRC) provided a high-frequency exciter, HEX, and a receiver for the exciter, REX. HEX and REX exploited the unique configuration for new two-point plasma wave propagation investigations. Don Wallis (NRC) provided the magnetic field observations, Al McNamara (NRC) Langmuir Probes, and Paul Kintner (Cornell U.) a dc-VLF electric field detector. Fokke Creutzberg contributed ground-based optical monitoring.

"The OEDIPUS-A payload was developed by Bristol Aerospace and launched into an active auroral display on January 30, 1989 from the Andoya Rocket Range in Norway. After reaching a few hundred km altitude the forward sub-payload was deployed along the magnetic field and reached a 1 km separation from the aft subpayload. The probe data showed how the tethered payload functioned as a large double electrostatic probe. Synchronized sweeps of the frequency range 0-5 MHz by the 2-W HEX on the upper end of the tether and its associated REX on the lower end produced signatures of quasi-electrostatic waves guided along field-aligned depletions of ambient density and of electromagnetic 'sheath' waves guided along the tether.

"OEDIPUS-A did not make its targeted parallel electric-field measurement because of the complicated spin state of the subpayloads. In the subsequent flight OEDIPUS C, space mechanics experts led by Frank Vigneron (CRC) joined the science team to design a double payload that deploys stably, and to provide instrumentation for observations of the tethered payload mechanics. Gordon James assumed responsibility as Project Scientist, Don Wallis provided fluxgate magnetometers, Langmuir probes and the Tether Current Monitor, David

Knudsen electron and ion spectrometers, and David Hardy (AFRL) energetic electron detectors. This OEDIPUS-C payload, also developed by Bristol Aerospace, was launched from Poker Flat Research Range on November 7, 1995, reached 800 km, and encountered a tenuous but active auroral plasma. This time, the 1-km tether and its two spinning subpayloads were successfully deployed, as seen in real time via a television camera mounted in the aft subpayload. Two-point density and plasma potential measurements revealed deep (>50%), narrow (20 m diameter) cylindrical density depletions aligned with the geomagnetic field. These data suggest that the depletions resulted from the expulsion of ions perpendicular to the magnetic field by intense confined plasma waves. The HEX-REX synchronized pair hosted several novel radio-science findings, in such areas as plane, ducted and tether-guided electromagnetic waves, sounder-accelerated electrons and optical and radio emissions that they excite, plus counter-intuitive properties of antennas in plasma enhancing their effectiveness. Thus, the disappointment in not reaching the goal that inspired its acronym has been mitigated by the scientific yield in other areas of OEDIPUS C, arguably the most productive sounding rocket mission led by Canadians. In 2008, analysis of data continues to extend the program bibliography of OEDIPUS C, and to show how the exercising of Canadian niche specialties can be fruitful in geospace science."

OEDIPUS C, a rocket payload launched on 7 November 1995 from Poker Flat Research Range in Alaska for the study of the electrodynamics of the auroral ionosphere. This artist's rendition shows the upward travelling payload one second after the separation of its two payload parts, beginning the deployment of a connecting tether to a final length of 1200 meters. Drawing by courtesy of Bristol Aerospace Limited provided with the assistance of Gordon James of CRC.

Brian Whalen's story continues again here.

Nozomi Thermal Plasma Analyser (TPA)

"The success of the Freja CPA and the good working relationship established during the Akebono SMS program prompted ISAS Japan to invite the NRC Space Plasmas team to provide an instrument, a variation of the CPA design, to fly on the Planet-B Mars Orbiter mission. The TPA (Thermal Plasma Analyser) proposal with Brian Whalen (PI) and collaborators from U. of Saskatchewan, U of Edmonton, York University and Western Ontario was accepted by the CSA and ISAS.

"The scientific objective of Planet-B (named Nozomi after launch) was to investigate the near-Mars plasma environment. Unlike Earth, Mars has no strong intrinsic magnetic field, so the solar wind impacts directly on the upper atmosphere of Mars and produces a number of unique plasma interactions and neutral atmospheric loss processes. Since the spacecraft would also spend a significant time in the unperturbed Solar Wind, studies of solar wind characteristics were also to be undertaken during the mission.

"The TPA design was similar to the CPA (on Freja) but included a time-of-flight capability to provide modest mass composition measurements of the Martian ionospheric ions. Images of ions and electrons in the thermal and suprathermal energy ranges (1 – 200 eV) over half a spacecraft spin would be combined to obtain the 3-dimensional velocity distribution."

The Japanese Nozomi spacecraft that travelled to Mars. Canada provided the Thermal Plasma Analyser (TPA) which is the white object at the left end of the boom that is folded against the left face, of the two spacecraft faces shown. The hinge is at the right hand end of the boom. Photo by courtesy of ISAS/JAXA.

"A number of theoretical studies were conducted at the University of Western Ontario to advance the knowledge of the science team on the Martian environment in preparation for the expected observational stage of the mission. These studies also assisted in the initial design and development of the instrument which was nearing completion by early 1995."

Termination of Space Plasmas studies at NRC

In the late 1980s the NRC administration began withdrawing support for long-term curiosity-oriented research. As mentioned previously, in 1993 the NRC decided that space research activities at the HIA would be terminated within three years and that the Canadian Space Agency would immediately assume responsibility for the future of these activities including those being conducted by the Space Plasma group. With the support of the space plasmas team, Whalen initiated informal discussions with The University of Victoria and the

University of Calgary about the prospects of relocating the Space Plasmas group activities to their institutions. The University of Victoria indicated interest and undertook initial discussions.

Simultaneously discussions also ensued between members of the Aeronomy and Space Physics Division of the Canadian Association of Physicists and the CSA about the future. One proposal discussed involved the section activities being transferred to the CSA. This proposal was opposed by a number of scientists including Brian Whalen, who maintained that the climate for productive research was now only to be found at Universities. In the fall of 1994 Whalen accepted a position as Adjunct Professor at the University of Victoria.

In early 1995 Whalen retired from the NRC, and the CSA then appointed Andrew Yau as PI for the Nozomi TPA. In the summer of 1995 some remaining members of the Space Plasmas group, with the support of the Canadian Space Agency, moved to the University of Calgary where two (Andrew Yau and David Knudsen) found permanent positions.

Continuation of the Space Plasmas Studies at the University of Calgary

With this change of location, the space plasma story is continued by Andrew Yau, at the University of Calgary.

Nozomi Thermal Plasma Analyser (TPA)

"*In July 1998, the Nozomi spacecraft (formerly Planet-B) was launched. However, after a minor failure in the propulsion system and a later encounter with an intense solar flare event, the Nozomi spacecraft was unable to achieve Mars orbit. The spacecraft passed Mars at about 1000 km distance and entered orbit around the Sun and was lost as a scientific satellite.*

Suprathermal Ion Imager (SII)

"*The instrument innovation, development and observational program which characterized the NRC Space Plasmas group activities has continued at the University of Calgary under the direction of David Knudsen . The SII instrument, which stems from the Freja CPA ion focusing design, utilizes a phosphor screen followed by a CCD to image ions and replaces the earlier charge amplifier based sensor of the CPA and TPA instruments. This approach provides a much improved 3,000 pixel resolution of images formed by the CPA focusing scheme. This instrument has been flown on five sounding rockets and has successfully measured the ionospheric electric fields using a precision ion drift technique pioneered by the Space Plasmas group. SII is currently being developed by COM DEV Ltd. to fly on all three of the ESA SWARM satellites, intended to provide the best ever survey of the Earth's geomagnetic field and its temporal evolution, improving our knowledge of the Earth's interior. The spacecraft are scheduled for launch in 2010.*

Imaging and Rapid scanning ion Mass spectrometer (IRM)

"*Development of the TPA paved the way for the TSA (Thermal and Suprathermal Analyser) led by Andrew Yau, which was flown on the SS520-2 rocket (also called GEODES-IC) in February, 2000. Subsequently, development of the IRM, a new toroidal deflection sys-*

tem which enables 3-D ion sampling from a non-spinning platform was initiated. A new time-of-flight gate will provide for improved mass resolution and the discrete anode imaging electronics has being improved to eliminate cross-talk present in the original TPA design.

The IRM (P. V. Amerl PI) and an electron version of SII (David Knudsen PI) will be flown on the CASSIOPE Enhanced Polar Outflow Probe (e-POP) satellite (Andrew Yau PS) which is dedicated to space weather research and will be the first satellite under the CSA's Canadian Small Satellite Bus Program."

With the design of the small hybrid satellite CASSIOPE, scheduled for launch in 2008, the CSA is pioneering in a new direction. The satellite will include the telecommunications instrument Cascade, implemented by MDA, which will provide the very first digital broadband courier service for commercial use, and the scientific payload ePOP (Enhanced Polar Outflow Probe), which will be used to study the ionosphere and is being prepared at the University of Calgary. The eight scientific instruments, one of which is an imager, will collect new data on space storms in the upper atmosphere and their effects on radio communications, GPS navigation, and other space-based technologies. Geospace storms generate huge electrical currents in the upper atmosphere's polar regions, and also produce enhanced displays of the aurora.

Rocket measurements of the Cosmic Background Radiation

As described earlier, Herbert Gush was an M.Sc. student of Alister Vallance Jones at the University of Saskatchewan, and after obtaining a Ph.D. in high pressure spectroscopy at the University of Toronto under Harry Welsh, he received an NRC post-doctoral fellowship to go to Pierre Jacquinot's laboratory in France—Laboratoire Aimé Cotton—within the Laboratoire de Bellevue, near Paris. During his two years there he developed an infrared Michelson interferometer that worked in a mode called "Fourier Transform Spectroscopy." This has a basic similarity to the WINDII instrument, but WINDII kept its mirrors in fixed locations, while this Michelson interferometer has a mirror that is moved through a few centimeters. If the spectrum of light being examined contained just a single wavelength, the record obtained would be just a simple wave pattern (a sinusoidal pattern), and the wavelength could be obtained accurately with a ruler, since a travel of 5 cm, say would produce a record of 250,000 waves and dividing the distance by this number would produce an accurate wavelength. If two wavelengths were present, they would be seen to be getting in and out of step, so the wave pattern would rise and fall in amplitude. But they could be separated from the repetition distance of the peaks and the wavelength of each determined. If there were one hundred waves the separation would seem impossible except for the French scientist, Joseph Fourier (1768-1830), who had developed a mathematical procedure, now called the Fourier Transform, for disentangling the constituent waves that make up a pattern. In fact he had astounded the mathematical world by stating that any arbitrary function could be represented by a superposition of waves. This was considered to be impossible, and since a fully convincing proof (in the sense of a mathematical proof) was not available then, this remarkable result was strongly resisted. Such a proof (which holds only under certain conditions) was demonstrated later, and this theorem later became the foundation of communications theory. Pierre Connes, Jacquinot's most innovative scientist had built an instrument to work in the laboratory, but Gush set out to build one that would measure the spectrum of the infrared airglow. Pierre's wife Janine Connes had become an expert in the Fourier transformation of the record (called the interferogram) that would produce the spectrum. This was 1957-1959 and it has to be remembered that computers were still rather primitive. In any case, the instrument was taken to the Observatoire de Haute-Provence in the south of France, and beautiful spectra of the infrared emission from the airglow, some 85 km overhead were obtained.

Gush returned to Canada to a professorship at the University of Toronto where he built a scanning Michelson interferometer to be carried by a high altitude balloon. The atmospheric group at CARDE was extremely interested in infrared emissions and the stratosphere; they supported the work and provided the balloon flights. From above the bulk of the atmosphere, around 45 km, the range of the spectrum opened up, and superb spectra were obtained. The student who worked on the project was Henry Buijs and he decided to put this new-found skill to commercial use by founding a company with Gary Vail, called Bomem, in Quebec City, which welcomed entrepreneurs. Bomem developed an instrument that would work on a bench in the laboratory and accurately measure spectra of samples for all kinds of research. Buijs conceived of a method of maintaining the parallelism of the two mirrors as the one mirror was moved as much as one meter, called the dynamic alignment method, while Vail designed the fast electronics needed to maintain the alignment as the mirror moved. The result was spectra of much higher resolution than obtainable with spectrometers and these became a worldwide hit. By now the mathematical transform was no longer a problem and a computer was built into the system (using a unique fast vector processing method developed by a third partner in the company – Jean-Noël Berubé) so that the user simply got a spectrum directly, with no effort.

After Herb Gush had moved to the University of British Columbia he was looking for new things to do. He decided to combine his expertise with Fourier transform spectrometers (Michelson interferometers) with rocket capability and study the Cosmic Background Radiation (CBR). This is radiation left over after the big bang that created our universe, that has by now cooled to about three degrees Kelvin, that is three degrees above absolute zero, the minimum attainable temperature. Zero degrees Kelvin is -273 degrees Celsius. All "warm" bodies emit radiation, easy to believe when the body is the sun, at 5,700 Kelvin (K), or a hot stove at 500 K. It is also true for the Earth's surface at 300 K, which is what the greenhouse effect and global warming is all about. It's still true at 3 Kelvin—its just that the amount of radiation is extremely small. As the temperature falls, the radiation becomes weaker and the wavelength of light emitted becomes longer. For the sun, the peak wavelength is green, about 500 nanometers, or 0.0005 millimeters. The CBR emission that Gush wanted to measure had a wavelength of about 2 millimeters, 4,000 times longer. The Michelson interferometer would work perfectly well at this long wavelength but there was a fundamental problem. If the detector looked at any object warmer than 3 K its radiation would swamp the radiation to be observed, and this includes the interferometer itself. The only solution is to cool the instrument to less than 3 K, which required liquid helium. Constructing a vacuum dewar to contain the refrigerant (called cryogen) and the instrument in a way that would survive the rocket launch was a very challenging undertaking. The latest highly successful flight, called COBRA (Cosmic Background Radiation), was launched from the White Sands Missile Range, NM January 20, 1990. Gush obtained the most accurate temperature of the CBR that had been obtained, at 2.7 degrees Kelvin, but because of technical and other delays in the program his results were published a few months after those from a NASA satellite called COBE (Cosmic Background Explorer). The results from COBE won the 2006 Nobel prize in physics for US scientists George Smoot and John Mather.

Measurements of Pollution In The Troposphere (MOPITT)

As Canada entered the new millennium, its space science program began to unfold in new directions. The Canadian Measurements Of Pollution In The Troposphere (MOPITT) instrument was launched on the NASA Terra spacecraft on December 19, 1999, just in time. The Principal Investigator, James R. Drummond, then at the University of Toronto but now at Dalhousie University, was born in the UK and received his Ph.D. at Oxford University, where the concept of pressure-modulated radiometry was invented and developed. Drummond came to Canada in 1979, to a position at the University of Toronto already well-experienced in this technique, in a broader sense called correlation spectroscopy. Optical spectroscopic techniques are based on the measurement of light at specific wavelengths, associated with particular molecular species. Different instruments make that association in different ways.

An instrument that measures at just one wavelength is called a photometer if it works in the visible region of the spectrum, and a radiometer if it works in the infrared or microwave region. Usually these have several channels, in order to make measurements "on" and "off" the molecular line, and to determine any background light underlying the molecular spectrum. Devices in which a scanning filter moves across a range of the spectrum are called spectrometers. With a CCD detector the device doesn't need to scan at all but can measure all wavelengths simultaneously; these are called spectrographs. Photographic film spectrographs operated in the same way, but are much less sensitive, harder to calibrate in terms of number of photons, and it is much harder to retrieve the data. In fact, the only way is to have the space vehicle return to Earth with the film, or throw it overboard and have it caught during descent (this was actually done in the early days).

Filters with narrow wavelength bands give a more precise association with the species but are difficult to make, especially in the infrared and the Oxford group (under John Houghton, now Sir John) recognized that an essentially perfect filter was the gas that was the object of study. If a cell containing that gas, say carbon dioxide (CO_2), was placed in the path viewed by an optical instrument it would absorb light at the wavelengths corresponding to the lines of carbon dioxide, so if light emitted by carbon dioxide molecules in the atmosphere passed through the cell, it would be partially (or totally) absorbed. If the carbon dioxide were then removed from the cell, the signal would go up, and the intensity of the light and thus the number of carbon dioxide molecules in the atmosphere could be measured. The cell couldn't be rapidly emptied and refilled so instead of removing the gas they changed the pressure in a cylinder, with a piston. The spectral shape of the absorbing line changes with pressure, changing the signal reaching the detector. It was called a Pressure Modulated Radiometer (PMR), as the modulation of the signal by the pressure changes could be interpreted in terms of the gas species concentration. There are different mechanisms that cause molecules to emit radiation but the simplest one is heat, so-called Black Body radiation, caused by the collisions of molecules that raise some to higher energy states, that drop down to lower levels, emitting the difference as light photons. Since the molecular motion is determined by its temperature, the temperature will in part determine how intense the light is, but it also depends on the concentration of the CO_2 in the atmosphere. The Oxford group used its PMR to measure atmospheric temperature; since carbon dioxide is a stable gas that is well-mixed in the atmosphere, with a known percentage, five one-hundredths of a percent, by mass—so a measurement of its radiation will give the temperature. Once the temperature is known, the unknown concentrations of other unknown species can be measured.

Drummond had a somewhat different idea. Instead of varying the pressure with a piston, a rather complex business, he rotated a disk having two different thicknesses inside the cell, modulating the absorption without changing the pressure. He proposed this technique for the NASA Terra mission, for which the announcement was made in 1989, and the proposal was accepted. The CSA agreed to support the development of the MOPITT instrument, with COM DEV of Cambridge, Ontario as the prime contractor, designed to measure not carbon dioxide, but carbon monoxide (CO) in the atmosphere. Carbon monoxide is a direct measure of human activity, and burning in general, such as forest fires including deliberate biomass burning. This is described in detail by Drummond (2005).

The instrument, however, is only a small part of the problem. Drummond was carving out a new direction in Canada, by proposing to make measurements in the troposphere, the roughly 10 km of atmosphere that is closest to the surface, where all of life exists and weather takes place, including clouds. In all the previous atmospheric measurements made by Canadian scientists described earlier all the measurements were made in the upper atmosphere. Anyone who has looked at images of the globe from space will have been struck by how much of the Earth's surface is covered with cloud. In order to view the troposphere, this cloud must be avoided. The best way to do this is to look straight down from the satellite between the clouds but this means looking at light emitted from the Earth's surface, including reflected sunlight. All of these factors become problems in the data analysis, all of which were overcome; and in fact the data are processed at the National Center for Atmospheric Research, in Boulder, Colorado.

In global maps produced by MOPITT, biomass burning in Southern Africa is a dominant feature, accounting for some 30-40% of the carbon monoxide worldwide. The carbon monoxide is carried by the winds; in the north the flow is across the Atlantic, and to the south it is eastward across the Indian Ocean and into the Pacific. Overall there is more carbon monoxide in the Northern hemisphere than in the Southern, and Southern China is a region of notable production. Although North America has a high concentration of industry, associated with combustion, the processes are better controlled, converting more of the monoxide to dioxide, although North American carbon monoxide is still evident. Individual forest fires can also be studied. MOPITT has laid a foundation for the future measurement of tropospheric pollution by Canadian space scientists. It is another example of a fruitful relationship between Canada and the US.

Ozone and the Odin mission

Another good example of international cooperation, but not involving the US, was the Canadian participation in the Swedish Odin mission. As noted earlier, Sweden has been a consistent collaborator with Canada. But the background to this involves the measurement of stratospheric ozone, which has a long and proud Canadian history (McElroy, 2005). E.H. Gowan, a student at the University of Alberta became a Rhodes Scholar and studied at Oxford University where G.M.B. Dobson was doing his pioneering work on ozone. Dobson had realized that he could measure stratospheric ozone from the ground by measuring the absorption of solar ultraviolet light in the atmosphere. To do this one would like to know the intensity of the sunlight (at ozone wavelengths) at the top of the atmosphere, and at the ground, but

Dobson conceived an easier method which was to look at two different wavelengths, one absorbed by ozone, the other not. There are other complications such as light scattered by atmospheric particles in the beam, but a total solution is possible by measuring the intensity of the light as the sun moves from its high point in the sky on a given day, down towards the horizon, as the change in the absorbing path gives the additional information required. His original instrument used film recording that was later replaced by a photoelectric detector in what he called a Féry Spectrometer. The Meteorological Service of Canada provided Gowan with the funds to bring one of these instruments back to Canada with him, and a program of ozone measurements was started. But the Second World War intervened, and a regular program of ozone measurements began only with the International Geophysical Year in 1957.

There was a major impact when Alan Brewer moved to the University of Toronto in the early sixties. Brewer had worked with Dobson at Oxford and together they had developed an understanding of the circulation of the atmosphere, in which ozone is produced at the equator where the sun is strongest, which then circulates to higher latitudes in the Northern and Southern hemisphere – this became known as the Brewer-Dobson circulation. Brewer set out to build a better Dobson spectrometer. His concept involved looking at several ultraviolet wavelengths, with a chopper mechanism exposing one wavelength to the detector at a time, in rapid succession. This would use the newly developed diffraction grating, as had been adopted by Hunten in Saskatoon, rather than the glass prism used by Dobson. This effort was joined by the Meteorological Service of Canada, including David Wardle, James Kerr and Tom McElroy, leading to a practical design that was manufactured by a company in Saskatoon, SciTech, (that has now changed hands), and these have been installed worldwide. The Meteorological Service of Canada is the repository for the world ozone database, and has put Canada at the forefront of this field. Canada was the first country to issue an "ultraviolet index" for its citizens, so that they could determine the tolerable exposure to the sun, with respect to skin cancer, on a given day. This story is told by McElroy (2005). Alan Brewer died on November 21, 2007 in Berkshire, England at the age of 92.

The Swedish auroral satellites, Viking and Freja have already been discussed. When their atmospheric scientists put forward a proposal for a satellite, it was felt that the group was too small to warrant it, so they teamed up with some astronomers. This may seem an unlikely combination, but molecules exist in space as well as in the Earth's atmosphere, and a promising new technology, sub-millimeter wave radiometry offered a way to measure both. This is essentially a radio receiver, but working at extremely high frequencies, a few hundred GHz (1 GHz = 1,000,000,000 cycles per second). This is a wavelength that overlaps the optical region, where Gush's CBR experiment used optical methods. The idea was that the Sub-Millimeter Receiver (SMR) would point at the sky for a period, collecting data for astronomers, and then at the Earth, providing information for atmospheric scientists. The SMR was a major effort and so France and Finland became partners. Canada also contributed to the design of the SMR and provided the cryocoolers for it. One of the Swedish scientists, Georg Witt, wanted a simple instrument to view the atmosphere in the visible region, and he came to Saskatoon to look for this second instrument for the Odin spacecraft. Ted Llewellyn was only too happy to provide it, and with support from the CSA, an instrument called the Optical Spectrograph and InfraRed Imaging System (OSIRIS) was developed and fabricated by Routes AstroEngineering of Ottawa and provided by Canada to Sweden for the mission. OSIRIS was designed to view the Earth's limb in the daytime and look at sunlight scattered from the atmosphere, measuring its entire visible spectrum using a CCD detector. The molecular species of interest, particularly ozone, would produce absorption lines in the spectrum from the passage of the sunlight in and out of the atmosphere. The instrument approach was simpler than in the infrared (the conventional method) although the scattered light analysis is more complicated. Odin was launched on February 20, 2001, and was highly successful, producing excellent ozone height profiles and leading to new information about the ozone hole. This idea of using visible region scattered light was so good that it has been used by instruments on the European Space Agency (ESA) satellite ENVISAT. The Scanning Imaging Absorption Spectrometer for Atmospheric CHartographY (SCIAMACHY) is a much more elaborate instrument that uses the same approach.

The Canadian OSIRIS instrument launched on the Swedish Odin spacecraft on February 20, 2001. Photo by courtesy of the University of Saskatchewan.

The OSIRIS instrument has provided more than ozone profiles and total ozone concentration maps at each altitude. The close collaboration with Sweden, France and Finland and the SMR team has also resulted in new measurements of the NO_2 height profiles and the chlorine fractionation that occurs each polar spring.

However, perhaps some of the major advances with Odin/OSIRIS have come in very unexpected ways. The optical spectrograph section of the instrument has an extended dynamic range and so can make measurements of the faint airglow. The resonant OH (A-X) emission that is excited by the incident sunlight provides an indirect measurement of the mesospheric water vapour content and profile. The OH concentration is related to water vapour, H_2O, through its chemistry. Direct confirmation of these measurements has been provided from observations of the rotational lines of the A-X emission that are formed through photolysis of water vapour by solar Ly-α (ultraviolet hydrogen emission from the sun). The OSIRIS instrument has also detected polar mesospheric clouds and has shown that the ice particles that occur at the coldest region on the planet grow in size as they fall through the upper mesosphere.

The imager section of OSIRIS has provided a novel new approach to atmospheric observation. The Infra Red Imager (IRI) channel has realized a tomographic capability that yields information on both horizontal and vertical structure in the atmosphere. Atmospheric tomography provides vertical slices of the atmosphere in a way similar to CAT scans of one's brain, by viewing the same feature from many different angles. A remarkable demonstration of the success of this approach was provided through the measurement of ozone destruction that occurred in the Halloween storm of 2003. This is a now famous geomagnetic storm named after its date. The IRI observations provided a time history of the depletion over the south polar region.

Thus the Odin/OSIRIS collaboration has provided an invaluable scientific return to the partners that far exceeds the sum of the parts. A measure of the success of the collaboration can be recognized through the acceptance of Odin as a Third Party Mission by ESA. The Canadian members of the Odin science team are Ted Llewellyn, Doug Degenstein, Nick Lloyd and Dick Gattinger of the University of Saskatchewan; Wayne Evans (Adjunct), Ian McDade, Craig Haley, Brian Solheim, John McConnell of York University; Chris Sioris of Environment Canada and Kim Strong of the University of Toronto. Sun Kwok of the Univesity of Calgary is involved with the astronomical measurements and Harvey Richardson of NRC made major contributions to the original OSIRIS design.

The SCISAT mission

WINDII, MOPITT and OSIRIS all achieved tremendous scientific advances, but all were "free rides" on foreign satellites. This is nevertheless a very effective and relatively inexpensive way to get the best science – to concentrate on building the very best instruments. Working with "foreign" scientists also helps to advance the science. Yet Canada's new agency felt it had matured to the point where it could achieve what the Alouette/ISIS program had done, to again build a scientific satellite in Canada. As part of the agreement to participate in the ISS, the CSA had been offered two free launches on the Pegasus launcher, which itself is launched from an aircraft. The CSA issued an announcement of opportunity and from the responses three groups were awarded contracts to support Phase A developments (conceptual design). From these a final selection was made for what was called SCISAT, its first autonomous scientific satellite. The successful proposal came from Peter Bernath at the University of Waterloo, and it involved a Canadian technology that had been a long time in development, as described in the earlier section, "Rocket measurements of the Cosmic Background Radiation."

Peter Bernath was a chemist, and a spectroscopist who was familiar with this technology through the use of Bomem Michelson interferometers and he became convinced that one could be flown on a satellite. The problem is that it would have to be rather small, compared to the laboratory versions, but they could bounce the light back and forth twice in order to increase the path. The mirror movement was also a problem but this was solved using a "double pendulum" with two arms oscillating about a common pivot, one mirror on each arm. As one mirror moved closer to the beamsplitter the other moved farther away, another doubling effect. To make this feasible the mirrors were replaced with "cube corners", prisms with three faces at right angles to one another, just as one has in the corner of a room. Such prisms return the light back in the direction from which it came regardless of their orientation, so alignment is no longer a problem. These are no different in principle from the "retroreflectors" used as highway markers that return the light from the headlights back to the driver's eyes. The application was similar to Marc Garneau's use of SPEAM, to point the instrument directly at the sun as it set behind the Earth as a result of the satellite motion. The ray from the sun passes through the atmosphere, at say 40 km above the surface, before continuing upwards to the satellite. The spectrum observed would reveal absorbed lines corresponding to the species present at 40 km, giving the concentrations of each. As the spacecraft moved, the solar ray would move down, say to 38 km, giving the species concentrations above this altitude, ultimately giving an altitude profile of species concentrations for every sunset and sunrise along the orbits for a given day. There are fifteen of these for a satellite at 600 km, giving thirty profiles. This is not many, but the incredible quality of the spectra more than made up for this. The technique is called solar occultation, and had been done before by Barney Farmer of the Jet Propulsion Laboratory (JPL), using a very large Michelson interferometer called ATMOS, as it was mounted on the space shuttle it provided only a limited amount of data. Solar occultation has been used by a number of investigators on free-flying satellites, but none with the resolution of SCISAT.

The SCISAT spacecraft in the laboratory prior to launch. Photo courtesy of the Canadian Space Agency.

The SCISAT instrument, which Bernath called ACE (Atmospheric Chemistry Experiment) was able to observe the following species: H_2O, O_3, N_2O, CO, CH_4, NO, NO_2, HNO_3, HF, HCl, N_2O_5, $ClONO_2$, CFC-11, CFC-12 (these two CFC constituents are what comes from spray cans), COF_2, HCFC-22, SF_6, HCN, CH_3Cl, CF_4, C_2H_2, N_2, CCl_4, ClO, HO_2NO_2, H_2O_2, HOCl, H_2CO, HCOOH, CFC-113 and HCFC142b, along with the isotopic species for the first five molecules.

In addition to the ACE-FTS (Fourier Transform Spectrometer), the SCISAT mission carried a second instrument, led by Tom McElroy of the Meteorological Service of Canada, called Measurements of Aerosol Extinction in the Stratosphere and Troposphere Retrieved by Occultation (MAESTRO). The focus was on the measurement of small particles in the atmosphere, called aerosols. In the troposphere, aerosols are associated with pollution and have a strong relationship to human health. The SCISAT spacecraft was built by Bristol Aerospace, no longer busy building rocket payloads, but with the same skills, and more. It was launched from a Pegasus vehicle on August 12, 2003 and was still operating in 2007.

SWIFT on Chinook

On February 14, 1994, Tony Raab of what was then CAL corporation, Ian Rowe of CRESTech (of the Ontario Centres of Excellence), Phil Merilees of Environment Canada and Gordon Shepherd met to discuss a possible follow-on to WINDII. What was discussed was how to measure winds lower in the atmosphere, in the stratosphere. There is no airglow emission here, so the Doppler shift of some other emission would have to be used. The only entity available is thermal radiation, from the heat of the molecules, which would have to come from some minor atmospheric constituent, of which ozone was ultimately identified. The technical problem is that the observations would have to be made in the "middle" infrared, at about 9 micrometers wavelength. But the benefit foreseen by Merilees is that winds at these altitudes, especially if extended downwards to about 15 km, would improve weather forecast models. This would be done by "assimilating" the satellite data into the forecast model within a few hours of being acquired. This was an attractive but also very challenging idea, which Merilees strongly supported. During the next few years the concept was developed and tests done on critical items, such as the Michelson interferometer and the filters, both of which would have to operate at low temperatures, around -130 °C. When the concept was ready it was presented to the European meteorological satellite organization, EUMETSAT, only to be told that they would have to go through a scientific "demonstration" mission before becoming an "operational" experiment.

At this point the Stratospheric Wind Interferometer For Transport studies, SWIFT, was born. The scientific objective was to be the measurement of winds in the stratosphere, along with the ozone concentration (since an ozone spectral line was to be used), allowing the transport of ozone to be studied. The CSA had up to this point not been able to find a way to accommodate SWIFT in a Canadian mission so suggested that the science team submit their proposal to a European Space Agency (ESA) competition. In this SWIFT was highly praised, but not given a mission. Instead, ESA identified a possible place on a Japanese satellite, GCOM (Global Change Observing Mission). This mission had space for an extra instrument so they held an international competition in which all parties made presentations on February 16, 2000, almost six years to the day after the original meeting. Canada was supported by ESA agency people and made a good presentation – they won the competition and the CSA had already agreed to support the mission if the SWIFT team had won.

This seemed to bring closure and so Ian McDade of York University took over the leadership of the project for this next phase. Unfortunately it turned out not to be the final phase as Japan did a reassessment of its projects and GCOM (climate change) was converted to a greenhouse gas mission (GOSAT). Soon afterwards, SWIFT was judged not to be consistent with these new mission objectives. By this time the CSA had developed a strong interest in SWIFT, and during a visit to York University on May 5, 2003 Marc Garneau, then president of the CSA gave his assurance that "SWIFT would fly." This meant a Canadian mission dedicated to SWIFT. This concept was developed and became what is now known as the Chinook mission. In July, 2004 the CSA invited applications for a low-cost secondary payload. At this time, Marianna and Gordon Shepherd were visitors at the research campus of Kyoto University in Uji, Japan and they proposed a collaborative experiment in which

Professor Toshitaka Tsuda would provide a GPS receiver. When the radio wave from a GPS satellite passes through the Earth's atmosphere before being received by another satellite, the wave is bent slightly, to an extent that can be determined because of the extremely high precision measurement of the satellite locations. In the stratosphere the bending is mostly due to temperature, which can be determined accurately, to within about 1 °C. In the troposphere the bending is due mostly to water vapour, which can be determined there. This proposal was accepted and Jim Whiteway of York University later became Principal Investigator. Finally the Chinook mission went to Treasury Board and was approved. During the bidding process there was only one bid, and it was too high, outside the pre-determined scope. So Chinook is currently in a state of budget reduction. Each mission has its own dynamic, and all can have twists and turns before they finally fly.

Ground-based Space Science

It was explained much earlier that prior to the advent of the space age what Balfour Currie and Frank Davies were doing at Chesterfield Inlet was "ground-based space science." Once rocket and satellite vehicles became available, these were the primary platforms used, and that is what has mostly been described so far. However, many of the ground-based observations pioneered by Forsyth, Hunten and Vallance Jones continued in many variants, but are too numerous to describe in this small volume. However, one example of how global space science can be done from the ground deserves to be described as an example, and that takes us back to Saskatoon. What follows is an account adapted from material provided by George Sofko, one of Forsyth's students and later a faculty member at the University of Saskatchewan. Jim Koehler has already been introduced through his work with Alex Kavadas in Chapter 5.

In 1990, Jim Koehler attended a meeting in Germany called by Ray Greenwald of the Applied Physics Laboratory (APL) of Johns Hopkins University (JHU) to discuss the expansion of the HF (High Frequency) Super Dual Auroral Radar Network (SuperDARN), which at that time consisted of their Goose Bay radar, the Halley Bay radar (run by the British Antarctic Survey) and a French radar at Shefferville, which acted as the "partner" radar for Goose Bay. The idea of SuperDARN was to use HF signals scattered from the F-region of the ionosphere, where virtually everything, including the plasma waves responsible for the coherent scattering of the radio waves, moves at a speed determined by the electric field strength multiplied (a vector multiplication) by that of the magnetic field. Since the magnetic field is well known, a measurement of the velocity would in effect be a measurement of the electric field. However, to measure the velocity vector in the F-region, two components are required, and so two radars had to be used to look at the scattering region simultaneously from two different directions – hence the word Dual in the acronym DARN.

In 1992 Koehler and Sofko proposed to the Canadian radar community to apply for funding for two HF SuperDARN radars at Saskatoon, one radar looking towards the east, one towards the west. The SuperDARN radars are single-beam radars operating in the 9 – 20 MHz range, but the single beam is swept horizontally through 16 successive positions separated by 3.25°, so that a complete 16-beam sweep looks at a wedge 52.0° in extent. For each beam azimuth, there are 75 range (range is distance along the radio beam) cells of 45 km extent, allowing for a range extent of 3,375 km along the beam (the first range is usually set to 180 km, so the radar measures from 180 to 3,555 km distance). When the 16 beams of the one radar and the 16 beams of the partner radar are taken into account, a radar pair can obtain echoes from the overlapping fields-of-view of the two radars over a large region, about 3.5 million square km in the ionosphere. By having a number of such pairs distributed around the Earth in the auroral zone, it was hoped to measure the complete global pattern of ionization movement, or equivalently the electric field and voltage patterns in the northern and southern hemispheres.

Because the Earth's magnetic field lines are highly conducting, the voltage pattern in the ionosphere is transferred along the magnetic field lines to the magnetosphere, where the deposition of energy from the solar wind and the connecting of the interplanetary magnetic field (IMF, also referred to as the solar magnetic field) and the Earth's outer magnetic field are primarily responsible for producing the electric field/voltage pattern. This ultimately is transmitted down from the magnetosphere to the ionosphere. Just as weather systems in the neutral atmosphere of the earth are driven by high and low pressure systems, so are the "space weather" systems in the charged particle environment of the magnetosphere and ionosphere driven by high and low voltage systems, and the latter are measured by the SuperDARN radars. In that sense, SuperDARN is a key to the measurement of 'space weather'', which has become very important because of the large number of satellites (about 800) that, through telecommunications, facilitate the transaction of millions of dollar of business transactions each day (international banking, telephone, TV, internet) and because of the increasing number of humans that are entering space, either as astronauts or space tourists.

The 1992 Canadian SuperDARN application to NSERC was partly successful. Money was awarded to build the east-looking radar at Saskatoon. At about the same time the US team, led by the APL group, obtained funding for the partner radar at Kapuskasing, Ontario, while the French team led by Dr. Jean-Paul Villain of Orléans obtained funding for the Stokkseyri Iceland radar that would be the partner for the Goose Bay radar. A "triumvirate" was formed between Canada, US and France to share the construction of the Saskatoon, Kapuskasing and Stokkseyri radars, with Canada providing the transmitters, France the phasing matrices, and the US the receivers and software for the controlling computers. As a result, the first truly successful SuperDARN pair was brought online in June, 1993, namely the Saskatoon-Kapuskasing pair. Later in 1993, the Stokkseyri radar built by the French team came online so that the Goose Bay-Stokkseyri pair was operational. It took seven more years for the Canadian team to raise NSERC funding for the Prince George, BC radar which in 2000 became the partner for the 1999 radar built at Kodiak, Alaska, radar under the leadership of Dr. Bill Bristow of the University of Alaska at Fairbanks.

By November 1 of 2007, the SuperDARN network had expanded greatly. The Canadian team had proposed the construction of two "PolarDARN" radars at Rankin Inlet (Nunavut) and Inuvik (NWT) to extend the SuperDARN ionospheric drift measurements into the polar cap region, because that is where the signature of magnetic reconnection will be seen frequently, particularly during the times when the IMF is strong. In the northern hemisphere, there were 13 radars, the two PolarDARN radars, nine auroral zone radars located at King Salmon (Alaska), Kodiak (Alaska) and Prince George (BC), Saskatoon (SK) and Kapuskasing (ON), Goose Bay (Labrador) and Stokkseyri (Iceland), Thykkvibaer (Iceland) and Hankasalmi (Finland), and two midlatitude radars at Wallops Island (Virginia) and near Rikebetsu (Hokkaido, Japan). In the southern hemisphere, the installation of radars is restricted by the presence of water, but there have been 7 radars installed. By the end of 2007, there will thus be 21 radars, and there are plans to extend that total to at least 32.

The SuperDARN network is now widely recognized for its ability to generate global-scale ionospheric convection (voltage) maps. As such it has become a very successful international scientific radar collaboration between ten nations – Canada, US, France, Great Britain, Japan, Italy, South Africa, Australia, China and Finland. Canada plays several very important roles in SuperDARN, and has been a key nation since the first truly successful SuperDARN pair – the Saskatoon and Kapuskasing radars – was installed in 1993. Nationally, the Canadian team is virtually Canada-wide, extending across five Canadian universities – the Universities of Alberta, Calgary, Saskatchewan, Western Ontario and New Brunswick – and three government groups – CSA Space Science, CRC (Paul Prikryl and Gordon James) and NRCan Geomagnetics.

Satellites do provide global coverage, but one satellite measures only at one place at one time. It visits a given site, like Saskatoon only twice per day, once for each side of the orbit, giving very poor time resolution. A ground-based instrument makes measurements continuously, but only at one place. But by employing the huge coverage of a single SuperDARN pair, a relatively small number of radars can be used to practically cover the globe, and thus continuously monitor the state of the magnetosphere. It's a long way from Nate Gerson's gift radar to Balfour Currie, but the connection is clear.

Another example is the chain of magnetometers established by Gordon Rostoker at the University of Alberta. This became an important element of the CANOPUS network, implemented by the CSA, now to become CARISMA (Canadian Array for Realtime InvestigationS of Magnetic Activity). This has been taken over by a new faculty member and Canada Research Chair at the University of Alberta, Ian Mann, who is extending it further. The combination of satellite and ground-based measurements is a powerful one and Canada has a lot of territory in the right location for these instruments.

Postscript

The move of the focus of space activity from Ottawa to Montréal was a dislocation because almost all of the individuals who had developed their space expertise there were not prepared to move so close to the end of their careers (the younger astronauts were an exception to this). This required a fresh start, with new people. On the other hand, the other participants on the national space scene just stayed where they were, in universities and industries across the country. As well, many of the initiatives and programs that were begun by the NRC were continued by the CSA, so there was considerable continuity. Following the "birth of the CSA" in 1989, Canada has enjoyed many successes in space, and has developed a new level of maturity. This is expected to continue in the future.

References:

Drummond, J.R., Measuring Pollution from Space, Physics in Canada, September, 2005.
Fowler, B. Bock, O. and Comfort, D., Is dual-task performance necessarily impaired in space? Human Factors, Volume 42, 318-326, 2000.
McElroy, C.T., The Canadian Ozone Story, Physics in Canada, September, 2005.
Shepherd, Gordon G., Spectral Imaging of the Atmosphere, Academic Press, 2002.

Chapter 8

Beyond the Earth

Very Long Baseline Interferometry – a Canadian invention

Two of Herbert Gush's contributions to space science have already been mentioned, both related to his pioneering work on Fourier Transform Spectroscopy (FTS) using scanning Michelson interferometers; his measurement of the infrared airglow spectrum from a high-altitude balloon and his measurement of the Cosmic Background Radiation from a Canadian rocket. Gush made one further major contribution that is the subject of this section. In the sixties, radio astronomy was in its infancy, and Gush, then a faculty member at the University of Toronto, realized that the Michelson interferometer could be applied to radio imaging of astronomical radio sources. He wrote down his idea and in 1964 discussed it with Y.L. (Allen) Yen of the University of Toronto, John Galt of the Dominion Radio Astrophysical Observatory (DRAO) in Penticton, British Columbia (that had installed a new telescope in 1960) and Alex Kavadas of the University of Saskatchewan who was working with the Prince Albert Radar Laboratory; in fact, it was Don Rose who had suggested PARL.

The idea was to replace the two Michelson mirrors with two radio telescopes situated a thousand or more kilometers apart. In the optical Michelson instrument the light beams returned from the mirrors are combined at a beamsplitter, allowing them to be compared. But how could radio beams received at telescopes thousands of kilometers apart be combined and compared? The answer was to record the signals on magnetic tape, along with extremely accurate clock signals. The two tapes could then be brought to the same computer and compared against their time records. The timing requirements were severe but Gush argued that the technology was now available in the form of atomic clocks, fast tape recorders and computers. The group expanded to include N. Broten of NRC and R. (Bob) Chisholm of Queen's University and in 1966 they took the concept to Don McKinley at NRC who committed $250,000 to it. This was a lot of money in those days, and could be produced then only because of the freedom from bureaucracy that existed at that time. Still they had to be careful with the funds. They purchased three used Ampex tape recorders from the CBC (who were purchasing new ones), at $25,000 each, and two rubidium clocks for $50,000. It took a long time to learn how to time-tag the time records, which included verbal counts for the longer intervals. During the preparations PARL closed down. The first comparison between DRAO and the Algonquin Radio Observatory (ARO) (in Algonquin Park, Ontario) failed, so they did a check between the relatively short baseline of ARO and an antenna at DRTE in Shirley Bay near Ottawa – this worked well. They traced the problem to a timing error and when this was corrected, on May 21, 1967, they saw the first "fringes", interference between the signals from DRAO and ARO some 3074 km apart, a wonderful accomplishment for Canada.

The story of this invention and implementation of the principle is told in a set of four articles for its 20th anniversary in the *Journal of the Royal Astronomical Society of Canada* by Locke, Gush, Broten and Galt (1988). It's interesting that Locke, Gush and Galt were all students of Harry Welsh at the University of Toronto, who was a spectroscopist, not an astronomer.

The 46 meter antenna of the Algonquin Radio Observatory, employed in the first successful Very Long Baseline Interferometry experiment. Photo by courtesy of Wayne Cannon.

The optical Michelson interferometers were used to derive the wavelength spectrum of the light, but the interferograms from the radio telescopes were spatial so the FTS conversion technique generated an image of the radio source, with a resolution the same as would be obtained with a single telescope thousands of kilometers in diameter. This was an enormous step forward for astronomy. The technique was quickly picked up by the US, who added the name "Very" in front of the more modest Canadian term "Long-Baseline Interferometry", and now everyone uses the acronym VLBI. Through international collaboration, signals from radio telescopes from around the world are combined in this way, but if multiple stations are used then there is redundant information. This means that if the positions of stations A and B on the Earth are accurately known (which is necessary in order to analyze the data) then the position of station C can be accurately determined. Thus the method became a technique not just for astronomy but for geodesy as well. All this was well before GPS appeared, and at that time it provided the only means of determining the accurate positions of isolated locations such as in Northern Canada. Using this technique it has been possible to measure the rebound of the Earth, still recovering from the disappearance of the last Canadian ice sheet.

The diameter of the Earth limited the ultimate length of the baseline for terrestrial measurements, but in the space framework one telescope could be located on the ground and the other on a satellite, at any desired distance. Both Russia and Japan initiated plans for this and Japan succeeded first, with its VSOP (VLBI Space Observatory Program), using the satellite HALCA (Highly Advanced Laboratory for Communications and Astronomy). Canadian Peter Dewdney of the DRAO led the Canadian team that included Russ Taylor of the University of Calgary and Wayne Cannon of York University. DRAO was one of three sites where the data were correlated using a specially constructed correlator designed by DRAO and funded by the CSA. The satellite was launched from Uchinoura on February 12, 1997 on the maiden flight of the ISAS M-V rocket and placed in a 6.3 hour orbit, with its apogee some 21,400 km above the Earth. After placement in orbit the mesh antenna 8 meters in diameter was unfurled. The primary object of study was super-massive black holes in active galactic nuclei and the mission, lasting 3 years, was spectacularly successful, producing images one million times more detailed than had been obtained with conventional ground-based optical telescopes. With so many collaborating ground-based telescopes there was a clear advantage in having the same recording equipment at all of them, to facilitate the comparison (correlation) of the data afterwards. A common receiver was developed by Wayne Cannon of York University, with the support of the Institute for Space and Terrestrial Sciences (ISTS), later CRESTech, both part of the Ontario Centres of Excellence program, and the Canadian Space Agency as well. This S-2 receiver became a world-wide standard.

The HALCA spacecraft of the Institute for Space and Aeronautical Sciences, Japan, before launch. The 8-meter diameter antenna was fabricated from gold-coated molybdenum wire, suspended from six masts. Photo by courtesy of ISAS/JAXA.

Space Astronomy

Astronomers have been studying objects in space for thousands of years, so what did the arrival of the space age mean to them? Space astronomy is defined as astronomical measurements made "from space", so this meant a major new window on the Universe was opened with access to space. There are three main advantages that astronomers gain by going to space. The first two relate to the Earth's atmosphere. First, the atmosphere is transparent in the visible region, but is opaque in the shorter wavelength part of the spectrum, the ultraviolet and X-rays, and in much of the infrared (the longer wavelengths). Basically, space opens up huge new windows on the universe, which are blocked by our atmosphere. Second, the "transparent" visible atmosphere is a problem because it is turbulent. Even on a wonderful site on a high mountain, telescopes experience a mild version of looking at the horizon on a hot afternoon and seeing it shimmer, or even seeing a mirage. The light waves bend slightly because of the variations in temperature and atmospheric density. With a large telescope, the rays entering on one side may be bent differently from those on the other side, blurring the image. The third advantage has to do with viewing time. An Earth-bound telescope spends roughly half its time in daylight, making continuous measurements impossible (except for six months at either pole). By getting away from the Earth in a suitable orbit, continuous measurements are made possible, increasing the continuity of observing time.

The most famous space telescope is the Hubble Space Telescope (HST), but there have been many other orbiting facilities for astronomy, beginning in the 1960s. Canadian scientists have been involved individually in HST but below are some space telescopes that have carried Canadian hardware. The assistance of John Hutchings of the Herzberg Institute of Astrophysics (HIA) of NRC is gratefully acknowledged in the preparation of these sections.

Far Ultraviolet Spectroscopic Explorer (FUSE)

The Far Ultraviolet Spectroscopic Explorer (FUSE) was a NASA telescope launched aboard a Delta II rocket on June 24, 1999 for a planned three years of operations. FUSE was designed to acquire ultraviolet spectra in a wavelength region inaccessible to Hubble, requiring different optical and detector technology. The FUSE wavelength region is very rich and contains many signatures of the interstellar medium, including ionized atomic oxygen (with not one, but five electrons removed) which traces hot gas. Principal science goals include the deuterium abundance of the universe, hot gas in the universe, and the gas that constitutes most of the baryons (protons and neutrons). The FUSE telescope had four segments, designed to respond to different parts of the far ultraviolet spectrum, with a different diffraction grating for each. FUSE was eventually turned off in October 2007 when the last spacecraft reaction wheel failed.

The FUSE satellite at NASA in August, 1998. Photo by courtesy of NASA.

The Fine Error Sensor (or FES) was basically the "eyes" of the satellite and it was provided by Canada. The FES worked in visible light, and imaged a region a little smaller than the moon. The FES could track on stars down to about 14th magnitude, which is about 5,000 to 10,000 times fainter than the stars that can be seen with the naked eye. The FES produced the only direct images provided by FUSE; the real task of FUSE was to observe the spectrum of astronomical objects in far-ultraviolet light invisible to ground-based telescopes.

There were two FES units, the second as a backup. Each consisted of a camera with a sensitive CCD detector. These cameras obtained visible-light images of a small patch of sky surrounding a science target. The FES determined the positions of several field stars in the neighborhood of the target, and provided this information to the satellite's control system in order to keep FUSE centred on the target to an accuracy of about 0.3 arcseconds during spectrograph exposures (one arc minute is 1/60 of a degree and one arc second is 1/60 of an arc minute). The CCDs in the FES units had 1024 x 1024 pixels, each of which was 24 micrometers square. In order to be sufficiently noise-free, they were cooled to approximately -60° C by means of a thermoelectric cooler unit. The excess heat generated by these units was removed from the spectrograph cavity by conduction along a heat strap to a large, flat radiator mounted on the side of the satellite. The satellite was positioned so that the active FES radiator was kept in the shade and thus radiated heat without absorbing it.

FES was called upon for a completely unplanned purpose when, in December 2004, the third of four onboard reaction wheels failed, depriving the satellite of stability and fine-pointing capacity. By using FES along with onboard magnetometers on FUSE it was found possible to measure drift rates after slews, allowing the spacecraft to operate long beyond its planned three-year lifetime.

The leader for the Canadian Science Team was John Hutchings. Aside from FUSE team investigations, all Canadian were guaranteed access to at least 5% of the time, and more than 50 Canadian astronomers conducted investigations, addressing questions such as: What were the conditions like in the first few minutes after the Big Bang? How are the chemical elements dispersed throughout galaxies, and how does this affect the way galaxies evolve? What are the properties of interstellar gas clouds out of which stars and our solar system form? By making a unique contribution to the spacecraft, Canadian astronomers are able to work with their international counterparts on these and other fundamental questions. At the time of writing, FUSE had produced 430 referenced research papers, some 20% of which have Canadian authors.

Herschel

ESA's Herschel mission was named after William Herschel and his wife Caroline. William was born in Germany but later moved to England and became famous for first observing the planet Uranus (it was named Herschel for a time). But his identification with this infrared mission also comes from the fact that he discovered infrared radiation. While using a thermometer to monitor the temperature in his laboratory he placed it just beyond the red light emerging from a prism and found an elevated temperature, due to the infrared (heat) radiation there. In terms of famous astronomers, this spacecraft could also have been named after his son, John Herschel, who mapped the southern skies from Capetown, named the seven then-known moons of Saturn, and also observed the four moons of Uranus. Herschel Island in the Canadian Arctic is named after him.

The spacecraft Herschel operates at the opposite end of the spectrum from FUSE, in the far infrared, at wavelengths between 60 and 670 micrometers. The 9 meter long 3,300 kg satellite (at launch) is planned to be launched in July, 2008 (the month of the 37th COSPAR Assembly in Montréal). With a 3.5 meter diameter telescope it will surpass Hubble, at 2.5 meters. Huge clouds of dust grains can absorb ultraviolet radiation emitted by the stars near them. The heated dust then cools by radiating at infrared wavelengths. To understand all the processes occurring in a galaxy, it is necessary to measure the total energy emitted at all wavelengths. With this capability Herschel will study the formation of galaxies in the early universe and their subsequent evolution, investigate the creation of stars and their interaction with the interstellar medium, observe the chemical composition of the atmospheres and surfaces of comets, planets and satellites and examine the molecular chemistry of the universe. Herschel contains two instruments with Canadian involvement, SPIRE and HIFI.

SPIRE is the Spectral and Photometric Imaging Receiver consisting of a three band imaging photometer (250, 360 and 520 μm) using highly advanced infrared detectors and an imaging Fourier transform spectrometer. CCD imaging detectors don't work at these wavelengths, so the arrays are made up of a grid of individual infrared detectors, 139 of them for the 250μm channel, for example. Each detector will record an interferogram that can be transformed to a spectrum, as described earlier for the scanning Michelson interferometers (FTS). Canada is contributing a Fourier Transform Spectrometer to characterize the instrument prior to flight, as well as data analysis software and personnel for the test teams. Canadian scientists will also participate in the SPIRE Instrument Control Centre. The Canadian SPIRE team is led by Professor David Naylor, of the University of Lethbridge where software is being written to allow researchers to process the enormous amount of data collected during observations.

The Heterodyne Instrument for the Far Infrared (HIFI) is dedicated to the study of interstellar chemistry. At these long wavelengths the transition is made from optical to radio wavelengths. Radio techniques provide much higher resolution as the detection is coherent. With its high spectroscopic resolution and sensitivity, the instrument can detect and analyze emissions from a variety of molecules. It uses far-infrared and submillimeter wavebands to observe molecules in various chemical processes and astrophysical environments. The instrument uses a local oscillator to produce its reference signal. This is mixed with the signal from the source object, and the result is a signal frequency much lower than the original and much easier to process with electronic circuits. The advanced local oscillator source unit being contributed by Canada has unprecedented precision and stability.

The recent development of power amplifiers at very high frequencies, beyond 100 GHz, now makes it possible to use a different approach to building the local oscillator (LO). A synthesizer can be built at approximately 10 GHz, the output of which can be amplified in power, and multiplied in frequency until the desired output for HIFI at up to 2.7 THz is reached, the maximum frequency at which HIFI will operate (One THz = 1 Tera Hertz which is one followed by twelve zeros—think mega mega). In order for this to work, the synthesizer output must be extremely clean in terms of phase noise and out-of-band noise. In spite of many challenges in building such an instrument, a team from COM DEV in Cambridge, Ontario successfully fabricated and delivered the Local Oscillator Source Unit to the project which, at the time of writing, has been successfully integrated into the HIFI instrument. The Principal Investigator for HIFI in Canada is Professor Michel Fich of the University of Waterloo.

In order to cool the detectors it is desirable to be as far from the warm Earth as possible, and in addition the light from the sun, moon and Earth are to be avoided. ESA has chosen a parking place in space called the L2 Lagrange point. In 1772 the French mathematician Joseph-Louis Lagrange discovered that there are five places in the sun-Earth system where the gravitational forces on a third body balance with the centripetal force required to keep that body in a location that maintains fixed distances from both the sun and the Earth (this is true of any two bodies, e.g. the sun and Jupiter). Two of the points (L4 and L5) lie along the Earth's orbit around the sun, 60° ahead of and behind the Earth. The others lie along a line passing through the sun and the Earth. L1 is between the Earth and sun, L3 is behind the sun as seen from the Earth and L2 is on the side of the Earth away from the sun. In this location Herschel will maintain an orbit radius larger than that of the Earth, but always keeping pace with it. In fact this location is unstable, so ESA plans to allow Herschel to oscillate around L2, and use propulsion to maintain the desired location.

The James Webb Space Telescope (JWST)

James Edwin Webb, born in 1906, was the second administrator of NASA, serving from February 14, 1961 to October 7, 1968. He is given credit for managing NASA through the period of the lunar landings, a truly great challenge.

This major facility-class space observatory, the James Webb Space Telescope (JWST) will be launched in 2013. It is a successor to the Hubble Space Telescope except that while Hubble makes observations mainly in the visible and ultraviolet light wavebands, the JWST will operate at infrared wavelengths. JWST is a joint mission of NASA, ESA, and the CSA. The mission concept is for a large 6.5 meter diameter telescope (Hubble has a diameter of 2.5 meters) located 1.5 million km from Earth at the L2 Lagrangian point as was described for Herschel.

Shielded from the Sun and Earth by a large deployable sunshade, the entire telescope assembly will be passively cooled to less than 50 degrees Kelvin (50 degrees above absolute zero, or -223 degrees Celsius). This gives JWST exceptional performance in the near-infrared and mid-infrared wavebands. The wavelength range for the instrumentation is 0.6 to 27 micrometers, not as far in the infrared as Herschel.

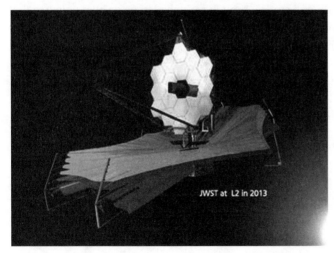

Drawing of the James Webb Space Telescope (JWST), to be launched to the L2 Lagrangian point in 2013. Image by courtesy of NASA.

The sensitivity of the telescope will be limited only by the natural zodiacal background (light scattered from electrons and dust in the solar system), and should exceed that of ground-based and other space-based observatories by factors of 10 to 100,000, depending on the wavelength and type of observation. The JWST observatory will have a 5- to 10-year lifetime but will be too far away to be serviced by astronauts.

Like Hubble, the JWST will be used by the astronomy community to observe targets that range from objects within our Solar System to the most remote galaxies, which are seen during their formation in the early universe. The science mission is centered on the quest to understand our origins, and specifically aimed at:

*observing the very first generation of stars to illuminate the dark universe when it was less than a billion years old;

*understanding the physical processes that have controlled the evolution of galaxies over cosmic time, and, in particular, identifying the processes that led to the assembly of galaxies within the first 4 billion years after the Big Bang;

*understanding the physical processes that control the formation and early evolution of stars in our own and other nearby galaxies;

*studying the formation and early evolution of proto-planetary disks, and characterizing the atmospheres of isolated planetary mass objects.

Because of wavelength redshifts, Doppler shifts to longer wavelength caused by high outward velocities in the distant universe, and dust that obscures star-forming regions, all of these fundamental questions are best addressed in the near- and mid-infrared wavebands.

Canada is providing hardware for JWST: the Fine Guidance Sensor and Tunable Filter Imager (FGS and TFI). The FGS is integral to the attitude control system of JWST, and consists of two fully redundant cameras that will report precise (3.5 milli-arcseconds) pointing information for JWST. Canadian expertise in this area has been established with the successful fine error sensors for the FUSE mission. Packaged with the FGS but functionally independent, the Tunable Filter Imager is a unique, narrow-band camera that can scan its wavelength of observation. It will also include a coronagraph, which suppresses the bright light from a star or quasar to allow detection and study of much fainter planets or host galaxies that surround them. The TFI will tune over wavelengths between 1.5 to 5.0 microns, with a gap in sensitivity between 2.7 and 3.0 microns.

The Canadian Project Scientist for JWST is Dr. John Hutchings of the HIA/NRC. Dr. René Doyon, of the Université de Montréal is the TFI principal investigator. The prime contractor for the FGS and TFI is COMDEV.

A full-scale model of the JWST built by Northrop Grumman, the prime contractor, is to be erected in Montréal for the COSPAR 2008 Assembly. A problem arose in that it was too large to fit in the Palais des Congrès where the Assembly is being held. It will therefore be placed in the Old Port of Montreal near the Montreal Science Centre.

ASTROSAT

ASTROSAT is a project of the Indian Space Research Organization (ISRO), that will be an observatory in low Earth orbit, equipped with four X-ray telescopes and two Ultraviolet telescopes, all pointing at the same place in the sky. The UV telescopes (UVIT) will have a one-half degree field of view and one arcsecond resolution. Canada is providing the photon-counting detectors for the UVIT, via CSA contract with Routes Astroengineering.

The observatory will be launched by ISRO in 2009, for a minimum 5-year lifetime. The unique feature of ASTROSAT is that it can obtain simultaneous observations from the visible to hard X-rays, using the full complement of instruments. It will also have an all-sky X-ray monitor to spot transient sources and then investigate them with the full suite of pointed instruments.

The Canadian PI is John Hutchings and there is a team of Canadian scientists who will work with the Indian teams. Canada is also guaranteed a minimum amount of competed open time on ASTROSAT.

BLAST

One can get above all but 999/1000 of the Earth's atmosphere in a high altitude balloon and this offers a way to get a large telescope aloft for a few days, at modest cost. Canada and three partner countries, the US, the UK and Mexico, conducted an unusual experiment during June, 2005 in the Arctic skies. Attached to a huge helium balloon flying at 38,000 metres, a 2,000-kg telescope called BLAST (Balloon-borne Large Aperture Sub-millimetre Telescope) viewed deep into the sky to study distant stars and galaxies. Launched from Kiruna, Sweden, on June 11, BLAST flew for six days before reaching Inuvik, on the Beaufort Sea in northern Canada. The two-metre telescope offers levels of sensitivity and resolution unmatched by any facility on Earth.

BLAST will provide a new view of the universe, according to Dr. Barth Netterfield, the Canadian science team leader and professor in the departments of Physics and Astronomy at the University of Toronto. BLAST will identify large numbers of distant star-forming galaxies, study the earliest stages of star and planet formation, and make high-resolution maps of diffuse galactic emission. Canada has provided the gondola, the pointing control system, the data acquisition system, the flight and ground station software, the power system, and overall system integration. Canadian partners in this project include the University of Toronto (Dr. Barth Netterfield), the University of British Columbia (Dr. Mark Halpern), and AMEC of Port Coquitlam, B.C. Canadian funding was provided by the CSA, the Natural Sciences and Engineering Research Council of Canada (NSERC), the Ontario Innovation Trust, and the University of Toronto.

International partners included the University of Pennsylvania, Brown University, the University of Miami, the Jet Propulsion Laboratory, Cardiff University, and the Instituto Nacional de Astrofisica of Mexico, with funding from NASA and the UK's Particle Physics and Astronomy Research Council (PPARC).

MOST – an all-Canadian astronomy mission

The first all-Canadian astronomy satellite arose from a proposal to launch a small (15 cm diameter) telescope to study small light oscillations in stars. Conceived by Slavek Rucinski and led by Jaymie Matthews of the University of British Columbia, and called Microvariability and Oscillations of Stars (MOST) this is classed as a "microsat" and was executed at low cost by having university scientists work closely with a small company, Dynacon, that had spun off from the University of Toronto Institute for Aerospace Studies (UTIAS). The satellite is "suitcase-sized", 65 x 65 x 30 cm, with a mass of 60 kg. It points accurately at stars using a system of miniature reaction wheels and magnetic torquers, pointing to within a few arc seconds. It was launched on a Russian three-stage *Rockot* (formerly an ICBM, no longer in service) on June 30, 2003 from Plesetsk, Russia, to an altitude of 820 km.

How can such a small telescope make a contribution to astronomy? The concept lies in the microvariability. As with the sun, the surface of a star oscillates as a result of seismic-type waves traveling in its interior. These result in variations in the light output which are very small, perhaps one-millionth of the total intensity, but periodic. To observe such tiny regular fluctuations needs a very long and weather-free observing period. An Earth-bound telescope can observe a star for roughly twelve hours before the onset of sunlight. Telescopes have been proposed for the South Pole, but that involves problems. MOST has been put into a sun-synchronous polar orbit where it can track the same star for as much as 60 days continuously - this allows very small variations to be detected. There is a second reason why the light output of a star might vary and that is because of a planet passing in front of it or behind it, blocking a small fraction of the light. So the same technique can be used to search for planets of distant stars, without actually seeing them. MOST continues to have an active and productive observing program at the time of writing, and has yielded some important scientific results. MOST is a fine example of ingenuity and execution; we may expect to see more microsatellites and even nanosats.

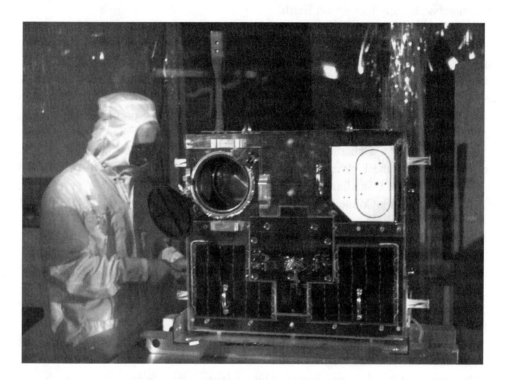

The Canadian microsatellite Microvariability and Oscillations of Stars (MOST) during testing at UTIAS. Photo by courtesy of the Canadian Space Agency.

Gravity Probe B mission to test Einstein's general theory of relativity

Wayne Cannon worked closely with Norbert Bartel, a Professor in the Department of Physics at York University. It was a good combination as Cannon was interested in the accurate determination of positions on the Earth, particularly in its movements; his laboratory in ISTS was known as the Space Geodynamics Laboratory. Bartel is a radio astronomer, with an interest in the radio sources and what could be revealed from them with the VLBI technique. But Bartel became involved in a US program to test Einstein's general theory of relativity.

The method was to place gyroscopes in space, completely isolated from the environment – that is, with no contact with the spacecraft. This was known as the Gravity Probe B (GP-B) mission. The changing position of the spin axis of the gyroscope would depend on whether the universe behaves according to Einstein's predictions or not. But against what reference should the gyroscope position be made? It should be the background universe of stars, and so GP-B would use a radio star as a reference. But this star might be moving against the background stars, so its motion would have to be tracked. Bartel agreed to undertake the tracking, using the VLBI technique.

On April 20, 2005, Bartel was asked to mark the one-year anniversary of the flight of Gravity Probe B (GP-B) by talking about the work he and his team have contributed to the experiment to prove Einstein's General Theory of Relativity. GP-B was mounted by California's Stanford University and NASA researchers. Bartel's team at York included researchers Michael Bietenholz and Ryan Ransom, and graduate student Jerusha Lederman.

GP-B was launched on April 20, 2004 and will continue collecting data until at least the spring of 2008, when the science portion of the mission will be complete. Researchers in Canada, the US and France, are contributing data to the final calculation which will eventually be released to the public. Here is Bartel's report:

"Nothing has seriously gone wrong yet, and that alone is exciting. The most difficult experiment ever done in space is working to near perfection—but that is about all we can say at this time. GP-B has been orbiting Earth now for one year, measuring the warping and twisting of space-time to figure out whether Albert Einstein was right or wrong. It had been 40 years in the making at NASA and Stanford University in collaboration with Lockheed Martin till the satellite could be launched exactly a year ago. It is just mind-boggling to think of the small spheres, called gyroscopes, as they freely fall around the Earth while the satellite is just literally built around them to shelter them and to provide the experimental platform for the experiment. No touching, no contact between the spheres and the satellite is allowed.

Do the gyroscopes move as predicted by Einstein?

"We do not know. We actually do not have the slightest inkling of an answer. The team at Stanford keeps extremely quiet as planned but does that mean they could perhaps already know the answer? They cannot know it. The final answer depends also on our work here at York which is done in collaboration with our colleagues at Harvard University. It is a double-blind experiment. At NASA and Stanford they measure the tiny motion of the gyroscopes relative to a star, and we at York and Harvard measure the motion of the star relative to the distant universe. No group is allowed to peek over the shoulder of the other so as to avoid any human bias, intentional or unintentional.

"I am already anticipating with some nervousness the dramatic moment when our two numbers will be combined to give the final result.

"A hundred years ago Einstein started a revolution in physics by redefining space and time. In commemoration of Einstein's miraculous year in 1905 UNESCO has declared this the World Year of Physics. Here we are sitting and looking at this data point and that data point trying to make absolutely sure that our analysis is correct. No error allowed. But the

larger picture is really that we are testing Einstein's Universe. Space-time is at the heart of Einstein's theory, and what can be more fundamental than space and time?"

Canada Joins the Phoenix Mission to Mars

Mars, the Red Planet, is one of the best known to the general public, owing in part to the postulated "canals" by early ground-based observers. Although it is now clear that these are not man-made canals, it is increasingly evident that strong natural forces shaped the surface of Mars. The adventures of the NASA rovers Spirit and Opportunity, which landed on Mars on January 4 and January 25, 2004 are well known. They explored significant regions of the Mars surface, but near the equator. This story will not be repeated here.

ESA was late in getting into planetary science but scored well with its Mars Express orbiter, so named because it was executed on a fast time line. The orbiter has the disadvantage of observing from a distance, but the advantage of global coverage. Mars Express was launched by a Soyuz-Fregat launcher from the Baikonur Cosmodrome in Kazakhstan in June 2003. On December 19, 2003, 5 days before orbit insertion, the Beagle 2 lander was successfully released towards the surface of the planet. On December 25, 2003 the orbiter underwent a successful orbit insertion manoeuvre and then reached its operational orbit. However, the Beagle 2 was never heard from, so apparently did not survive its landing on the Martian surface.

The following material is taken from the Mars Express website:

http://sci.esa.int/science-e/www/object/index.cfm?fobjectid=31026

"Large outflow channels, each of which could have been formed only by the massive release of water over a short period of time, scar four regions: Chryse-Acidalia, Elysium Planitia, the eastern Hellas Basin and the Amazonis Planitia. Several of these channels drain into the northern plains, lending support to the existence of an ancient ocean over most of the northern hemisphere. However, there is other evidence for flowing water in earlier times.

"The history of water on Mars as revealed in these three different types of evidence, suggests a dramatic change in the climate about 3.8 billion years ago. The atmosphere today is too cold to support liquid water on the surface for long and too thin to support ice - any ice that does form will quickly sublimate into water vapour. But before 3.8 billion years ago many scientists think that liquid water must have existed on the surface for quite some time."

Early Background

The Canadian story is told by Allan Carswell, with slight editing.

This first Canadian science mission to the surface of Mars was not the result of a logical decision by an executive officer but rather the culmination of a long series of events stretching over almost four decades. In the beginning these activities had no connection at all with the eventual development of a Canadian meteorological station, MET, for studies of the

Martian atmosphere as part of the NASA 2007 mission, "Phoenix." Things actually began in the 60's with the establishment at York University in Toronto of an advanced lidar (laser radar) capability made possible because of the advent of the laser. It is of value to briefly trace these events as an interesting case history showing how the long term development of basic "bread and butter" scientific and technical skills can provide opportunities for unforeseen but very significant advances. In addition to the basic skills however such advances require the application of a dedicated drive towards a well-defined goal.

The beginning can probably be set in the summer of 1968, when Allan Carswell joined the faculty of the Department of Physics at York University. He had arrived at York from a position as Director of the Optical and Microwave Physics Laboratory at the RCA Victor Research Laboratories in Montreal where he had been working on the development of lasers and microwave systems for a variety of applications.

Carswell was a graduate of physicist Harry Welsh of the University of Toronto, who trained many graduate students. Herbert Gush, Gordon Shepherd, John Galt and Vic Gaizauskas (the latter two became members of the Herzberg Institute of Astrophysics) were there at the same time in the late 50s. After a post-doctoral fellowship in Holland Carswell came back to a position at RCA Victor in Montreal, where the ISIS satellites were built. At this time Morrel Bachynski was leading a first-rate research team there, demonstrating for the first time that excellent research could be done in Canadian industry. In this stimulating environment Allan Carswell chose to work on lasers, a new capability that had just been discovered.

The Science Faculty of York University had opened its doors in 1965 with a strong emphasis on atmospheric physics and chemistry by Ralph Nicholls and Harold Schiff respectively. Carswell was anxious to collaborate with these research activities. As a result he decided to initiate research on the study of the atmosphere using lidar techniques. Lidar is the optical analog of radar in that it employs a pulsed laser rather than a microwave source to probe the atmosphere by studying the reflected signals. The time delay provides information on the location of the probed volume and the analysis of the optical signal (e.g. intensity, wavelength, polarization, etc.) provides detailed information on the characteristics of the atmosphere.

An extensive array of lidar facilities was developed by the York lidar group. These included van-mounted mobile lidars for tropospheric studies as well as lidar stratospheric observatories in the Canadian Arctic at Eureka (the MSC ASTROLAB facililty) and on the York campus in Toronto, both as part of the international Network for the Detection of Stratospheric Change (NDSC). Since that time in the 60's lidar has become one of the instruments of choice for atmospheric measurements.

The lidar activities at York benefited greatly from the establishment, in 1987, of the Institute for Space and Terrestrial Science (ISTS) as one of the first Ontario Centres of Excellence. The headquarters of ISTS was located on the York campus and Carswell served as the founding Director of the ISTS Lidar Laboratory. Several years later ISTS was reconfigured as CRESTech (Centre for Research in Earth and Space Technology) where the lidar work continued.

As lidar technology and measurement capabilities expanded, Carswell recognized that many of the lidar capabilities could have significant commercial applications. As a result Carswell along with his wife, Helen, decided to start a company capable of pursuing some of these applications. Thus Optech Incorporated was founded in 1974 as a spare-time activity. Early work at Optech was based on atmospheric lidar systems but this was soon expanded to include lidar capabilities for underwater remote sensing applications. This led to the development of down-looking airborne systems for measurement of water depth (bathymetry). This airborne work has subsequently grown to include lidars for accurate digital terrain mapping and a variety of other three dimensional imaging capabilities. Optech now has a staff of over two hundred with facilities in the US as well as Toronto. It is the leading provider of airborne lidar survey systems with customers all over the world. Lidar systems for space applications have also been developed at Optech and these have involved uses for satellite remote sensing as well as systems for spacecraft rendezvousing and docking, precision landing and terrain mapping. In 2005 Optech provided the first Canadian lidar in space as part of the XSS11 program of the US Air force.

MITCH

In 1998 Carswell retired from his position at York and moved full-time to Optech. In the summer of 1999 he was approached by the Jet Propulsion Laboratory of the California Institute of Technology (JPL) and formally by the CSA to see if Canada could provide a lidar system to be incorporated in a proposal to NASA in response to an Announcement of Opportunity (AO) for the Mars 2003 lander. This AO was part of the NASA HEDS (Human Exploration and Development of Space) activities and the focus was on investigating the hazards to be encountered by astronauts and equipment on the surface of Mars. The purpose of this proposal was to develop an instrument group called MITCH (Mars Investigation of Total Climatological Hazards).

Canadian involvement in the MITCH proposal was a direct result of the previous development by Optech of a lidar system for JPL. In 1998 Optech had developed a highly successful engineering prototype lidar for JPL for use on the 2003 Deep Space 4/ Champollion comet-landing mission. Although this mission was cancelled, the lidar system had been very well received by the groups at JPL. In particular, the JPL MITCH team headed by Peter Smith of the University of Arizona recognized that Optech's Canadian-developed technology had the capability of meeting the needs of the MITCH lidar. JPL approached CSA and Optech to ascertain if this technology could be included as part of the MITCH mission. Optech determined that technically this was indeed feasible. The CSA agreed to fund this undertaking on the condition "that there is a significant interest in MITCH scientific investigations in the Canadian scientific community as confirmed by a panel of scientists involved in Space Exploration." The Canadian scientific interest in this opportunity was quickly established and the Canadian activities on the MITCH proposal moved ahead rapidly.

It is interesting to note that at that time in 1999 the NASA program had very ambitious and optimistic concepts for early human missions to Mars. It is also worth pointing out that the Canadian lidar played a key role in the proposal, since it was the principal instrument for remote detection and tracking of the dust storms and dust devils. Based on existing terrestrial experience it was felt that lidar could reliably detect incoming storms and dust devils with sufficient warning to allow shut-down of sensitive instruments and protection of personnel. These views are reflected in the following quotation from the MITCH proposal.

PHOENIX Begins

The focus of the MATADOR work changed abruptly in 2002 when NASA came out with the AO under the new Scout mission concept for the 2007 Mars mission. NASA's Scout selection process was the first fully competed opportunity for scientific missions to the Red Planet. Twenty-five teams assembled and submitted proposals in response to this Scout opportunity. Among these was the Phoenix proposal headed by Peter Smith. Although the Phoenix proposal did not closely mirror the work of MATADOR it did involve a number of the key components and research teams from MATADOR. The focus was no longer on dust and dust devils, but on the search for water and "habitable" zones on the surface of Mars where the conditions for past or present life could exist.

As the Phoenix proposal to NASA took shape Peter Smith, invited the Canadian team to participate in Phoenix by providing a lidar for studies of the Martian atmosphere. The lidar was to be used to directly observe dust plumes, ice clouds and fog layers and the depth and dust content of the Martian troposphere. From this distribution of dust and ice particles, scientists can make important inferences about how energy flows within the polar atmosphere, important information for understanding Martian weather. These particles serve as tracers to reveal the formation, duration, and movement of clouds, fog, and dust plumes, improving scientific understanding of Mars' atmospheric processes. Lidar measurements of the convective layer depth also allow the estimation of the altitude at which atmospheric convection deposits tracer species.

The CSA responded positively to the Scout opportunity and agreed to support Canadian teams proposing to collaborate with US-led proposals. Nine such Canadian collaborations were submitted (out of the 25 proposals received by NASA). As part of the nine, the CSA agreed to fund a Canadian-developed lidar for inclusion in the Phoenix proposal. Canadian work then proceeded apace to provide the lidar component of the Phoenix proposal. It is interesting to note that this Canadian contribution to the Phoenix proposal was undertaken mainly by staff scientists and engineers from Optech and MDA, since the focus was heavily directed to demonstrating that a lidar could provide the range of capabilities needed to satisfy the Phoenix requirements that were significantly different from those of the MITCH program. Allan Carswell, Arkady Ulitsky and Bob Richards headed up the Optech proposal efforts. At MDA Chris Sallaberger, David Hiemstra, Mike Daly and Andy Kerr led the work. These activities included not only the development of the concept and the quantitative capabilities of the lidar, but also the defence of the lidar part of the proposal before NASA adjudication and referee committees.

The need for Phoenix to include an advanced Martian weather station was recognized. As a result the Phoenix proposal to NASA included a weather station, MET, whose primary goal is to study the hydrological cycle at the landing site. Its secondary goal is to study the cycle of dust and other tracers and how they are affected by weather. MET incorporated an array of sensors for measurements of the temperature, pressure and winds at the landing site. Included in MET was the Canadian lidar and initially this was to be the only contribution of the CSA. As the proposal neared completion however, the US team recognized that inclusion of the full MET station would probably exceed the NASA stipulated total mission cost cap of $325 million. In response to this Peter Smith asked if CSA could expand its contribution to include the other MET sensors as well. This request came very close to the August

2002 submission deadline for the proposal and after many last minute discussions and deliberations CSA agreed to provide the essential MET component of the proposal.

The final Phoenix proposal proved to be highly successful, since in December of 2002, in the first step of a two-step process, NASA selected four out of the twenty-five proposals for a six month detailed implementation study as candidates for the 2007 "Scout" mission in the agency's Mars Exploration Program. Phoenix was one of the four. These four teams were informed that their science was accepted, but that they had to clearly demonstrate the full implementation capabilities of their proposed mission through a detailed mission-concept study. The results of these studies were due for submission by July 2003 and one of these would be selected for full development as the first Mars Scout mission to Mars in 2007.

The MET instrument with the lidar beam operating. Photo by courtesy of the Canadian Space Agency.

CSA funded the Canadian participation in the Phoenix implementation study as well as a similar study for a key Canadian contribution to the MARVEL proposal that was also part of the final four. As with the earlier Phoenix proposal this study involved mainly the industrial instrumentation teams at MDA and Optech, since NASA was already satisfied with the excellence of the Phoenix science in the proposal. The study involved six months of high intensity work to quantify the detailed properties of the proposed instruments and convince NASA of their mission performance capabilities.

During preparation of the Phoenix proposal and the implementation study the details of the Canadian contribution varied quite substantially. In fact there were two channels of proposal activity. The first was the work associated with responding to the NASA competition. The second was the need for the preparation of a proposal from the Canadian team to the CSA to provide the funds needed to go forward as part of Phoenix. This CSA proposal required the development of Canadian science and engineering plans with detailed schedules and cost estimates extending over the full time period from 2003 to the end of the primary surface mission in late 2008.

The study reports of the four contenders were submitted in July and on August 4, 2003 outstanding news arrived when NASA announced that Phoenix had been selected. The Phoenix Mission is the first of NASA's Mars Scout class and is led by Peter Smith of the University of Arizona's Lunar and Planetary Laboratory as Principal Investigator responsible for all aspects of the mission. At JPL Barry Goldstein serves as the project manager with Leslie Tamppari as the project scientist. Lockheed Martin Space Systems is responsible for the spacecraft flight systems. The CSA provides the overall management of the Canadian Phoenix activities. CSA personnel also play active roles in the Canadian science and engineering teams leading this first significant involvement of Canada in a science mission on Mars.

Allan Carswell of Optech and York University was designated as the PI of the Canadian Phoenix team and lidar Instrument Co-I of the NASA Phoenix Science Team. For the first Phoenix phase Optech was the prime contractor to CSA for the science team activities. This somewhat unusual situation reflected the lidar focus of the Canadian involvement. The initial Canadian Science Team included atmospheric scientists Diane Michelangeli and Peter Taylor from York, Tom Duck an atmospheric lidar expert from Dalhousie University in Halifax and, from University of Alberta, and Warren Finley and Carlos Lange, specialists in heat and mass transfer, computational fluid dynamics and multiphase flows.

Subsequently Jim Whiteway joined the faculty at York in September and because of his advanced lidar skills he was immediately co-opted as a member of the Canadian science team. Duck and Whiteway are former students of Carswell and earlier had been members of the York and CRESTech lidar groups. Finley later withdrew and was replaced by Dave Fisher from the Geological Survey of Canada. Fisher's expertise includes ice cores and paleo-climate history in Canada's Arctic and has enabled him to contribute significantly to studies of polar ice caps on Mars. In addition to these members the team included a number of post-doctoral fellows and research scientists associated with the faculty members. These Canadian Scientists brought a range of expertise well-suited to the needs of the overall Phoenix program.

As Phoenix proceeded, the Canadian PI changed several times. Carswell served as Canadian PI until March of 2004, when he withdrew for personal reasons. He still retains his role on the Phoenix team as Co-Investigator, Meteorology. Diane Michelangeli became PI but her situation was complicated by continuing ill health. During much of the following period Peter Taylor took on the responsibilities of the PI, along with his role as Co-lead of the Phoenix Atmospheric Science Theme Group (ASTG), a position shared with Nilton Renno of the University of Michigan. In the fall of 2006 Diane relinquished the PI position for health reasons and Whiteway assumed the role of PI of the Canadian science team.

The whole Phoenix team was greatly saddened by the news of Diane's death on August 30, 2007. The successful launch of MET on August 3rd as part of the Phoenix mission was a clear testament to her outstanding capabilities. Although Diane did not live to see the arrival of the MET data, her contributions to the mission will continue through the scientific skills of her research team at York. Diane's seminal work on the clouds and aerosols on Mars will provide important contributions to the analysis and interpretation of the MET measurements.

The Canadian industrial team is led by MDA as the prime contractor to CSA. Andy Kerr is the Phoenix program manager, with Mike Daly as the chief engineer. MDA is the world's leading space robotics company committed to offering hi-tech, sophisticated robotic and engineering solutions for space and terrestrial applications. Tasks for the development, testing and integration of temperature, pressure and wind sensors were added to the responsibilities of MDA with the incorporation of the full set of MET instruments. Optech has a leading role in the lidar development for the mission as the principal subcontractor to MDA. Passat Limited later joined the team as manufacturer of the laser for the lidar. Passat is a Toronto-based company specializing in the development of solid-state lasers and related nonlinear optical components. The full roster of the Canadian Phoenix Team is given in the York web site http://www.yorku.ca/dvm/Phoenix

Phoenix is intended by NASA to be a low-cost, low-risk mission, designed to land in the high northern latitudes of Mars, and follow up on Mars Odyssey's discovery of near-surface water ice in such regions. The Phoenix mission will land at a safe site inside the Martian Arctic Circle, to investigate the discovery of the Odyssey team. This zone is attractive because near-surface ice and higher surface pressure creates conditions where water may be periodically stable as orbital dynamics change the regional climate. Phoenix will land in terrain suspected of harbouring as much as 80 percent water ice by volume within 30 centimeters of the surface, and conduct the first subsurface analysis of icebearing materials to assess the astrobiological potential of the planet.

The lander for Phoenix was built and was being tested to fly as part of the 2001 Mars Surveyor Program, but the program was cancelled after the Mars Polar Lander was lost in December 1999. Since then, the 2001 lander had been stored in a clean room at Lockheed Martin in Denver, managed by NASA's new Mars Exploration Program as a flight asset. This lander was incorporated in the Phoenix proposal and as a result the name "Phoenix" was chosen since the spacecraft was arising from the "ashes" of the cancelled 2001 mission. Of interest is the fact that the beautiful and colourful logo of the mission depicting the mythological Phoenix bird against the backdrop of the planet Mars was designed by a young CSA engineer, Isabelle Tremblay, who was a key part of the management team for the Canadian contribution.

The principal goals of Phoenix are to seek answers to the questions: 1) can the Martian Arctic support life; (2) what is the history of water at the landing site and 3) how is the Martian climate affected by polar dynamics? As well as the Canadian MET, the Phoenix lander incorporates advanced instrumentation from JPL, several US universities and a number of European research centers. The University of Arizona hosts the Phoenix Science Operations Center (SOC) in Tucson where the science and engineering teams will command the lander once it is on the surface of Mars. Full details of the mission as it progresses are available at the web site www.phoenix.lpl.arizona.edu.

The Phoenix mission tests the hypothesis that a habitable zone can be sustained during the wetter epochs on Mars. In June 2008, Phoenix will reach the northern plains between latitudes of 65 and 75° N and operate for up to 150 sols (Mars "days" equal to 24 hours and 39 minutes on Earth) during the northern summer. A stereo camera scans the environment and obtains multi-spectral images of the terrain and the atmosphere around the lander. A robotic arm excavates a trench and gathers samples for analyzing the geology and chemistry of the regolith. A mass spectrometer sensitive to minute quantities of organic molecules baked out of heated samples enables assessment of the habitability of the soil in this ice-rich region. By analyzing constituents that go into solution, wet chemistry chambers test the conditions created during potential wet epochs. A microscope station examines the grain properties of the soils. Climate studies provide information on the present-day environment, including the transport of water vapor during the polar summer. Past climates are studied from records left in the soil layers.

The role of the MET station is to provide information for characterizing Mars' present climate and climatic processes. In particular, combined data from the lidar and the temperature, pressure and wind sensors will provide a comprehensive characterization of Martian weather at the landing site. The Canadian lidar will be the first lidar on the surface of Mars and will have the capability of probing the Martian atmosphere to an altitude of about 20 km to provide much new information on the Martian boundary layer. The structure and evolution of the planetary boundary layer (PBL) is key to understanding the surface-atmosphere interactions, particularly the exchange of volatiles.

The lidar-derived information on the formation and movement of clouds, fogs, and dust plumes will add valuable new information on the climate of Mars and will enhance the ability to model the key atmospheric processes. The MET lidar will also provide a quantitative record of the spatial and temporal optical properties of the atmosphere over an extended vertical region. This will include information on diurnal and seasonal changes, particularly with respect to the location and opacity of cloud, fog and dust layers that form and dissipate at various altitudes during a sol. Such information will contribute directly to the overarching theme to "follow the water."

In May of 2004 the mass of the overall Phoenix instrumentation had increased to beyond the allowed level. Mass reductions were mandated and MET had to shed about 2 kg. The only way to achieve this and retain the core MET capabilities was to de-scope the pan-tilt capability of the lidar. As a result the final Phoenix lidar is a fixed vertically pointing system. It operates simultaneously at wavelengths of 1064 and 532 nanometers with a pulse repetition rate of 100 Hz. The dual wavelength information offers some capability for ice/dust discrimination in the atmospheric signals. The final version of the total MET system has a mass of 6 kg and a power consumption of about 30 Watts. The lidar science activities are led by Jim Whiteway and Tom Duck assisted by Allan Carswell along with the lidar teams at York and Dalhousie. The York and Dalhousie groups play the leading role in the overall characterization, calibration and testing of the lidar system in support of the instrumentation work at MDA and Optech.

The very cold temperatures of the Martian Arctic will be measured with thin wire thermocouples, a technology that has been used successfully on meteorological stations for both the Viking and Pathfinder missions. Three of these thermocouple sensors are located on a one meter vertical mast to provide a profile of the temperature variation with height. Peter Taylor's team at York led the overall development and testing of the temperature and pressure sensors conducted in the CRESS Space Instrumentation Laboratory by Stephen Brown as well as the modeling of the Martian atmosphere. This work incorporates into the Martian boundary layer model a dust model that takes into account the particle size distribution. In addition Taylor's group is developing a model that can incorporate both heat and water transfer in the soil into the Martian atmospheric modeling.

Michelangeli's group at York is focusing attention on a Mars Microphysical Model

(MMM) that will provide predictions for times that are most favorable for the detection of ice clouds and surface fogs based on near-real time analysis of MET temperature, pressure, wind and lidar data. A goal is to update the MMM to include heterogeneous chemistry of the Martian atmosphere and to develop methods to integrate the data from the full range of MET sensors.

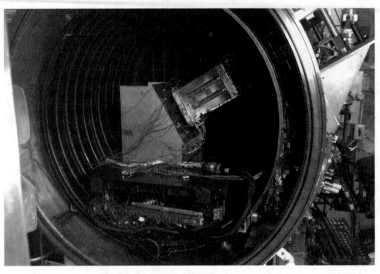

The MET instrument in a thermal vacuum test chamber at the David Florida Laboratory of the CSA. Photo courtesy of the Canadian Space Agency.

The MET wind measurements are being made with a telltale instrument (like a ribbon, blowing in the wind). Limitations of mass, power and data capacity ruled out the use of very sophisticated anemometers and the team opted for the elegant simplicity of the telltale. The calibrated movement of the telltale is observed and recorded by the Phoenix camera to provide quantitative information of the wind speed and direction. Carlos Lange from University of Alberta leads the Canadian wind measurement activities including the complex tasks associated with describing the flow over the lander and its effects on the temperature and pressure sensors. The work of his team includes small scale flow modeling to develop a 3-D model of the flow over the lander, to permit the correlation of the Phoenix in-situ measurements with ambient field values of temperature, pressure, velocity and humidity in the lander vicinity.

Phoenix had a perfect launch at 05:26 EDT on Saturday, August 4, 2007 on a Delta II launch vehicle from Cape Canaveral Air Force Station, Florida. It is scheduled to touch down on the flat arctic plains of Vastitas Borealis, near Mars' north pole, on May 25, 2008. Phoenix holds promise of providing a wealth of new information on the North Polar Region of Mars and on the potential for habitability in this region. The teams are looking forward with great optimism to the challenging and rewarding activities once the lander is on the surface.

Phoenix is still a work in progress but it can now be said that the mission is well along the path from A to B. To date the Canadian progress along the way is the result of the contributions of several hundred engineers and scientists. The full Phoenix mission has no doubt involved about ten times this number of contributors. The path on the way to Mars over the last decade has had its many twists and turns and a few big bumps. However, the direction of the journey has always been driven by several unwavering forces. The first is Peter Smith's recognition of the high science value of having a lidar on the surface of Mars and his wish to have the Canadian team provide it. The second is the continuing dedication of Allan Carswell to provide the know-how in response to Peter's invitation. Finally, there is the creative resilience of CSA to respond supportively throughout the vagaries of the journey. In particular, the enthusiastic support of Marc Garneau, the CSA President when the mission was selected, and the Space Science Program through David Kendall (Director General, Space Sciences) and Alain Berinstain (Director, Planetary Exploration and Space Astronomy) has ensured that CSA could provide the necessary resources even while often constrained by stringent fiscal and financial limitations. We all look forward now to the arrival at point B in the spring of 2008 when the stream of science information begins to flow from the Phoenix systems.

Postscript:

Space is indeed vast in its territory and its scope. Although Canadian space science began on the Earth and is still very much involved with its surface and its near-environment, other Canadians have been reaching farther and farther out into space. The study of the Earth from space will continue, and will continue to be enhanced, as this is where humanity lives. However, the greater scientific surprises will likely come from distances much farther from Earth.

References:

Broten, N.W., Early days of Canadian long-baseline interferometry: Reflections and reminiscences, Journal of the Royal Astronomical Society of Canada, Volume 82, page 233, 1988.

Galt, John. Beginnings of Long-Baseline Interferometry in Canada: A Perspective from Penticton, Journal of the Royal Astronomical Society of Canada, Volume 82, page 242, 1988.

Gush, H.P., Beginnings of VLBI in Canada, Journal of the Royal Astronomical Society of Canada, Volume 82, page 221, 1988.

Locke, J.L., The Twentieth Anniversary of Long-Baseline Interferometry, Journal of the Royal Astronomical Society of Canada, volume 82, page 219, 1988.

Chapter 9

Canada's Future in Space

Co-operation and competition in space

Working together has been a characteristic from the early days of the space age. The US has been very generous in sharing its space opportunities and space technologies with other countries, and Canada has been a benefactor of that. However, this was not done solely through benevolence; NASA followed this practice with the intention that those partners who learned of the benefits of the NASA systems would become customers, and this did happen in the short term. Eventually the country partners would decide it was better to become independent, even with the cost involved in doing so. From this point on they became competitors with NASA. This constant balance between co-operation and competition continues to mark the international character of space. These ideas, reflecting conditions in 1985, especially in the US Canadian context were discussed and recorded in a book edited by John Kirton (1986).

Early examples of this are in the fabrication of satellite launch vehicles. Originally only the USSR and the USA had this capability, and Europe made use of launch vehicles from both these countries. When ESA eventually decided to develop the Ariane it got no further help from the US and was not even able to buy US goods or services for this development. However, they succeeded and have had a highly successful commercial launch vehicle since 1979. In a somewhat different approach, the US assisted Japan with the development of its "N" vehicle, on the condition that it would be used only to launch Japanese payloads. Japan later developed its "H" vehicle without the use of US technology, so that it would be able to market it worldwide. Today China and India have substantial launch capability. This duplication and competition between countries increases the capability and level of space activity worldwide. This leads all countries to seek more and more applications.

The US has certainly had major benefit from its co-operative enterprises. One can argue that the space shuttle would never have been accepted by the US government without the contributions by Europe of Spacelab and Canada of Canadarm. Similar arguments, stronger ones, apply to the International Space Station, despite all the difficulties involved.

Communication satellites provide another example. Writing in Kirton (1986), David A. Golden, Chairman of the Board of Telesat Canada was then somewhat pessimistic. At that time operators were facing increased launch costs (Space shuttle or Ariane), increased insurance costs, increased competition from optical fibres, where the technology was advancing much more rapidly than expected, and increased bandwidth in terrestrial microwave communications. For Canada, the manufacturing of satellites is made more expensive because there is no parallel military industry to share the cost of development of these systems. Although the original Canadian space program was predicated on the development of a Canadian company that could provide its communications satellites, this has simply not turned out to be viable, despite a valiant effort. It would have been possible only if sales to other countries had been significant, which did not turn out to be the case.

However, the communications satellite business did stabilize and is still highly successful and important to Canada. In other countries it is even more important. In India there are many problems in implementing fibre optics systems while it is relatively easy for India to build and launch its own communication satellites. The demand has grown so rapidly that India is running out of space in geosynchronous orbit. They have 9 communications satellites in orbit at the moment; with six slots they plan to put three satellites in each slot, parking the satellites just 12 km apart, a remarkable accomplishment.

Canada has collaborated with many other countries, especially in space science as described in earlier chapters. Canada collaborated with the USSR and later Russia in search and rescue satellites as well as through participation in the Mir space station.

The Canadian space industry

Generally speaking, space activity can be divided into space science and space applications. In Canada space science is a relatively small part (about 15%), which does not allow for spectacular space science activities. In the US there are extensive space activities outside NASA. The military is one, and applications such as Earth Observations and Weather Satellites are transferred to other organizations in government, such as the National Oceanic and Atmospheric Administration (NOAA). The Canadian space program has always been based on the desire to support the national industry, with considerable success in terms of the export market that Canadian companies have developed.

Canada has always felt the need for at least one space industry that can act as prime contractor for major missions. While that didn't work out for communications satellites it has proved feasible until recently for the Canadarm activity and RADARSAT. Going back in history the prime contractor was intended to be RCA, but later became Spar Aerospace, and after some other maneuvers became MDA (MacDonald Dettwiler and Associates Ltd). But on January 8, 2008 MDA announced that it was selling, for $1.325 billion, its satellite and space division businesses to Alliant Techsystems (ATK) of Minnesota, whose web banner reads, "An advanced weapon and space systems company." ATK is a major space company, leading the market in solid rocket motors, including the shuttle solid boosters. They are also the main contractor for the first stage of the new Ares-1 vehicle that will replace the space shuttle, but they also manufacture ammunition. The plan, according to ATK is that the 1,900 jobs involved, including 800 engineers, will stay in Canada. John MacDonald, a co-founder of MDA, established 40 years ago, was quoted in the Financial Post as saying: "If the company wants to fulfill its destiny, it has to do this. None of us likes it, but the market for the space side of MDA isn't really in Canada. But we're all disappointed." After noting some similarity to the Avro Arrow, MacDonald went on to say that, "But I think from a business point of view, it's the right choice. It's unfortunate, but I mean – isn't that the story of Canada?" The sale is expected to close in March, 2008, pending approval. This raises a host of questions for Industry Canada, and the CSA. A significant complication is that in order to foster privatization of Earth observations, the Government of Canada had transferred ownership of RADATSAT-2 to MDA, so that this highly advanced radar satellite will now be owned by ATK. It is ironic that after the US decided to withdraw its support for the project and the CSA decided to go to it alone, at a cost of about $500 million of Canadian taxpayers' money, that it is now owned by the US.

Critics say that the sale to a weapons manufacturer is in conflict with the primary mandate of the CSA, which is, "To promote the peaceful use and development of space." Marc Garneau said that "If this deal goes through, it's a sad day for Canada. There's no question that [MDA] is the jewel in the crown in terms of Canadian space capability, so I am very concerned." Some MDA senior employees have resigned because of the military aspect. Supporters of the sale say that the 1,900 positions in Canada will have a more secure future because of the ATK access to the military market in the US. The fundamental question for Canada is, "Why are we creating Canadian space companies that are too large to exist on the Canadian market alone, and that can only survive with foreign ownership." The answer seems to be that it is the small and medium size companies that should be fostered, as they have done very well with their export sales. The CSA retains the right to data access from RADARSAT-2 and so Canada may not suffer in the short term. Canadarm2 will still be called Canadarm2. If the supporters are correct, then the robotics and radar technologies will remain in Canada, ATK will be a good Canadian corporate citizen and business will flourish. Only time will tell.

A smaller company that has done well in the space communications subsystems business is COM DEV of Cambridge, Ontario. Another company that has specialized in space science instruments was Canadian Astronautics Limited, later renamed CAL Corporation before being purchased by EMS of the US. More recently its Ottawa office that has been responsible for science instruments (the other was the Spar Aerospace component in Montreal), was purchased by COM DEV and has become known as COM DEV Ottawa. This organization built the ultraviolet auroral imagers and the WINDII instrument. Earlier some CAL employees split off from their parent company and formed a smaller company called Routes AstroEngineering and they built the OSIRIS instrument. Bristol Aerospace in Winnipeg is still a significant player that manages to sell Black Brant rockets to NASA, and built the satellite "bus" for SCISAT. There are a number of small companies such as Spectral Applied Research in Concord, Ontario (just north of Toronto) and Resonance Limited in Barrie. Having built the FTS Michelson interferometer for SCISAT, ABB Bomem are now well established in the space business as well.

Arthur Collin, secretary to Canada's Ministry of State for Science and Technology as well as the chief science advisor to the Government of Canada in 1985 (Kirton, 1986), discussed the Canadian long-term space plan at that time. The three projects on the table were 1) MSAT; 2) RADARSAT and 3) the International Space Station. It was felt that to accomplish (i.e. be funded for) all three would be a remarkable accomplishment. In spite of economic difficulties at that time, this happened. The greatest impact was expected to be that from the space station, saying that "when it is launched in the early 1990s, the space station will be one of the most complex and visible international achievements in history." Although Canada's role in the construction of the space station has been significant owing to the Mobile Servicing System (incorporating Canadarm2), the delays in construction mean that the benefits of utilization are still in the future.

Canada's co-operative approach to space activity has allowed a highly successful program at the least cost. But the approaches are different with different countries. With ESA the relationship is highly structured, permitting Canada to be involved in the planning process. With the US there has never been a general agreement; rather, Canada has provided specific components to particular projects, limiting the scope of the involvement to that project, as with Canadarm and Canadarm2. The reverse has also been true such as for RADARSAT-1 where Canada built the entire spacecraft and it was launched by the US However, for

RADARSAT-2 Canada has had to go it alone, raising the question as to whether the control of one's program is worth the additional cost. Canada answered this in the affirmative at that time. With science projects, the normal approach has been the provision of an instrument to the collaborator that incorporates it into their payload. This is an extremely effective way to do science at the least cost, as was accomplished with Sweden and Japan as well as with the US. Generally the platform provider does not ask for a financial contribution to the bus costs, as in such bilateral agreements no funds are exchanged, only responsibilities. But there are other ways to share the financial burden and for the Swedish Viking satellite Canada provided some support in the form of Black Brant rockets. In 2004 the CSA launched SCISAT, the first Canadian-built scientific satellite launched since ISIS II in 1971, and is currently planning future Canadian small scientific satellites. However, even SCISAT was launched by the US, on a Pegasus vehicle, a part of the compensation for Canada's contribution to the ISS.

David Williamson, senior fellow in science and technology at the Center for Strategic and International Studies, Georgetown University, Washington, D.C., and formerly with NASA argues (Kirton, 1986) that the US does not have a real space policy, in spite of its capacity. What it is, is a "collection of industrial policies, commercial policies, scientific policies, trade policies, anti-trade policies, military policies and alliance policies" that is very difficult to translate into a space policy. These different rationales for the space program doesn't stand up to hard academic scrutiny. This reflects the interdisciplinary nature of space research and the number of jurisdictions that it cuts across. Effectively, "the United States is in the space business because it can afford it. Space is pretty much icing on the economic cake." If there was retrenchment, the US space program would suffer. Yet in spite of what Williamson says, the US public accepts a space program that spends more per citizen than any other country.

This pessimistic view would appear to make it difficult for Canada to formulate an effective policy. But Canada has some advantages. The relationship between industry and the CSA is more collaborative in Canada than in the US, and this has been borne out by experience. But as noted earlier, the results cannot always be predicted.

How Does one Formulate a Program?

As described in earlier chapters, space science in Canada grew rapidly through immediate responses to external forces existing at that time. The responses evolved into clearer objectives over the decades to the program that exists today. But one should not assume that having proven its worth, a space program can simply continue doing the same things as before. At the COSPAR Assembly in Montreal in July, 2008, the Fiftieth Anniversary of the organization will be celebrated – with four talks: one a look backward, but the other three looking forward. Forward is the right way to look. The USA tends to lead the way, with the International Space Station (ISS) in the near future and landing humans on Mars in the more distant future. But even the US cannot achieve these objectives alone; they have, and will continue to have international collaborations. Projecting ahead a full fifty years to 2058 is difficult. A human presence on Mars may just barely be accomplished by then. The driver for Mars is the human desire to explore, supported by the public will to see its scientists make novel discoveries. Life on Mars is a great driver. But this is a conditional driver, conditional on the required budget and the available resources. There is no necessity to get there in twenty years, or fifty years; another decade or two does not present a problem.

Others might take a more aggressive point of view. There clearly is available water on Mars, so it is feasible to sustain life there, to colonize it, so to speak. It could be made truly habitable by creating an oxygen atmosphere, as the Earth did, while Mars did not. This process is considered as having two steps. Pioneering studies on this were carried out by a Canadian, from York University, Robert Haynes, who passed away in late 1998, not long after he became President of the Royal Society of Canada. Haynes coined the word "ecopoeisis" from two Greek roots, "an abode" and "fabrication" and it is now accepted by workers in the field. To achieve this on Mars would involve 1) increasing the global surface temperature by about 60 °C; 2) increasing the atmospheric pressure; 3) the presence of liquid water and 4) the reduction of surface ultraviolet and cosmic ray radiation (this comes naturally through increasing the pressure of the atmosphere). Reaching this stage would allow anaerobic life to develop (life that does not require the presence of oxygen). It is argued that just a little (engineered) warming through releasing stored CO_2, or placing large solar mirrors in space might be enough to initiate the required global warming. This could be done in as little as 200 years. The next level is called terra-forming, which provides an oxygen atmosphere that supports aerobic life. Photosynthesis is the route to this; the concentration of oxygen needed is not necessarily as great as on Earth for many life forms (terraforming involves more than human life!). This process could take thousands of years. One might ask why the creation of new planetary atmospheres should be considered when our own is not fully under control. Will a corner have been turned in our response to Earth's warming by 2058?

Indeed the study of Earth's atmosphere from space is becoming a priority because of global warming, as it was earlier because of ozone depletion. That latter challenge, which was met admirably by scientists and policy-makers world-wide is still not over, as while the point of recovery is thought to be at hand, it is not yet confirmed by observations. Canadians are fully aware of the challenge of global warming as they watch the Arctic change perceptibly on a year-by-year basis, with new species being observed there never seen before by Arctic dwellers, as well as the prospects of ship traffic through the Arctic Ocean. It is also evident at mid-latitudes. A taxi-driver in Rome commented that they had no winter last year, only as cold as 14 °C. These are effects of which the public is very aware. Some parts of Canada may benefit from global warming, but for most there is apprehension, as detailed predictions are so limited. There is no doubt that the monitoring of our atmosphere, shared with other countries, and the surface of Canada, will be a priority and that the problem will not be solved in fifty years. A great improvement in instruments and in the sophistication of the data analysis methods can be expected, as well as an enhanced capability of making predictions with computer models. Atmospheric modeling will play an increasingly greater role.

ESA places great importance in having a balanced plan. It allocates considerable resources to the observation of Earth and its atmosphere, with the orbiting ENVISAT just one example of this. Currently they are considering a suite of missions: BIOMASS, TRAQ (TRopospheric composition and Air Quality), PREMIER (PRocess Exploration through Measurements of Infrared and millimetre-wave Emitted Radiation), FLEX (FLuorescence EXplorer) to observe global photosynthesis, A-SCOPE (Advanced Space Carbon and Climate Observation of Planet Earth) and CoReH2O (Cold Regions Hydrology High-resolution Observatory) to make detailed observations of key snow, ice and water cycle characteristics.

However, within the ESA exploration also shares the spotlight. SOHO (Solar and Heliospheric Observatory), the Huygens probe that successfully landed on Titan, Saturn's largest moon; the Mars Express already described; Rosetta, now on its ten-year journey to Comet 67 P/Churyumov-Gerasimenko are all examples of exploratory missions covering a wide range of objects in and outside the solar system. The BepiColombo mission to Mercury, in collaboration with JAXA consists of planetary and magnetospheric orbiters that will explore the origin and influence of the magnetic field of that planet; it is to be launched in 2013. Fifty proposals have been received for ESA's Cosmic Vision, to unfold between 2015 and 2025. The ESA Columbus module for the ISS was launched on Atlantis on February 7, 2008, which will mark the beginning of European activity on the space station. The US has perhaps concentrated more on major facilities for space exploration such as the well-known Hubble space telescope and its successor, the James Webb Space Telescope, a huge telescope that will be sensitive enough to reach far enough back in time to examine the first galaxies that formed in the early universe. It will be launched in 2013. But in late 2008 ESA will launch the Herschel spacecraft, as described earlier.

Smaller countries cannot afford to have a balanced program in the sense of covering a bit of everything. They must look for niche areas, and at the same time collaborate with large organizations like ESA and NASA in order to be a part of some major endeavours. Both China and India have chosen the moon as their stepping stone to more distant space, as the US did earlier, only to abandon it—now to go back again. In the reverse of what most countries did, India began with its application satellites (communications, weather and Earth observations) and is now expanding its scientific missions, currently ASTROSAT which includes collaboration with Canada. China has an extensive array of communications, weather and Earth observations satellites as well as an astronaut program with considerable ramifications for the future. Japan has followed a more independent and ambitious track, covering a broad range, with some notable successes such as Akebono, YOHKOH (a solar mission with US and UK collaborators), HALCA and now Hinode (which means sunrise), a successor to YOHKOH with NASA and ESA participation.

Canada's space future will continue to be firmly associated with NASA. This brought major successes in the past and likely will continue to do so in the future. Working collaboratively with NASA, and its other partners will always continue to be the most effective way to work at the forefront. Making sophisticated measurements from space is just too complicated for a single small country to manage on its own. Current Canadian collaborations with NASA are MOPITT (Terra), ISS, Phoenix, JWST, Cloudsat and THEMIS (Time History of Events and Macroscale Interactions during Substorms). THEMIS is a suite of five spacecraft that are lined up along the anti-sun line in the Earth's magnetotail in order to understand what causes an auroral substorm (event). This is being supported through ground based imagers provided by Eric Donovan at the University of Calgary and magnetometers provided by Ian Mann at the University of Alberta.

With a Canadian instrument on the Phoenix mission to Mars, launched in August, 2007 and due to touch down in May, 2008, Canada's appetite for Mars has been stimulated. But an all-Canadian mission, while in the minds of some, seems still to be over the horizon.

Approaches by other countries

ESA's approach is more organized and more predictable in the long term. Canada is an Associate Member of ESA. As such it contributes to general revenues an amount proportional to its GNP, but all of this money apart from an administrative and infrastructure component flows back to Canadian companies through contracts for projects. In addition to this base cost, Canada can then subscribe to any optional program (all its programs other than space science) at any level it wishes. The total cost is between $25M - $30M annually. The rationale for joining ESA was to strengthen other areas of space activities, especially satellite communications and Earth observation, with a view to strengthening Canadian industrial capacity and international share in these commercial and semi-commercial activities. The ESA approach to selecting its future missions is completely transparent, but still extremely competitive. Canada is involved in the ESA Swarm mission, but only because to ESA it is not space science, but Earth Observations.

SWIFT went the ESA route, under the Earth Observations program, and was highly praised, but not selected for a mission. However, ESA encouragement helped make the Canadian Chinook mission possible, the one that will carry SWIFT. Swarm, a constellation of satellites that will study the Earth's magnetic field in extreme detail, was selected and will carry instruments from the University of Calgary, led by David Knudsen. John Miller, of York University, participated in the SPECTRA (Surface Processes and Ecosystem Changes Through Response Analysis) project, but it was not finally selected for a mission. John McConnell of York University has participated in atmospheric studies for ESA (see the earlier list). Although there are many good reasons why participation in ESA missions would be beneficial to Canadian space science, the main impact to date has been on industry.

Countries that wish to be labeled "space-faring" nations need, in addition to a collaboration, some independent autonomous programs. A good example of this is France, with its Centre National d'Etudes Spatiales (CNES) that conducts its own significant missions in addition to having an active participation in ESA. After the USA, France in the past has spent more money per capita on space than any other country (current data are hard to find but this may no longer be true). This is almost certainly due to its space pioneers, in particular Jacques Blamont from the Service d'Aeronomie of the Centre National de la Recherche Scientifique (CNRS), who promoted space research in France from its earliest days. He was a contemporary of Hunten and Vallance Jones, and he had a particularly good working relationship with Hunten, and jointly with Tom Donahue of the US, who sought Canadian support for the Apollo-Soyuz linkup. Like Canada, France does not have its own launch capability. This sets these two countries apart from Russia, Japan, China and India, that do have this capability and so put their national programs first and their international collaborations second. It is perhaps because of this that China and India began with their applications missions first, with the scientific missions following later. Japan was unusual in having two independent space organizations, one for science called the Institute for Space and Astronautical Sciences (ISAS), originally part of Tokyo University, as described earlier in connection with Obayashi and Jacobs, and another for applications called NASDA (National Space Development Agency). These organizations even had their own independent satellite launch vehicles. However, more recently these two bodies have been joined into one, called JAXA (Japan Aerospace Exploration Agency).

It is interesting that the UK, which was a leader in the pioneering days of space, led by Sir Harrie Massey, does not have a space agency, and a rather small national program, coordinated by the British National Space Centre (BNSC) executed primarily through ESA. Although the BNSC is considered to be underfunded, in fact it does not have a budget for space missions and payloads – the funding relies on money from other government sources, such as the granting councils (PPARC, etc.) and therefore is not very effective – there have nevertheless been innovative experiments produced. The University of Surrey space group with their Surrey Space Centre (SSC) has a world-wide reputation for building satellites in a short period of time. Their commercial arm is Surrey Satellite Technology Ltd, which has built many microsats (up to 100 kg) and minisats (400 kg). In 2000 they launched a demonstration nanosat (defined as less than 10 kg) called SNAP (Surrey Nanosatellite Applications Platform) that had a mass of 8.5 kg and featured 3-axis stabilization, propulsion and GPS autonomous control of the orbit, for station-keeping, for example. This is needed if one wants to pass over the same location each day. Colin Pillinger of the Open University led a low-cost mission to Mars, called Beagle 2. However it disappeared mysteriously during its landing, and contact was never established; another mission defeated by the Red Planet. The Royal Society of the UK is currently calling for a UK Space Agency to replace the BNSC. The UK last year contributed just 7% of ESA's total budget compared with 25% from France and 20% from Germany which means its returns are in the same proportion. Britain has not been the prime contractor on a major European mission since Giotto in 1986.

The Scandinavian countries have a long-standing interest in space owing to their proximity to, and long-standing interest in, the aurora. Norway maintains the Andoya Rocket Range, with a capability of launching high altitude rocket flights; many Black Brant rockets have been launched from here since the closing of the Churchill Research Range. The early pioneers were Størmer, Vegard and Birkeland; Leiv Harang was one of the more modern pioneers who overlapped with the space age. However, space activity in Norway falls under NDRE, the Norwegian Defence Research Establishment. Sweden also operates an ESA rocket range called ESRANGE, near Kiruna just inside the Arctic Circle in northern Sweden. They also operate a satellite ground station there; at high latitudes data can be downloaded from polar satellites on every orbit. Sweden has shown remarkable organizational ability and foresight in concentrating economic, scientific, environmental and educational resources on Kiruna. Kiruna was originally an iron ore mining town, with a long-standing Kiruna Geophysical Observatory, established by Bengt Hultqvist, one of Sweden's space pioneers. Another pioneer was Hannes Alfvén, a Nobel prize-winning founder of space plasma physics, whose ideas about currents flowing along the Earth's magnetic field lines bolstered those of the Norwegian Kristian Birkeland and eventually won out against Sidney Chapman's objections. His successor at the Royal Institute of Technology in Stockholm was Carl-Gunne Fälthammar, who joined with Hultqvist to create the Viking and Freja satellites, with Canadian participation as described earlier. The Odin mission followed, led by astronomers and Georg Witt on the atmospheric side, from Stockholm University. Partners were Finland, France and Canada. The Swedish space program is run by the Swedish National Space Board (SNSB), which provides coordination and funding, but is not an actual space agency. Denmark has a lower profile, but concentrated on specific areas, like Earth's magnetic field measurements. There is a Danish Space Research Institute.

Brazil is working hard to get into the space club, including their own launch capability. Its space agency is called INPE (Instituto Nacional de Pesquisas Espaciais). A disastrous launch vehicle accident in the laboratory has slowed their progress, but they are continuing to press forward. Taiwan has a small but very successful space program operated by its National Space Organization (NSPO), in which they have carefully selected missions that have unique characteristics or otherwise attract worldwide attention. The latest, FORMOSAT 3, also called COSMIC (Constellation Observing System for Meteorology, Ionosphere and Climate) is a constellation of six GPS radio occultation satellites whose results are being utilized by weather forecast models worldwide. South Korean President Kim Dae Jung has recently established a five-year initiative to design, build and launch a commercial space cargo rocket, and to use the rocket to create a commercial launch business in South Korea. It is not clear how much room is available in the commercial launch business so this is a challenging initiative. Interest in space activity is not waning, by any measure.

What of Russia, the country that started the space age? Despite the political perturbations it has undergone, Russia has maintained a strong space capability. Their ability to launch and operate the space station Mir, then bring it down successfully into the Pacific Ocean, and to support the building and servicing of the ISS is remarkable. The have also provided inexpensive launches to many other countries, drawing on their ballistic missile reserve, scheduled for depletion. Scientific expertise is concentrated in IKI (Space Research Institute, in English). At one time IKI was led by Roald Sagdeev, one of whose many projects was the Apollo-Soyuz mission. He is now a Distinguished Professor of Physics at the University of Maryland. He is married to Susan Eisenhower, granddaughter of the former President Eisenhower. IKI is now under the direction of Professor Lev Zelenyi. There have been few space science missions in the last two decades. The most noteworthy were the VEGA-1 and VEGA-2 missions that delivered balloons and landers to Venus in 1985 on their way to a rendezvous with Halley's comet in 1986. A very ambitious VLBI project called RadioAstron that will carry a 10-meter diameter radio telescope has required many years of effort and is planned for launch in late 2008. But the future looks brighter for Russian space science.

Canada's Future

After 50 years, it is an appropriate time for Canada to orient itself for the next fifty years in space. It will want to continue its collaborations with NASA, but the role is not as yet defined. Currently Canada is acquiring data from MOPITT on Terra and from Cloudsat, while the Phoenix mission is on its way to Mars, the James Webb Telescope is under construction, and the ISS is approaching completion. Taking advantage of the ISS research facilities, when they become available to Canada, will be very expensive and challenging. But Canada cannot afford to be left out of some major NASA and ESA missions.

As has happened before, a new big idea may come along. One possibility is already developing. It has long been known that when a mercury container is rotated about a vertical axis the surface of the liquid mercury takes up a parabolic shape, the same shape as is required for the big primary mirror of an astronomical telescope. Ermanno Borra of Université Laval in Quebec City was the first to construct useful telescopes this way, of a few meters in diameter, costing a factor of ten or more less than a glass telescope mirror of the same size. The problem is that it can only look upwards, as it cannot be tilted. This is not a problem for lidar systems (light radar) probing the upper atmosphere, as they only need to look vertically. Borra implemented one such lidar system, called the Purple Crow Lidar at the University of Western Ontario under the leadership of Robert Sica, assisted by Stoyan Sargoytchev of York University and it has performed well. Astronomers have also found other ways to use it, as at UBC by Paul Hickson. However, Borra has since found a way to tilt a liquid mirror. In a tilted mirror an element of liquid on the high side of the tilt will tend to move downwards towards the axis and one half revolution later when it is on the low side it will tend to move away from the spin axis. If the liquid is sluggish (viscous) then it won't move significantly during this half-revolution and in the long term the two motions cancel, so the tilted surface actually remains stable. The problem is to find the right viscous material, but there now exist Metallic Liquid Like Films (MELLF) that meet the requirements. But large astronomical telescopes have another problem: turbulence in the atmosphere means that incoming light from the star on one side of the mirror is bent, by the atmosphere, differently from that on the other side, blurring the image. Although astronomical telescope sites are always chosen at high altitude, where the air is thin and as still as possible, this remains a fundamental problem. An alternative solution that has been adopted is to make the mirror of many segments, each independently controlled in orientation by a computer system in a way that a change in the angle of the segment compensates for the deflection of the atmosphere for that segment at that moment and this does improve image quality. But as described in *Macleans* magazine, July 9, 2007, Borra aims to solve that problem completely by putting a liquid mirror on the moon, which has no troublesome atmosphere—a mirror of some 20 to 100 meters in diameter. Mercury wouldn't work here anyway because of the low temperature, but Borra has found a salt solution that can be coated with vapourized silver that meets the requirement for temperature (and hopefully viscosity, although the article doesn't say that).

Another activity that is difficult for scientists to comprehend is the Virgin Galactic offer to fly passengers to 110 km above the Earth, and then glide back, taking 2.5 hours altogether. In the *Globe and Mail* newspaper of July 18, 2007 it was announced that John Criswick is one of six Canadians that have made a deposit. John received an M.Sc. degree from York University in 1988, applied to be an astronaut and was not selected. Since then he has done well enough in cell-phone technology to afford the $200,000 price tag. Compared with conventional rockets this is a relatively inexpensive way to get packages into space, so who knows where this will take us in the future?

Canada will also maintain an autonomous space science component. After ISIS II was launched in 1971, Canada returned to its own satellites only with SCISAT, launched in 2003. The next Canadian satellite is to be Chinook, to be launched around 2013. One expects on the applications side that the RADARSAT series will continue, with enhancements, perhaps including optical observations, although this view is currently clouded by the ATK acquisition of RADARSAT-2. In the near future, Cassiope with the ePOP (enhanced Polar Outflow Probe) suite of 8 scientific instruments, is the next Canadian small satellite mission, intended to be launched in 2008. This is a dual communications and magnetospheric mission. The first component is the experimental Cascade payload to demonstrate the world's first commercial space-based digital courier service. Similar to the operations of a traditional courier company, Cascade will pick up a data parcel, anywhere from 50 to 500 Gigabytes and deliver it to a recipient somewhere else on the globe by the next day. MDA is the prime contractor for this, and will operate the satellite. For the second component, ePOP, six of the eight instruments on board are to be manufactured in Canada, one in Japan and one in the US.The purpose of this CSA-financed project is to help researchers gain a better understanding of how our space environment is influenced by solar variability. Design and manufacture of the ePOP has been carried out by a team from the Institute for Space Research at the University of Calgary, directed by Dr. Andrew Yau. The probe will be able to study meteorological phenomena in the transitional region between the atmosphere and the magnetosphere, where the solar wind interacts with the Earth's magnetic field and creates, in particular, the Northern Lights. The ePOP science team is made up of researchers from ten Canadian universities, two foreign universities, and research agencies. The instrument will carry a series of plasma imagers, radio receiver instruments, magnetometers and a CCD camera to make in-situ measurements at resolutions never previously achieved. It will also do an almost real-time tomographic scan of the ionosphere in conjunction with Earth-based radars.

But launches every seven years are not frequent enough to satisfy the needs of individual researchers or the national program. There are very strong arguments for starting a small scale component, based on balloons, rockets and small satellites that could involve students and be built at universities, in order to restore in part the pace of the sixties. Steps to achieve that are in progress.

It may be time to revisit the question of Canada's launch capability. Bristol Aerospace has the capability to do this, but despite the recommendation of the John Chapman report, this has never been implemented. Nanosats are normally launched as piggy-backed on other satellites, but this is very restrictive. After Marc Garneau's resignation as President of the CSA a series of acting Presidents followed, before Laurier (Larry) Boisvert was appointed President in April, 2007. Boisvert had spent 34 years with Telesat, the final 13 as President and CEO during which Telesat achieved dramatic growth and became an international competitor in satellite services. He was prepared to establish a new direction for the CSA as well, involving some of these considerations above, but also reducing the level of bureaucracy to something closer to that of the private sector. However, he resigned in December, 2007, saying that a President with a longer tenure would be required to meet this challenge. Guy Bujold, formerly Assistant Deputy Minister at Industry Canada was appointed interim President on January 1, 2008.

Canada's contributions in the future will depend on its continuing to find niche areas. This has been true from the beginning of the space age and will continue to be so. Expertise in imaging from space was established with Viking, and continued with Freja, Interball and Image, and will be re-established with ePOP. Expertise in Doppler Michelson interferometers led to WINDII on UARS, and SWIFT on Chinook. An opportunity to put a small wind instrument on a Mars mission was missed, but this opportunity may recur. Expertise with diffraction grating spectrographs as in OSIRIS, MAESTRO and ARGUS (to be launched on CanX-2 in late 2007) will continue to be important. Lidar from space will become an important tool, but at present is in the category of large expensive missions. Canada's short-range lidar expertise on Mars, developed by Optech and MDR is a wonderful way to begin. RADARSAT expertise is also important.

Canada's ideal space future would consist of an environmental program with both collaborative and autonomous elements, having both science and applications components, an exploration program and an active small payloads program.

The benefits of space research

This story has been about fifty years of exploration, motivation, imagination, challenge, initiative and effort. The resulting success of these early undertakings certainly encouraged Canada to continue at top speed. But what has been the benefit of all that activity, and can it be justified in terms of dollar cost? There was an immediate benefit to those who participated, the scientists, engineers and technologists. But the knowledge gained by those scientists of the space about them, from the Earth's surface to the edge of the universe is valuable not just for them, but for their institutions, their country and the world at large. Canada needs knowledgeable space scientists, as leaders, as educators and policy-makers. It receives the benefit of knowledge gained by all the other countries involved in space activity because Canadian space scientists are able to absorb it, are capable of working within international teams and of presenting their work at international conferences such as COSPAR. Thus the knowledge gained by Canadians has its own inherent value, but it is also the access to international involvement. Similarly the engineers and technologists are a national resource. Their skills and capability form the basis of space activity including design, fabrication and the export market, but these skills may be transferred to areas other than space. Space activity can be expensive, as it may require large numbers of people with wide-ranging skills for a given project, but the dollars spent do not fly into space with the rocket; they stay on Earth with the employees, who not only provide these benefits for the population, but inject their dollars back into the economy. As transmitted by the media and by the public through computer networks, this knowledge is shared and appreciated by all Canadians.

Canadians use satellites all the time, while scarcely being aware of it. When one calls a service provider for help with software, the advisor may as well be in Bangalore as in Kenata. The Alouette/ISIS satellites were intended to study the ionosphere, to solve the problems of northern communications, but they became auroral observatories, unlocking many of the secrets of that elusive phenomenon while at the same time, the solution to northern communications problems emerged—as communications satellites. Thus the Alouette/ISIS strategy evolved to the creation of a Canadian company able to build (and export) communications satellites. This strategy didn't work as intended, but Canadian companies are actively involved in communications, and Canadian citizens are constant users of satellite communications, in making long-distance phone calls, watching television and using the Internet. Canadian hikers and drivers use the Global Positioning System routinely without visualizing the satellites overhead that make it possible.

The global society is a concept that is already accepted in principle, and exercised in practice to the level that technology allows. But space and computer technologies will continue to make this global contact more realistic and immediate. If you have a laptop, a telephone is not required. Hand-held communications devices can now work world-wide. The conference calls (telcons) of today will eventually became virtual conference rooms, allowing colleagues to not only listen to each other and watch their presentations, but watch them too, for their facial responses and hand gestures. Meeting and business travel will then begin to reduce, hopefully in time to respond to the twin crises of reduced petroleum production and greenhouse gas emissions, as well as providing reduced stress from travel.

There are levels of space utilization which are so buried that they are almost invisible. Weather forecasts have been improving – why? Because of the increased use of satellite data. More parameters are measured, over a greater altitude range, and ingested into more powerful forecast models; the improvements have been significant and the limits are far from being reached. Higher spatial resolution and faster computers will eventually bring long-range weather forecasts. There is a second kind of "weather" that is less familiar, "space weather", the weather of the solar wind, the magnetosphere, the ionosphere and the aurora. This weather can include "killer particles" that can disable expensive communications satellites, with enormous consequences on communications traffic (communications satellites in geostationary orbit are particularly vulnerable since they are in the radiation belts). These solar storms can also produce electric currents in the Earth, as were measured by Currie and Davies at Chesterfield inlet, that damage electrical power transmission systems and corrode pipelines. Storm warnings allow protective measures to be implemented.

In terms of global imaging, if a camper wants to view a campsite in one of Canada's beautiful parks before booking it, this is possible using satellite images, freely available to the public. The same capability is available to a real estate agent, looking for locations world-wide. RADARSAT was seen as a way of allowing ships to navigate through the Arctic ice floes; now it is seen as a way of monitoring the rapidly receding polar ice extent. The fertil-izing equipment used by farmers will soon be able to adjust the level according to satellite data – this is moving from the research to the operations stage. These are just a few examples of the benefits of space research. These comments have not even touched upon medical ben-efits and those of improved fabrics and other materials (others are bar codes and smoke detec-tors). NASA has identified 30,000 commercial applications from space technology; many can be found on the web, including the well-known teflon.

Space travel is a dream that will ultimately be realized and Canada's astronauts have contributed to the beginning of it. But it is these other hour-by-hour and day-by-day activities that all Canadians experience that have the most impact on our society. Canada is a beautiful country and Canadians would like it to stay that way (or preferably as it once was). But this will not be the result of doing nothing, there are too many other global forces to be counter-acted. Space technologies can help in monitoring and meeting the challenges of global cli-mate change.To do this, the influences and the consequences must be monitored on a global scale, and this is one of the major problems being addressed by space agencies worldwide. What needs to be monitored? Almost everything. Can that be done? Not entirely, but the scope is gradually increasing and the key parameters are gradually coming under routine observation. It is this Global Watch that is the key to preserving our future environment and creating the quality of life to which all aspire, worldwide. To achieve this is the greatest gift that space research can provide.

How to end this discourse without making a firm prediction? With a recent inspiring event in Canada's space history, namely Steve MacLean's STS-115 mission of September 9-21, 2006. MacLean wrote an account of this spectacularly exciting flight for *YorkU* the mag-azine of York University. The account is introduced by the Editor, and Steve's story follows. Both have kindly granted permission to reproduce this story here.

Steven MacLean and the International Space Station

"It took a few tries, but on Sept. 9, 2006, York alumnus Steve MacLean and five other astronauts lifted off from the Kennedy Space Center in Florida aboard the space shuttle Atlantis. Their destination was the International Space Station, to install a giant truss carry-ing two solar arrays that would double the power available to the station. Mission specialist MacLean (BSc '77, PhD '83) of the Canadian Space Agency and his NASA colleagues – com-mander Brent Jett, pilot Chris Ferguson and mission specialists Heidemarie Stefanyshyn-Piper, Joe Tanner and Dan Burbank – spent 12 days aloft performing the first work on the space station in four years. But for MacLean, making his second shuttle mission since 1992, the long-awaited highlight was his first-ever spacewalk, known to NASA as Extra Vehicular Activity, or EVA. Here, exclusively for YorkU, *MacLean writes about the beauty of that - experience as well as his much-publicized tribulations with unruly bolts."*

Camping Out in Space

"It was 6:30 pm, the evening before my space walk. Dan and I were in the airlock. I had a moderate headache because the carbon dioxide levels in the space station were slightly high. I put on my oxygen mask to begin the process to purge nitrogen from my blood. During training, I did not like this mask. Because I have a narrow face, I had to wear it uncomfortably tight, so that oxygen would not leak out. Just this tightness could evolve into a slight headache so I expected worse today. But not two minutes after I put the mask on, my carbon dioxide headache disappeared. I was breathing pure oxygen. I now thought this mask was the greatest invention.

"Joe and Heide helped Dan and me close the hatch between the airlock and the rest of station. We were now camping out in outer space, isolated from the rest of the crew until morning. After completing the slow depressurization of the airlock, we took off our masks, stretched out our sleeping bags and ate supper. I had shrimp cocktail followed by beef tips and green beans. I washed it all down with a protein-laced chocolate energy shake. We reviewed our procedures for the next day, took an aspirin to thin out our blood, put on our diapers and went to bed at 20:30 GMT. The potentially lethal nitrogen would continue to slowly purge from our bloodstream.

"The best sleep I had on station was that night. The carbon dioxide levels were low in the isolated airlock and the temperature was quite cool. I woke up refreshed at 3:45am, about 45 minutes before the official wake-up time. At 4am we opened the hatch to the station. I floated through the lab to the shuttle for the morning hygiene break, trailing the 90-foot oxygen hose. After I finished, I floated back to the airlock, passing Dan in the lab as he headed towards the shuttle. Back in the airlock I ate a power bar, some dried apricots and two more energy shakes and put on my heart-monitoring electrodes.

"Once Dan returned we closed the airlock hatch. Brent and Thomas Reiter, a European Space Agency astronaut living on station, joined us to help us with suit-up, a three-hour process. Then, four hours after wake-up, Dan and I were in the crew lock, the inner hatch was closed and we were at vacuum. So cool. Very soon we would get the 'go.'

"This is it, what we had trained for in years of preparation – basic EVA course, Advanced Skills course, 40 seven-hour dives in the big pool, all the contingency runs that we did. Which scenario would play out? I had confidence in the suit. I had worn it so much; it was a part of me. I was comfortable with the task at hand. But what would be my physiology? How would I react...how would I react? I knew that most spacewalkers made their mistakes in the first few minutes. When I was a Capcom, communicating with astronauts from the ground, I had seen a colleague disconnect his tether unknowingly, and another who had tumbled away from his handholds, and on videos I had seen a former astronaut disappear over the nose of the shuttle, only to be catapulted back a minute later by the recoil of his tether. Though rarely mentioned, it happens. Would it happen to me? Would I think I was falling towards the Earth? Would I grip the handholds too tightly?

"Brent broke into my thoughts with, 'Dan, Steve, I know you guys will do well. Trust your training. You have a "go" to open the hatch.'

"'Oh Boy, This is High'

"I looked down through my feet, through the opening in the hatch, at the Earth – the deep blue dotted by scattered white. Oh boy, this is high! It looks so much higher than from inside...I have to stay focused...but this is so cool...look at that speed of the blue and white...this is so cool...I have to stay focused.

"I verified my safety tether, picked up my white transfer bag and slowly floated out of the airlock. I did not look at the Earth. I locally tethered to the airlock ring and checked the manual isolation valve and the hand controller lock on my Buck Rogers backpack. They were both good. I did not look at the Earth. I closed the thermal cover on the airlock, disconnected my local tether and translated – moved – on to the airlock tool box. Here I tested the inertia of my suit. In succession I yawed, rolled and pitched to see how much force or perhaps how little force each movement required, while simultaneously maintaining overall stability. I did not look at the Earth.

Steve MacLean during the STS-115 mission, working at the far end of the truss, just below where Canadarm2 is attached to the ISS. Israel may be seen on the Earth far below. Photo by courtesy of Steve MacLean and the Canadian Space Agency.

"I reached above me for the Ceta Spur, a ladder that extends up from near the airlock tool box at a 45-degree angle to the front face of the main truss. It did not appear to be at 45 degrees. Its orientation was distinctly horizontal and I felt like I was on my back, holding on to the Ceta Spur from underneath. I knew about the different perceptions of orientation. There is no up or down in space. As I crawled along the spur I played with these perceptions. From that initial position I flipped the entire station. Now I perceived that I was on top of the station floating above the Ceta Spur. Then I flipped the Ceta ladder 90 degrees back so it was vertical. Now I perceived myself to be slowly floating vertically up a ladder. Very

comfortable at being able to decide on any orientation, I reached past the top of the ladder for the front face of the main truss.

"I tethered locally to the front face, let go with one hand, turned and for the first time since I left the airlock, seriously looked at the Earth. Virgin feelings of....awe...and...wonder... embraced me...and the fleeting thought...I come from there...but....I was falling. There was no doubt I felt like I was falling towards the Earth. I looked along the truss towards the Earth. And sure enough the station was falling with me. We were both falling at the same rate of curvature as the Earth, and I could distinctly feel it and see it and yet I was comfortable. Stabilizing myself, I let go with my second hand. I could not see the station; the Earth dominated my field of view. I felt apart...an external observer...with each breath, I inhaled the beauty of the Earth from afar...and yet...floating, my back to the truss, I was in a co-orbit with the station, I was my own spaceship, a small human satellite circling the Earth. The Earth, the moon and me...I really felt a part of it all.

The Big Chill

"Ten minutes into the walk, the physiology and stability tests behind me, I started the long translation out to the end of the truss, when I heard Joe's voice in my helmet. 'Steve, two minutes to sunset.' It was a reminder to check tethers and tools, verify the integrity of your suit, put up your visor and turn on your heater gloves.

"In space, we fall into night instantly. It would last 30 minutes, followed by 60 minutes of daylight. I stowed my transfer bag on the front face of the truss and translated up to the zenith side to start on my first launch lock. Our main task was to free up the 10-foot-diameter rotary joint between the P3 and P4 truss. We had 16 launch locks and six launch restraints to remove. It would take half the EVA for me to remove my locks distributed on the zenith, aft and nadir portions of the truss. The plan was to rotate the truss just after our EVA and then deploy the panels in the morning. Once deployed, the P4 solar panels could follow the sun, double the power available to the station and provide a redundant channel. I was glad that this redundant power channel was making it so much safer for my friends on the station's longer-term crew.

"All of a sudden I realized I was cold. I cranked up the heat on my suit to maximum. Minutes later my legs were still freezing. The last week before flight the thermal engineers had told Dan and me that we would be the 'coldest spacewalkers ever' on the zenith side of the station at night. Their comment seemed to be an aside at the time, so I did not worry about it. Besides these calculations are often a factor of two in error. But now I thought....the truss shadows me from the Earth, my body is radiating to deep space and the station itself is in a very cold attitude...those guys were right! I had not been this cold since I was stuck in a snowstorm near Mt. Everest. I needed the sun to come up.

"I did not delay, taking only glimpses of the beauty below. I finished on the zenith, worked around to the aft and by the time I was on the nadir side it was night again. I was so comfortable on the nadir side at night. The Earth was at my back silently radiating its warmth. The striking contrast with the other side briefly inspired poetic notions that Mother Earth was actually comforting me.

Those Bolts

"Still on the nadir side, I was putting the cover back on my second last launch lock when I realized a cover bolt was missing. Keeping to communication protocol, the expletive "damn" only echoed in my mind. This was serious. If the bolt floated inside towards the rotary joint's race ring it could jam the mechanism and prevent the solar arrays from moving. If I was lucky and the bolt drifted away, it would burn up on re-entry within three days without consequence. I was pretty sure that the bolt must have drifted away and Houston agreed, but for the rest of the EVA I was vigilant, with one eye looking for that bolt. Since the flight, many of my friends have said that they found the bolt in their driveway. I now have several bolts of various sizes on my fireplace mantle at home.

"I finished the last launch lock and went back to the aft to start on the launch restraints. I set up the ratchet wrench and braced myself to break the torque on the first bolt. It needed more than I thought. Keeping the same position I put more force on the wrench and the socket tool broke. At the same time, on the opposite side Dan had a restraint bolt that he could not break the torque on. After some discussion with the ground, Joe said, 'Steve, we would like you to go to the Z1 port tool box to get the nine-inch socket and the cheater bar, and it is now two minutes to sunset.' I thought, 'Oh no, I have to translate through the rat's nest at night.' The rat's nest is a plethora of cables, T-clamps and connectors that your tether and suit can easily catch on. Translating through there during the day requires caution but translating through there at night is tedious and slow. I was on my way back with the cheater bar and the socket just as the sun was coming up.

"Dan was pushing on the end of the cheater bar, which extended the socket handle to give us more force, and I was halfway up. All the other launch restraints were now removed but this bolt was seized. If we did not succeed, the rotary joint could not be activated. I had just over three hours of oxygen left. I was not coming inside until this bolt was gone.

"I recalled a time over 35 years ago when my great uncle, my father and I used an iron pipe as a cheater bar down on our ancestral farm in Nova Scotia. We were repairing a slapboard barn when one of the foundation bolts showed us the same stubbornness. Back then we were worried about side loading, stripping or breaking the bolt head, and we were worried about that now. Dan and I stayed in step, continually discussing the bolt's progress, a vivid echo of the discussion 35 years ago. There was one difference. A slip then meant bruised hands or at worst a dirty scrape. A slip here in this harsh, pristine environment could mean an unforgiving hole in your gloves.

"Afterwards, Dan would tell me that with every stroke he had been very careful not to slip. At 15 minutes my right hand was pretty beat up inside the glove and I knew that at least two of my fingernails would delaminate. Dan's hands were worse. After 30 difficult minutes we were successful. As we celebrated, I realized I had no feeling along the edge of my hand up to the top of my index finger. It would be six weeks before that feeling would start to return. But, I could go back inside now.

Absorbing It All

"So far, I had allowed myself only periodically to steal glimpses of the Earth, the *atmosphere and the universe as I focused on getting the job done. Now with the main task complete and still a few hours left, I felt I could lengthen my glimpses and ponder the view...what a privilege.*

"Dan cheered, 'Steve, you have to look behind you!' As I turned, directly below was the Manicouagan crater. You could see most of Quebec in a single view. I slowly lifted my eyes to see the Maritimes extending towards the horizon, stopped there for a second to consider the royal blue thinness of the atmosphere, and then continued my scan up into the universe. No words can describe your emotions when your mind is filled with wonder like that, where each vivid second stimulates an eternity of intellectual wandering. Minutes later it was the Mediterranean centred on Italy, followed by the surprising silent beauty of the Middle East...and...somehow there is ample time for your mind to absorb it all.

"Each time I looked at an area on the Earth I had been to, my mind would instantly shift to when I was there. I had travelled extensively through Europe and had taken the time off to backpack in Africa and throughout Asia, including China, India and Nepal. The memories were so vivid in their detail. It was as if I had beamed down to the surface to relive each local adventure in parallel with this, the ultimate adventure.

"We finished clearing the front face when the inevitable words echoed in my helmet, 'Steve, after you help Dan pack his transfer bag, you can pick up yours, and I know you guys do not want to hear this but it will then be time to head on in to the airlock.' I translated back to the Ceta Spur, where it all began, and there I took one last, thoughtful, lingering look at the Earth and our universe.

"And then it was time. Time to go home, back to the safety of the crew lock...a tough, challenging spacewalk behind us...nowhere near routine...one stubborn bolt and one lost bolt...but bottom line, the solar array rotary joint would work, we would deploy the arrays the next day and the next truss element could be added. I was coming home feeling good and proud that we had nailed it...leaving a remarkable world of awe and wonder, merged with a deep understanding of humility.

"We closed the outer hatch at 7 hours 11 minutes, opened the inner hatch to all smiles...and I felt safe."

Postscript

The story just told is only one of many Canadian successes in space over the fifty years of COSPAR. This larger story is one in which individuals played the leading role, from within the organizing framework that was available to them. In terms of overall administrative structure, Canada has perhaps had a better space program than it deserves. If so, the credit goes to these individuals, who are really the heroes of this half-century. However, at the end of this period it is evident that the Canadian effort is becoming better integrated and hopefully more effective. The next fifty years should be even more rewarding.

References:

Kirton, John, Canada, the United States, and Space, Canadian Studies Program, Columbia University, 1986.

LIST OF ACRONYMS

ACE – Atmospheric Chemistry Experiment
ACOA – Atlantic Canada Opportunities Agency
ACSR – Associate Committee on Space Research (of NRC)
ACTORS – Atlantic Canada Thin ORganic Semiconductors
AECL – Atomic Energy of Canada Limited
AES – Atmospheric Environment Service
AF – Air Force
AFCRL – Air Force Cambridge Research Laboratories
AFOSR – Air Force Office of Scientific Research
AGU – American Geophysical Union
AIM – Aeronomy of Ice in the Mesosphere
AIT – Advanced Information Technologies
AMISR – Advanced Modular Incoherent Scatter Radar
AO – Announcement of Opportunity
APL – Applied Physics Laboratory (of Johns Hopkins University)
ARDC – Air Research and Development Command
ARF – Aquatic Research Facility
ARIES – Auroral Rocket and Imager Excitation Study
ARO – Algonquin Radio Observatory
A-SCOPE – Advanced Space Carbon and climate Observations of Planet Earth
ASP – Auroral Scanning Photometer (on the ISIS-II spacecraft)
ASTD – Atmospheric Science Theme Group (of the Phoenix mission)
BLAST – Balloon-borne Large Aperture Sub-millimeter Telescope
BNSC – British National Space Centre
CAP – Canadian Association of Physicists
CAPCOM – Capsule Communicator
CAPE – CAnadian Protein crystallization Experiment
CARDE – Canadian Armament Research and Development Establishment
CARISMA – Canadian Array for Realtime InvestigationS of Magnetic Activity
CBR – Cosmic Background Radiation
CCD – Charge Coupled Device
CCSS – Canada Centre for Space Science
CEP – Cylindrical Electrostatic Probe (on the Alouette/ISIS spacecraft)
CETEX – Contamination by ExtraTerrestrial EXploration
CFC – Chlorofluorocarbon
CFCAS – Canadian Foundation for Climate and Atmospheric Science
CFI – Canadian Foundation for Innovation
CFZF – Commercial Float Zone Furnace
CNES – Centre National d'Etudes Spatiales
CNRS – Centre National de la Recherche Scientifique
CoReH2O – Cold Regions Hydrology High-resolution Observatory
COSMIC – Constellation Observing System for Meteorology, Ionosphere and Climate
CRC – Communications Research Centre (of Canada)
CRL – Communications Research Laboratories (of Japan)
CRR – Churchill Research Range
COBE – Cosmic Background Explorer
COBRA – Cosmic Background Radiation
COSPAR – Committee on Space Research
CPA – Cold Plasma Analyser
CRESS – Centre for Research in Earth and Space Science (at York University)
CRESTech – Centre for Research in Earth and Space Technology)
CSA – Canadian Space Agency
CSAGI – Comité Spécial de l'Année Géophysique Internationale
CTS – Communications Technology Satellite
DAO – Dominion Astronomical Observatory
DARA – Deutsche Agentur fur Raumfahrtangelegenheiten (German Space Agency)
DASP – Division of Aeronomy and Space Physics (of CAP)
DEW – Distant Early Warning
DND – Department of National Defence (of Canada)
DOC – Department of Communications (of Canada)
DOT – Department of Transport (of Canada)
DOVAP – DOppler Velocity And Position indicator
DRAO – Dominion Radio Astrophysical Observatory
DRB – Defence Research Board (of Canada)
DRNL – Defence Research Northern Laboratory
DRTE – Defence Research Telecommunications Establishment
DREV – Defence Research Establishment Valcartier
EC – Environment Canada
EEE – Explosives Experimental Establishment
EIMS – Energetic Ion Mass Spectrometer
EL – Electronics Laboratory
ePOP – Enhanced Polar Outflow Probe
EROS – Earth Resources Orbiting Satellite
ERTS – Earth Resources Technology Satellite
ESA – European Space Agency
ESRO – European Space Research Organization
ETON – Energy Transfer in Oxygen Nightglow
FES – Fine Error Sensor
FGS – Fine Guidance Sensor
FLEX – FLuorescence EXplorer
FSL – Fluid Science Laboratory
FTS – Fourier Transform Spectrometer
FUSE – Far Ultraviolet Spectroscopic Explorer
GAS – Get Away Special
GCOM – Global Change Observing Mission
GLONASS – GLObal NAvigation Satellite System
GNSS – Global Navigation Satellite System
GP-B – Gravity Probe B mission
GPS – Global Positioning System
GSA – Geological Society of America
GSC – Geological Survey of Canada
HAD – High Altitude Diagnostic
HALCA – Highly Advanced Laboratory for Communications and Astronomy
HARP – High Altitude Research Project
HEDS – Human Exploration and Development of Space
HF – High Frequency (3 to 30 MegaHertz, where Mega = one million)
HIA – Herzberg Institute of Astrophysics
HIFI – Heterodyne Instrument for the Far Infrared
HST – Hubble Space Telescope
IAU – International Union of Astronomy
ICBM – Intercontinental Continental Ballistic Missile

ICS – Interdepartmental Committee for Space (of Canada)

ICSU – International Council for Science (earlier the International Council of Scientific Unions)

IGY – International Geophysical Year

IKI – Space Research Institute (of Russia)

IMF – Interplanetary Magnetic Field

IML – International Microgravity Laboratory

IMS – Ion Mass Spectrometer (on the ISIS spacecraft)

INPE – Instituto Nacional de Pesquisas Espaciais (Brazilian Space Agency)

IPY – International Polar Year

IRM – Imaging and Rapid scanning ion Mass spectrometer

ISAS – Institute of Space and Aeronautical Sciences (of Japan)

ISAS – Institute of Space and Atmospheric Studies (of the University of Saskatchewan)

ISI – Intra Space International

ISIS – International Satellites for Ionospheric Studies

ISRO – Indian Space Research Organization

ISS – International Space Station

ISTP – International Solar Terrestrial Physics

ISTS – Institute for Space and Terrestrial Sciences

IUGG – International Union of Geodesy and Geophysics

IUTAM – International Union of Theoretical and Applied Mathematics

JAXA – Japan Aerospace eXploration Agency

JHU – Johns Hopkins University

JPL – Jet Propulsion Laboratory

JSC – Johnson Space Center

JWST – James Webb Space Telescope

LEO – Low Earth Orbit

LFC – Large Format Camera

LO – Local Oscillator

LIMS – Limb Infrared Monitor of the Stratosphere

LMS – Life and Microgravity Spacelab

MAD DAWG – MArs Dust Devil Advanced Warning Gizmo

MAESTRO – Measurements of Aerosol Extinction in the Stratosphere and Troposphere Retrieved by Occultation

MANTRA – Middle Atmosphere Nitrogen Trend Assessment

MATADOR – Martian ATmosphere And Dust in the Optical and Radio

MC – Mission Controller

MCP – Micro Channel Plate

MDA – McDonald Dettwiler Associates

MELLF – MEtallic Liquid Like Films

MIM – Microgravity vibration Isolation Mount

MITCH – Mars Investigation of Total Climatological Hazards

MMM – Mars Microphysical Model

MOPITT – Measurements Of Pollution In The Troposphere

MOST – Microvariability and Oscillations of STars

MPE – Max Planck Institute for Extraterrestrial Physics

MSC – Meteorological Service of Canada

MSC – Mobile Servicing Centre

MSS – Mobile Servicing System

MVIS – Microgravity Vibration Isolation System

NAE – National Aeronautics Establishment

NANOGAS – Nanogas Get Away Special

NASA – National Aeronautics and Space Administration

NASDA – NAtional Space Development Agency (of Japan)

NATO – North Atlantic Treaty Organization

NDSC – Network for the Detection of Stratospheric Change

NGDC – National Geophysical Data Center

nm – nanometers

NOAA – National Oceanic and Atmospheric Administration (of the US)

NRC – National Research Council (of Canada)

NRL – Naval Research Laboratory (of the United States)

NSBF – National Scientific Balloon Facility

NSERC – National Sciences and Engineering Research Council (of Canada)

NSF – National Science Foundation (of the United States)

NSPO – National Space Organization (of Taiwan)

NWT – North West Territories

OAR – Office of Aerospace Research

OASIS – Oxygen Aeronomy Studies In Situ

OEDIPUS – Observations of Electric-field Distributions in the Ionospheric Plasma – a Unified Study

OGLOW – Orbiter Glow

ORRF – Ordnance Rocket Research Facility (of the US army)

OSIRIS – Optical Spectrograph and InfraRed Imaging System

OSR (Office of Scientific Research)

OSRD – Office of Scientific Research and Development

PARL – Prince Albert Radar Laboratory

PBL – Planetary Boundary Layer

PEARL – Polar Environment Atmospheric Research Laboratory

PI – Principal Investigator

PMC – Polar Mesospheric Clouds

PMDIS – Perceptual Motor Deficits In Space

PMR – Pressure Modulated Radiometer

PNL – Pacific Naval Laboratory

PPARC – Particle Physics and Astronomy Research Council (of the UK)

PREMIER – PRocess Exploration through Measurements of Infrared and millimetre-wave Emitted Radiation

PS – Project Scientist

PSAC – President's Science Advisory Committee

PTOC – Payload Telescience Operations Centre

PTV – Propellant Test Vehicle

PVT – Physical Vapour Transport

RCN – Royal Canadian Navy

REED – Radio and Electrical Engineering Division (of NRC)

REL – Research Enterprises Limited

RF – Radio Frequency

RLP – Red Line Photometer (on the ISIS-II spacecraft)

RMC – Royal Military College

RPA – Retarding Potential Analyzer (on the ISIS spacecraft)

RPL – Radio Physics Laboratory (earlier the Radio Propagation Laboratory)

RRL – Radio Research Laboratories (of Japan)

RSI – RADARSAT International

RSO – Range Safety Officer

SAC – Science Advisory Committee

SAGE – Stratospheric Aerosol and Gas Experiment

SAR – Synthetic Aperture Radar

SCIAMACHY – SCanning Imaging Absorption SpectroMeter for Atmospheric CHartographY

SEC – Space Engineering Division (of the University of Saskatchewan)

SCCO – Soret Coefficient in Crude Oil

SCOSTEP – Scientific Committee for Solar Terrestrial Physics

SEAC – South East Asia Command

SII – Suprathermal Ion Imager

SIR – Shuttle Imaging Radar

SMR – Sub-Millimeter Radiometer

SNAP – Surrey Nanosatellite Applications Platform

SNSB – Swedish National Space Board

SOC – Science Operations Center (for the Phoenix mission)

SOHO – SOlar and Heliospheric Observatory

SPAR – Special Products and Applied Research (of De Havilland)

SPEAM – Sun Photometer Earth Atmosphere Measurements

SPECTRA – Surface Processes and Ecosystem Changes Through Response Analysis

SPIRE - Spectral and Photometric Imaging Receiver

SPS – Soft Particle Spectrometer (on the ISIS spacecraft)

SRDE – Scientific Research and Development Establishment

SRFB – Space Research Facilities Branch (of NRC)

SRI – Stanford Research International

SSC – Surrey Space Centre

SSCO – Space Science Coordination Office (of NRC)

SST – Super Sonic Transport

STEM – Storable Tubular Extendible Member

STS – Space Transportation System

SuperDARN – Super Dual Auroral Radar Network

SVS – Space Vision System

TC – Test Conductor

TCTS – Trans Canada Telephone System

TDI – Time Delay and Integrate

TFI – Tunable Filter Imager

THEMIS – Time History of Events and Macroscale Interactions during Substorms

TIA – Transverse Ion Acceleration

TPA – Thermal Plasma Analyser

TRAQ – TRopospheric composition and Air Quality

TSA – Thermal and Suprathermal Analyser

UARS – Upper Atmosphere Research Satellite

UBC – University of British Columbia

µm - micrometers

URSI – Union Radio Scientifique Internationale

USAF – United States Air Force

UTIAS – University of Toronto Institute for Aerospace Studies

UVI – Ultra Violet Imager (on the Viking satellite)

UVIT – Ultra Violet Imaging Telescope (on the ASTROSAT satellite)

UWO – University of Western Ontario

VLBI – Very Long Baseline Interferometry

VLF – Very Low Frequency (3 to 30 kiloHertz, kilo = one thousand)

VM – Vehicle Manager

VSOP – VLBI Space Observatory Program)

VUV – Vacuum Ultra Violet

WAMDII – Wide Angle Michelson Doppler Imaging Interferometer

WAMI – Wide Angle Michelson Interferometer

WASP – WINDII Advanced Studies Program

WINDII – WIND Imaging Interferometer (on the UARS satellite)

WINDO – Winds IN-situ and Doppler Oxygen

WINTERS – Winds and Temperatures by Remote Sensing

WISP – Waves In Space Plasmas

INDEX

A

ACTORS (Atlantic Canada Thin Organic Semiconductors) 199, 273
Aerial 26, 45, 47-48, 150-152
Aerobee rocket 22-23, 100-101, 108, 110, 140, 142-144, 146
AFCRL (Air Force Cambridge Research Laboratories) 7, 63, 69, 134, 273
AFOSR (Air Force Office of Scientific Research) 63, 273
Airglow 35, 42, 72, 74-75, 91, 100, 122-123, 171, 180-181, 212, 218, 222, 225, 229
AIT (Advanced Information Technologies) 174-175, 273
Akasofu, Syun-Ichi 93-94
Akebono (satellite) 9, 184-185, 214, 260
Alaska 65, 83-84, 91, 93, 117, 133, 143, 152, 170, 207, 214, 227
Alberta 21, 29, 44, 51, 78-79, 83, 88, 127, 159, 195, 220, 228, 249, 252, 260
Alfven, Hannes 81, 139, 147, 187, 262
Algonquin Radio Observatory 52, 229-230, 273
All-sky 59, 90-92, 94-97, 100, 109, 150, 236
Alouette I 33, 83, 102, 128-130, 132-133, 137-138, 140, 159
Alouette II 133, 138
Amundsen, Roald 25
Anger, Clifford 69, 94, 128, 135-136, 150, 182-183, 212
Ångström, Anders 42
Anik (satellite) 13, 131, 156-157
Antarctic 33, 41, 57, 78-79, 86, 88, 92, 178, 208, 226
Antenna 52, 62, 66-67, 83, 109, 129, 134, 168, 205-207, 229-231
Apollo-Soyuz 9, 170, 261, 263
Appleton, Edward 26, 45-47, 171
Arcas rocket 83, 144, 146, 150
Arctic 19, 25-26, 33-35, 39, 49, 56-57, 63, 85-86, 94, 97, 99, 104, 108-109, 123, 141, 151, 156-157, 164, 168, 170, 178, 189, 204-205, 207-208, 233, 237, 241, 249-252, 259, 262, 267
ARF (Aquatic Research Facility) 9, 195, 201, 273
Associate Committee on Space Research 21, 28, 49, 159, 163, 273
Astrobee rocket 144, 146
ASTROLAB 178, 241
Astronaut 9, 14, 83, 113, 159, 173, 178-179, 181, 193-198, 201-202, 204, 264, 268
ASTROSAT 10, 236-237, 260, 276
Atmosphere 8, 13-14, 18, 21, 23, 26, 35, 40, 42-43, 45-46, 51, 56, 60, 72-75, 80, 83, 85-86, 92, 96, 98-101, 103-108, 110-111, 115, 118, 121-127, 132, 136, 139-140, 146-147, 150, 155-156, 161, 171, 174, 176, 178-180, 189, 208-212, 215, 217-228, 231, 237, 240-241, 243, 245-247, 250-252, 259, 263-264, 272, 275-276
Atmospheric electricity 35, 45
Atomic energy 30, 49, 89, 117, 128, 163, 273
Aurora 7-8, 13, 20, 22, 25-26, 35-36, 39-42, 52, 54-56, 59, 61-62, 65-69, 71-74, 79-81, 83, 87-88, 90-97, 100, 103, 105-111, 118, 122-124, 133, 135-137, 145-146, 148-149, 161, 164, 166-168, 176, 181-183, 187, 209, 213, 217, 262, 266
Auroral camera 74, 105, 180
Auroral oval 41-42, 94, 137-138, 183, 187
Auroral photography 39-40, 65
Auroral radar 7, 61, 65-66, 68-69, 74, 90, 92, 118, 226, 276
Auroral Scanning Photometer 128, 135-137, 273
Auroral spectrum 71, 107, 110, 124
Avro Arrow 17, 256
Axford, W. Ian 117, 138, 154

B

Bachynski, Morrel 68, 73, 241
Ballard, B.G. (Guy) 49
Ballistics 45, 50-51
Balloon 9, 40, 83, 92, 100, 123-124, 128, 162, 177-178, 218, 229, 237, 273, 275
Banting, Frederick 48

Barringer Research 172
Barrington, Ron 116, 138, 159-160, 176
Bates, David 71, 80-81
Bennett, Richard Bedford 30-31, 47, 170
Berkner, Lloyd 55, 85
Bernath, Peter 223-224
Birkeland, Kristian 107, 262
Black, John 94
Black Brant rocket 83, 108, 125, 140-146, 164-166, 257-258, 262
Blackett, P.M.S. 162-163
BLAST (Balloon-borne Large Aperture Sub-millimetre Telescope) 10, 237, 273
BNSC (British National Space Centre) 261-262, 273
Bolts 267, 271
Bomem 218, 223, 257
Bondar, Roberta 178, 181, 194
Borden, Sir Robert 28
Borra, Ermanno 263-264
Boulding, Dave 137, 139
Bowles, Ken 83
Boyle, Robert 29, 44
Brace, Larry 133, 136
Brazil 131, 262
Brazilsat 157
Brewer, Alan 221
British Columbia 21, 28, 52, 55, 77-79, 87, 121, 195-196, 201, 207, 218, 229, 237, 276
Broadfoot, Lyle 74
Brock University 94
Buijs, Henry 218
Bull, Gerald Vincent 114-115, 127
Burrows, Ron 136, 138-139, 184

C

CAL Corporation 173-174, 183, 225, 257
Calgary 9, 99, 120, 127-128, 136-137, 154, 177, 182-183, 185, 187-188, 195, 212, 216-217, 223, 228, 231, 244-245, 260-261, 264
Cambridge 7, 18, 29, 43, 45, 54, 63-64, 71-72, 76, 79, 81-82, 91, 105, 107, 113, 117, 134, 220, 234, 257, 273
Camera 20, 39, 45, 54, 59, 62, 67-68, 74, 90-97, 100, 105, 109, 179-180, 182-183, 188, 211, 214, 232, 236, 243, 250, 252, 274
Canada Centre for Remote Sensing 8, 150, 154, 192
Canada Centre for Space Science 162, 185, 191, 273
Canadian Association of Physicists 83, 159, 216, 273
CANEX-2 194-195
CanX-2 265
CAPE (Canadian Protein Crystallization Experiment) 9, 202-203, 273
Cape Parry 148, 164-169
CARDE (Canadian Armament Research and Development Establishment) 21, 50-51, 74, 100-102, 114, 118, 121, 123, 127, 140-142, 144, 177, 212, 218, 273
Carnegie Institute 33, 35, 56
Carswell, Allan 11, 127, 240-242, 244-247, 249, 251, 253
CASSIOPE 14, 217, 264
CCD 182-183, 211, 216, 219, 222, 232, 234, 265, 273
Chapman, John 11, 19, 21-22, 24, 57, 60, 89, 113, 120, 131-132, 152-156, 158, 193, 202, 265.
Chapman, Sidney 56, 72, 80-81, 85, 139, 147, 262.
Chapman report 113, 120, 132, 154-156, 265
Chesterfield Inlet 7, 25-26, 33-36, 38-42, 53, 73, 75, 96-97, 100, 108, 226, 267
China 176, 187, 199, 220, 228, 255, 260-261, 272
Chinese records 106
Chinook 9, 225-226, 261, 264-265
Churchill 8, 19-20, 22-25, 35, 43, 53, 56, 59, 83, 90-93, 95-101, 107-110, 113, 116, 118, 120, 124, 135, 140-144, 147-148, 154, 161-165, 167, 170, 177, 188, 262, 273
Clyde River 56
CNES (Centre National d'Etudes Spatiales) 174, 213, 261, 273
Cogger, Leroy 11, 128, 137, 183, 188, 212
COM DEV 175, 185, 205, 207, 216, 220, 234, 257

Communications Research Centre 57, 59, 67, 101, 130-131, 156, 273
Connes, Pierre 209, 217
Corona 103
Cosmic Background Radiation 9, 217-218, 223, 229, 273
Cosmic rays 8, 20, 23-24, 26, 45, 49, 79, 87, 89-90, 98-99, 105, 117, 160, 163
COSPAR 20, 22-28, 30-32, 34, 36, 38, 40, 42-44, 46, 48, 50, 52, 54, 56, 58, 60, 62, 64, 66, 68, 70, 72, 74-76, 78, 80, 82, 84, 86, 88, 90, 92, 94, 96, 98, 100, 102, 104, 106, 108, 110-114, 116-118, 120, 122, 124, 126, 128, 130, 132, 134, 136, 138, 140, 142, 144, 146, 148, 150, 152, 154, 156, 158, 160, 162, 164, 166, 168, 170, 172, 174-176, 178, 180, 182, 184, 186, 188, 190, 192, 194, 196, 198, 200, 202, 204, 206, 208, 210, 212, 214, 216, 218, 220, 222, 224, 226, 228, 230, 232-234, 236, 238, 240, 242, 244, 246, 248, 250, 252, 254, 256, 258, 260, 262, 264, 266, 268, 270, 272-274, 276
Covington, Arthur 52, 88, 98
CRC 58, 60, 84, 86, 131, 134-135, 137-138, 141, 154, 156-159, 162, 173, 181, 188, 192, 213-214, 228, 273
Creutzberg, Fokke 11, 149-150, 167, 169, 213
Cronyn, Hume 28-29
CSA (Canadian Space Agency) 9, 11, 14, 53, 77, 84, 120, 126, 132, 154, 158, 160, 180-181, 184, 191-196, 198-205-209, 213-214, 216-217, 220-221, 223-225, 228, 231, 235-238, 242, 245-250, 252-253, 256-258, 264-265, 267, 269, 273
CSAGI (Comité Spécial de l'Année Géophysique Internationale) 85-86, 90-91, 273
Currie, Balfour 7-8, 11, 25-26, 33-42, 53-54, 59-62, 64-66, 68-69, 72-73, 75, 80, 84, 88, 90-91, 94-95, 97, 100, 104, 107-108, 118, 120-121, 126, 226, 228, 267
Cylindrical Electrostatic Probe 136, 273

D

Davies, Frank 7, 11, 25-26, 33-34, 37, 39-42, 45, 53, 55-60, 64-66, 73, 88, 90, 97, 116, 118, 126, 131, 226, 267
De Havilland Aircraft 25, 114, 129, 134, 155, 276
De Mairan, Jean Jacques 106
Deep River 99
Defence Research Board 8, 20-22, 26, 51, 61, 113-114, 273
Diefenbaker, John G. 17, 49, 83
Distant Early Warning 61, 97, 164, 273
Dobson, G.M.B. 92, 220-221
Dominion Radio Astrophysical Observatory 52, 60, 98, 161, 229, 273
Donahue, Tom 171, 261
Doppler 23, 68, 72, 100, 107, 166, 168, 172-173, 206, 210, 225, 236, 265, 273, 276
Doré, Roland 193
DRB (Defence Research Board) 21, 26, 51, 58-60, 65-66, 68, 72, 81-82, 84, 88-90, 92, 96-97, 113-114, 116-118, 124, 126-129, 151, 154, 156, 192, 273
D-region 56
DREV (Defence Research Establishment Valcartier) 9, 176-177, 244, 274
DRTE (Defence Research Telecommunications Establishment) 7-8, 21, 24, 26-27, 51-52, 57-60, 65-68, 82-84, 88-90, 97, 101, 107, 115-116, 126, 128-131, 133-135, 138, 140-141, 154-156, 193, 229, 273, 274

E

Earth current 40
East Quoddy 163
Eaton, Sir John 48, 80
Eclipse 25, 42, 46-47, 52, 60, 83, 103, 163
Eggleston,Wilfred 27, 31
EIMS (Energetic Ion Mass Spectrometer) 173, 274
Eisenhower, Dwight 17, 31, 83, 86, 263
Electric fields 8, 119, 146-148, 169-170, 213, 216
Elvey, Christian 91
e-OSTEO 9, 202
E-region 46-47, 56
ESA (European Space Agency) 77, 132, 157-158, 170, 193, 196, 198, 200, 202, 204, 208, 216, 222-223, 225, 233-235, 240, 255, 257, 259-263, 268, 273
Evans, Mac 181,193, 205
Evans, Wayne 11, 123-125, 164, 166, 177-178, 180, 189, 212, 223

F

Farley, Donald 69
Feldstein, Yasha 42, 94
Fia, Al 142
Fine Error Sensor 232, 274
Fisher, Dave 249
Fleming, Sir Sandford 106
Forsyth, Peter 11, 21, 33, 48, 54, 61-62, 65-69, 74, 80, 82, 84, 86, 88, 90, 95, 101, 115-116, 118, 127, 155, 159-161, 169, 179, 226
Fort Sik-Sik 38-40
Foster, George 28
Foster, J.S. 69
Fourier Transform Spectroscopy 121, 217, 229
Fowler, Barry 11, 197, 203-4, 228
Franklin, John 25
Franklin squirrel 38
Franklni, Colin 130
Fraunhofer 70
F-region 26, 46-47, 59, 115, 138, 212, 226
Freja mission 9, 128, 186-188, 214-216, 221, 262, 265
FUSE 10, 232-233, 236, 274

G

Garneau, Marc 14, 173, 178-181, 193-195, 197, 199, 224-225, 253, 257, 265
Gattinger, Dick 72, 122, 212, 223
Gault, Bill 125, 212
Gauss, Carl Friedrich 106
GEODESIC 217
Geomagnetism 7, 64, 76-77, 81, 87-90, 92
Gerson, Nate 7, 11, 61-67, 69, 75-77, 80-81, 84, 86, 147, 228
Gold 34, 57, 82, 89, 105, 116, 211, 231
Gravity Probe B 10, 173, 238-239, 274
Greeks 106
Gregory, John 126,
Gregory, Alan 151, 153
Gun 45, 52, 73, 107, 114-115, 127, 160, 169
Gush, Herb 72-73, 118, 121, 217-218, 221, 229, 241, 253

H

Hadfield, Chris 195, 197
Halley, Edmund 106, 226, 263
Hampson, John 123, 177, 189
Harang, Leiv 61, 262
Harang, Ové 209
Harrington, Ertle Leslie 34, 72
Hartz, Ted R. 88, 116, 128, 134, 154, 158-159, 189
Heikkila, Walter 101, 116, 134, 136-137, 140
Henderson, John T. 7, 45-49, 60, 83, 128
Hermes 131-132, 158, 174-175, 192, 205
Herschel mission 10, 233-235, 260
Hersom, Rudy 212
Herzberg, Gerhard 54-55, 65, 71-72, 81, 116, 122, 125, 161
Herzberg, Luise 116
HIA (Herzberg Institute of Astrophysics) 60, 89, 161, 163, 173, 183-184, 215, 232, 236, 241, 273
Hines, Colin 7, 81-82, 84, 87, 115-117, 138, 154, 212
Hirao, Kunio 77, 148, 164, 166
Hoover, Herbert 34, 57
Huancayo 55, 68, 84, 91
Hudson's Bay 36, 56, 99
Hunten, Donald M. 8, 11, 69-75, 80, 86, 88, 90-91, 95-96, 100, 109, 118, 122-126, 176, 183, 221, 226, 261
Hutchings, John 11, 232-233, 236-237
Huygens spacecraft 74, 259
Hydroxyl 100, 121

I

IAU (International Union of Astronomy) 85, 274
ICBM 51, 237, 274
ICSU (International Council of Scientific Unions) 18, 21, 85, 115, 274
IGY (International Geophysical Year) 7-8, 17, 19-20, 22-24, 42-43, 45, 49, 59, 68, 75, 76, 82, 85-93, 95-105, 107-111, 113, 115, 124, 127, 140, 221, 274
IGY Advisory Committee 88
Incoherent scatter 83-84, 273
India 131, 176, 255-256, 260-261, 272
Intensity recorder 59, 90-91, 95-97
Interball 9, 186-188, 265
International Space Station 9-10, 13-14, 158, 181, 191, 193, 195, 197-199, 208-209, 255, 257-258, 267, 274
Ion Mass Spectrometer 9, 134, 136, 170, 173, 184, 216, 274
Ionosonde 25, 46-47, 55-56, 58-59, 68, 128
Ionosphere 20, 22-26, 45-46, 55-57, 59, 62, 64-65, 68-69, 72, 82-83, 88-90, 98, 100, 103, 115-118, 128-130, 133, 137, 140-141, 147-148, 150, 152, 161, 164, 166, 169, 184-186, 205, 214, 217, 226-227, 262, 265-266, 273
IPY (International Polar Year) 25, 34, 42, 53, 87, 274
ISAS (Institute of Space and Aeronautical Sciences) 77, 184, 186, 214-215, 231, 261, 274
ISAS (Institute of Space and Atmospheric Studies) 118, 126, 274
ISIS I satellite 134, 136, 138
ISIS II satellite 94, 128, 134-139, 156, 164, 167, 182-183, 188, 206, 212, 258, 264
ISS (International Space Station) 9, 13, 181, 191, 193, 195-197, 199, 203-204, 208-209, 223, 258, 260, 263, 274
IUGG (International Union of Geodesy and Geophysics) 18-19, 85, 88, 115, 274

J

Jackson, John 134, 139
Jacobs, Jack 7, 31, 64, 75-79, 87-88, 261
James Webb Space Telescope 10, 235, 260, 274
Javelin rocket 129, 143, 145-146
JAXA (Japanese Space Agency) 209, 215, 231, 260-261, 274
Jelly, Doris 11, 116, 131, 175, 189
Johnson, Dennis 119-121, 186
Johnson Space Center 172, 194-195, 246, 274

K

Kavadas, Alex 11, 118-121, 146-147, 173, 182, 226, 229
Keldysh (Academician) 17-18
Kendall, David 11, 180, 189, 253
Kennedy, John F. 17
Kennedy Space Center 267
Kerwin, Larkin 191, 193
King, Mackenzie 29, 47
Kirton, John 120, 154, 255, 257-258, 272
Kites 35, 37, 40
Koehler, Jim 11, 120, 146-147, 186, 226
Kornelson, Ernie 68, 73
Korolev, Sergei 17
Kruschev, Nikita 18

L

Lagrangian point 235
Lapp, Phil 25, 155, 205
Lethbridge University 234
Lidar 241-253, 263, 265
Life Sciences 9, 160, 194-197, 201
Llewellyn, Edward J. 11, 123, 126, 171, 212, 221, 223
Lockwood, Glen 83, 138
Loran, 59, 63
Lovell, Clegg and Ellyett 61
Lowe, Robert 11, 123, 127, 159, 212
Lunik 116

M

Macallum, A.B. 28
MacDonald, Dettwiler and Associates 158
Mackenzie, Chalmers J. 19, 30, 48-49, 52
MacLean, Steve 10-11, 178, 181, 194-195, 197, 267, 269
MAD DAWG 243-245, 275
Magnetic field 13, 23-25, 37, 40-41, 43, 46, 56, 69, 77-78, 80, 82, 92, 96, 98-99, 103, 105-107, 116, 136-139, 146, 148-150, 163, 169, 186-187, 213-215, 226-227, 260-262, 264, 274
Magnetometer 23, 59, 97, 109, 139, 163
Magnetosphere 8, 41-42, 79, 82, 103, 105, 116-117, 133, 137-139, 147-149, 156, 161, 165, 184-185, 227-228, 264, 266
Manchester University 43, 162-163
Manson, Alan 126
Marconi Group 26, 57, 59
Mars 10, 13-14, 182, 214-216, 240, 242-243, 245, 247-253, 258-260, 262-263, 265, 275
Mars Express 240, 260
Massey, Harrie 20, 80-81, 261
MATADOR 10, 245-247, 275
McConnell, John (Jack) 127, 223, 244, 261
McDade, Ian 127, 171, 223, 225
McDiarmid, Ian 11, 14, 27, 101, 138-140, 147, 162-163, 174, 176, 184
McElroy, Tom 220-221, 224, 228, 244
McEwen, Don 11, 97, 107, 138, 159, 164, 166, 186
McGill University 21, 28-30, 33-35, 43, 45-46, 58, 61, 67, 69, 80, 89, 114, 118, 127, 131, 163, 194, 196
McKinley, Don W.R. 21-22, 61, 68, 88-89, 229
McLennan, John 28, 35, 118
McNamara, Al 11, 68, 88, 91, 96, 164, 213
McNaughton, Andrew George 30, 43, 45, 47-50
McVeigh, Stuart 35, 39-40, 53
MDA (McDonald Dettwiler Associates) 159, 205, 207, 217, 247-248, 250-251, 256-257, 264, 275
Meek, Jack 56-59, 90-93, 126, 131
Meighan, Arthur 29
Meinel, Aden 72, 74, 80, 90-92, 100, 107, 121, 124
Mercury 154, 260, 263-264
Mesosphere 100, 103-104, 126, 223, 273
MET 18, 33, 44, 49, 62, 79, 87, 114, 117, 119, 131-132, 134, 137, 140, 159, 174, 205, 225, 241, 247-252, 259
Meteorological Service 33, 35-36, 53, 179, 196, 221, 224, 275
Meteorology 7, 21, 25, 54, 79, 88, 92, 160, 243, 249, 262, 273
Michelangeli, Diane 244, 246, 249, 252
Michelson interferometer 128, 173, 209-211, 217-218, 223-225, 229-230, 234, 257, 265, 276
Microgravity 9, 179, 194-196, 198-202, 274-275
Millman, Peter 61, 80, 88-89
MITCH (Mars Investigation of Total Climatological Hazards) 10, 242-245, 247, 275
Money, 28, 35, 65-66, 69, 71, 76, 135-136, 143, 175, 193, 227, 229, 261-262
Money, Ken 178, 194
MOPITT (Measurements of Pollution In The Troposphere) 9, 219-220, 223, 260, 263, 275
Morley, Larry 11, 150-151, 205
Morrow, Bill 11, 171-172
MOST 10, 14, 237-239 275
MSAT 181-182, 205
Muldrew, Don 11, 138

N

NANOGAS (Nanocrystal Get Away Special) 199-200, 275
NASA 8, 18, 22, 24, 74, 100, 104, 128-130, 132-134, 136, 138, 140-141, 145, 151-154, 157-159, 169, 172-175, 177-178, 180-181, 183, 191, 193-196, 201-202, 205, 207-210, 213, 218-220, 232, 235, 237, 239-242, 245-250, 255-258, 260, 263, 267, 275
NASDA 261, 275
National Research Council 7-8, 11, 19, 21-22, 27-29, 31, 34, 42-45, 54-55, 61, 77-78, 87, 111, 160-161, 164, 275

Newton, Sir Isaac 70, 198
Nicholls, Ralph 11, 72, 80-81, 84, 107, 127, 189, 241
Nike rocket 22-23, 100, 108-109, 142-144, 146
Nishida, Atsuhiro 11, 26, 77, 138
Nobel prize 45, 55, 70, 147, 161, 163, 195, 219, 262
Noctilucent clouds 103-104, 212
Norway 61, 73, 96, 107, 131, 170, 209, 213, 262
Noxon, John 122-123
Nozomi 9, 186-187, 214-216
NRC 7-9, 19, 26, 28-31, 35, 43-52, 54-56, 58, 60, 65-68, 71,
    80, 84, 87-90, 92-93, 96, 98, 101, 113-114, 117, 119-
    120, 126, 129, 132, 135-136, 138-141, 145-149, 154,
    158-170, 173, 176, 178, 180-181, 185-187, 191-194,
    209, 212-217, 223, 228-229, 232, 236, 273, 275-276
Nuclear Winter 9, 176-177, 189

O

Obayashi, Tatsuzo 76-77, 261
Odin mission 9, 220-223, 262
OEDIPUS 9, 147, 188, 213-214, 275
OGLOW 180-181, 189, 275
Optech 242, 244, 246-251, 265
OSIRIS 221-223, 257, 265, 275
Oxygen 35, 42, 81, 95, 122-124, 135, 138, 171-172, 212,
    232, 259, 268, 271, 274-276
Ozone 9, 14, 81, 92, 100, 103, 122, 124, 176-178, 180, 189,
    220-223, 225, 228, 259

P

Paghis, Irvine 116-117, 128, 154, 158, 189
Parallactic 36, 39, 53, 73, 96, 108
Parker, Eugene 116
Patterson, Gordon 21, 25, 33, 35, 155
Payette, Julie 196, 209
Pearl Harbour 66
Personal reflection 8, 107
Petrel rocket 171
Petrie, William 55, 60, 65, 68-69, 72, 88, 90, 124
Phoenix mission 10, 53, 240-241, 243, 247-253, 260, 263,
    273, 276
Photography 26, 39-40, 45, 54, 65, 108, 150-153
Photomultiplier 70-71, 73, 95, 100, 110, 182-183
Plasma resonance 138
PMDIS 9, 203-204, 275
POLAIRE 161-162, 174, 182
Polar mesospheric clouds 103, 223, 275
Pope Pius 86
Pressure Modulated Radiometer 219, 275
Prince Albert Radar Laboratory 7, 82, 126, 152, 229, 275
Propellants 24, 45, 50-51, 60, 141-142
PTV 24, 101, 141, 275

Q

Queen's University 34, 43-44, 47, 57, 81, 193, 229

R

Radar 7, 13, 23, 30, 46, 48-49, 52, 54, 58, 61-62, 65-69, 74,
    79-80, 82-84, 90-92, 100, 109-110, 117-118, 126, 146,
    151-152, 164, 166, 168, 179, 204-205, 208, 226-229,
    241, 256-257, 263, 273, 275-276
RADARSAT 9, 13, 179, 181, 193, 204-208, 256-258, 264-
    265, 267, 276
Radio and Electrical Engineering Division 51, 87, 119, 145,
    161, 275
Radio Physics Laboratory 60, 81-82, 90, 131, 276
Ratcliffe, Jack 81-82, 115-117
Red Lake 164
Red Line Photometer 135-136, 164, 275
REED 51-52, 88-89, 92-93, 119, 145-146, 161-162, 275
Research Enterprises Limited 48, 275
Resolute Bay 49, 84, 90-93, 96-99, 141, 150
Resonance Ltd 172-173

Rettie, Dick 120, 145
Rigden, John 17, 31
Roach, Franklin 91
Rocket 8-9, 13, 20-24, 26, 43-45, 51, 77, 83, 99-102, 104,
    108-110, 114, 116, 118-121, 123, 125, 127-129, 132,
    134-136, 140-150, 154, 158, 163, 165-167, 169-172,
    182, 186, 188, 200, 207, 212-214, 216-218, 223, 225-
    226, 229, 231-232, 256, 262-263, 266, 273, 275
Roosevelt, Franklin D. 62
Rose, Donald Charles 7, 11, 19-22, 25-29, 31, 33, 42-51, 60,
    64, 79, 83, 88-91, 99, 101, 106, 114, 127-128, 132, 135,
    139-142, 145, 162-163, 205, 229
Rosetta mission 260
Rostoker, Gordon 11, 77-78, 159, 228
Rowland, Henry Augustus 70
Russia 13, 88, 93, 188, 195, 200, 208, 231, 237, 256, 261,
    263, 274
Rutherford, Ernest 29, 43, 71, 163

S

Saint Hubert 191, 207
Saskatchewan 8, 13, 21, 25, 27-30, 33-36, 38, 40-41, 48, 53-
    55, 59-62, 65-68, 72, 83-84, 88, 90, 92-93, 95-97, 100,
    107-108, 117, 120, 125-128, 146, 154, 159, 164-166,
    177-178, 186, 207, 209, 212, 214, 217, 222-223, 226,
    228-229, 274, 276
Saskatoon 7, 29, 33, 41, 45, 53-55, 61, 65-69, 71-75, 80, 83,
    90-96, 98, 108, 116, 118, 120, 122-127, 173, 177-178,
    196, 200, 209, 221, 226-228
Satellite 9, 13-14, 17-18, 24, 27, 33, 60, 69, 83-87, 89, 94,
    102, 104-105, 114, 127-132, 135, 137, 139, 147, 149-
    162, 170, 174, 177, 179, 181-184, 186-188, 200, 202-
    203, 205, 207, 209-211, 216-218, 220-226, 228, 231-
    233, 237, 239, 242, 255-258, 261-262, 264-267, 270,
    273-274, 276
SCCO 9, 200, 276
Schiff, Harold 80, 127, 241
Science Council 114, 155
SCISAT 9, 14, 188, 223-225, 257-258, 264
Scott, Jim 56-58, 60, 88, 116, 131
Scott, Robert Falcon 78-79
SED 8, 119-121, 126, 146, 149, 170, 173, 175, 177, 184-185
Shackleton, Ernest Henry 35, 79
Shemansky,Don 72, 74, 121
Shepherd, Gordon 11, 13, 73-75, 118, 127, 134-136, 159,
    161, 164-165, 167, 173-174, 182, 209, 212, 225, 228,
    241, 244
Shepherd, Marianna 212
Shirley Bay 21, 59-60, 67, 82, 158, 229
Shrum, Gordon 28, 35, 122
Shuttle 8-9, 13-14, 153, 158, 160, 170, 173-174, 178-181,
    194-197, 199, 201-203, 208-210, 213, 224, 255-256,
    267-268, 276
Singlet delta 8, 121-123
SMS 9, 170, 173, 184-186, 214
Sodium 23, 42, 74-75, 100-101, 124-126, 144
Sofko, George 11, 226
Soft Particle Spectrometer 134, 164, 276
SOHO 259, 276
Solandt, Omand 113-114, 155
Solar 8, 14, 25-26, 40, 42, 46, 52-53, 56, 60, 70, 75, 83, 85,
    87-89, 91-92, 98, 103-105, 107, 116-117, 125-126, 128-
    129, 133, 137, 146, 149, 158, 163-164, 166, 175-176,
    179, 182, 185, 197, 207, 212, 215-216, 220, 223-224,
    227, 233, 235, 259-260, 264, 266-267, 270-272, 274,
    276
Solar flux 52
Solar maximum 146
Solar wind 8, 103, 105, 116, 137, 149, 215, 227, 264, 266
Space Science Coordination Office 160, 276
Spacecraft 13, 18, 116, 129-130, 132-137, 139-140, 156,
    158, 161, 164, 171-173, 175, 182, 185-188, 194, 196-
    198, 200, 205-208, 211, 215-216, 219, 221-222, 224-
    225, 231-233, 239, 242, 249-250, 257, 260, 273-276
Spar Aerospace 129, 157, 159, 181, 205, 207, 256-257
SPEAM 179-181, 189, 224, 276
Spectrometer 9, 70-75, 92, 96-97, 109-110, 121, 123-124,

134, 136, 151, 164, 170, 172-173, 184-185, 216, 221-
222, 224, 234, 251, 274, 276
Sputnik 7, 17-18, 31, 49, 63, 83, 86-87, 104, 193
Steacie, Edgar William Richard 49
Stoffregen, Willy 59, 61
Storey, Owen 97, 115-116
Størmer camera 35, 67, 73-74, 80-81, 105, 262
STRATOPROBE 9, 124, 177-178
STS-115 197, 267, 269
Submarines 26, 57, 78
Sulphur Mountain 49, 98-99, 127-128
Sun photometer 179, 276
Sunset 42, 103, 180, 224, 270-271
Sunspot 52, 59, 103-104, 133
SuperDARN 226-228, 276
Surrey Space Centre 262, 276
Sussex Drive 29, 44-45, 48, 191, 193
Swedish 9, 106, 138, 176, 182-183, 186, 220-222, 258, 262,
276
SWIFT 9, 92, 225, 261, 265

T

Taylor, Russ 231
Taylor, Peter 244, 246, 249, 251
Temperature 18, 23, 35, 37, 40, 70, 73-74, 76, 80, 92, 97,
100, 102-104, 106, 109, 133, 148, 166, 176, 184, 186-
187, 200, 209, 218-220, 226, 231, 233, 243, 247, 250-
252, 259, 264, 268
Temple of science 29-30, 44, 48
Terra 219-220, 259-260, 263
Terraforming 259
The Hague 18-20
Thirsk, Bob 178, 195-196, 209
Thistle, Mel 27, 31
Thuillier, Gerard 174, 213
Tokyo 77, 166, 172, 185, 261
Topside sounder 24, 102, 128-129, 131, 133-134, 136, 156,
173, 205
Toronto 18-19, 21, 25, 28, 31, 35-37, 42, 48, 53, 75-77, 79-
81, 88-89, 92, 106, 113-114, 117-118, 121, 124-127,
129, 133, 154-156, 170, 172, 178, 194-196, 200, 212,
217-219, 221, 223, 229, 237, 241-242, 250, 257, 272,
276
Tory, Henry Marshall 29-30, 43-44, 49
TPA 9, 214-217, 276
Trent University 212
Truman, Harry S. 17
Tryggvason, Bjarni 178, 196, 198

U

UARS 9, 174-175, 209, 211, 265, 276
Uffen, Bob 154
UK 13, 26, 34, 45, 48, 50-51, 99, 114, 130-131, 140, 142,
147, 163, 196, 205, 219, 237, 260-262, 275
Ultra Violet Imager 176, 182-183, 276
Universities, (see also individual names, Alberta, Calgary,
etc.) 7-9, 13-14, 18-19, 21, 25-31, 33-36, 38, 40-41, 43-
44, 46-49, 53-55, 58-65, 67-69, 71-72, 74-86, 88-97,

99-100, 105-109, 113-122, 124-127
URSI 85, 115, 276
US Air Force 18, 84, 113, 123, 126, 140, 143, 242
US Army 18, 100, 108-109, 140, 142-143, 275
USAF 63-66, 72, 80, 82-83, 124, 126, 144, 276

V

Valcartier 9, 21, 49-50, 101, 114, 176-177, 244, 274
Vallance Jones, Alister 11, 71-73, 91, 95, 118, 121-124, 126,
159, 182-183, 217, 226, 261
Van Allen, James (radiation belts) 8, 18, 25-26, 85, 104-
105, 107, 109, 116, 132, 138
Vandenberg 132, 207
Vanguard 18, 105
Vawter, Floyd 66-67
Vegard, Lars 69-74, 80, 86, 262
Velvet Glove missile 51, 142
Very Long Baseline Interferometry 10, 229-230, 276
Viking 9, 128, 180, 182-185, 187, 221, 251, 258, 262, 265,
276
Von Braun, Wernher 105, 173

W

WAMDII 9, 173-176, 180, 182, 210, 276
War 7, 18, 26-28, 30-31, 33, 35, 37, 39, 41-43, 45, 47-57,
59-63, 65-66, 68-70, 76-78, 80, 84-85, 87-88, 113-114,
118-119, 127, 141-142, 150, 152, 163, 172, 176-177,
189, 221
Watanabe, Tomiya 77, 186
Waterloo 223, 234, 244
Welsh, Harry 217, 229, 241
Western Ontario 7, 21, 69, 80-81, 107, 120, 127, 131, 154-
155, 159, 169, 194, 196, 212, 214-215, 228, 263, 276
Weyprecht, Karl 34
Whalen, Brian 11, 121, 147-148, 150, 164, 166, 168-170,
173, 184-186, 213-216
Whiteway, Jim 226, 249, 251
Wiens, Rudy 212
Williams, K. 120
Williams Lake, 121
Williams, Dave 196-197, 203-204, 209
Wilson, J.Tuzo. 18-19, 31, 75-76, 84, 87-89
Wilson, Brian 21, 99, 127-128
Wilson, Margaret 139
Wilson, Peter 53
WINDII 9, 125, 174-175, 209-213, 217, 223, 225, 257, 265,
276
WISP 173-174, 276
Wright, Sir Charles Seymour 7, 48, 77-79

Y

Yau, Andrew 11, 184-186, 216-217, 245, 264
York University 63, 65, 80, 125, 127, 134, 136, 149, 154,
159, 164-166, 170-173, 177-179, 194, 197, 203, 212,
214, 223, 225-226, 231, 238, 241, 243-244, 246, 249,
259, 261, 263-264, 267, 273
Young, Bob 127, 170-173